John Timbs

The Yearbook of Facts in Science and Art

Exhibiting the most Important Discoveries and Improvements of the Past Year

John Timbs

The Yearbook of Facts in Science and Art
Exhibiting the most Important Discoveries and Improvements of the Past Year

ISBN/EAN: 9783742808271

Manufactured in Europe, USA, Canada, Australia, Japa

Cover: Foto ©Thomas Meinert / pixelio.de

Manufactured and distributed by brebook publishing software
(www.brebook.com)

John Timbs

The Yearbook of Facts in Science and Art

Joseph Whitworth

THE

YEAR-BOOK OF FACTS

IN

Science and Art:

EXHIBITING

THE MOST IMPORTANT DISCOVERIES AND IMPROVEMENTS OF THE PAST YEAR

IN MECHANICS AND THE USEFUL ARTS ; NATURAL PHILOSOPHY ;
ELECTRICITY ; CHEMISTRY ; ZOOLOGY AND BOTANY ; GEOLOGY
AND MINERALOGY ; METEOROLOGY AND ASTRONOMY.

By JOHN TIMBS,

AUTHOR OF "CURIOSITIES OF SCIENCE," "THINGS NOT GENERALLY KNOWN," ETC.

"Every new development of intelligence is a source of pure enjoyment."
BABBAGE.

The Great Melbourne Telescope.—(See p. 102.)

LONDON:

LOCKWOOD & CO., 7, STATIONERS' HALL COURT.

MDCCCLXIX.

MR. JOSEPH WHITWORTH, C.E., F.R.S.

(*With a Portrait.*)

THIS eminent Mechanician was born at Stockport, on the 21st December, 1803. He received his early education from his father, who kept a day and boarding school. At the age of twelve years, he attended Mr. Vint's Academy, at Idle, near Leeds, where he remained for eighteen months. He then went to an uncle in Derbyshire, who was a cotton-spinner, with whom he spent four years and a half, and learned to work the various machines employed in that branch of industry. At the age of eighteen he went to Manchester, and made himself practically acquainted with the manufacture of cotton machinery in the works of Messrs. Crighton and Co., and others. In 1826 he went to London, and worked for some time at Messrs. Maudslay's, and at Messrs. Holtzapfel's. He was also with Mr. Clement, who was engaged in the construction of Mr. Babbage's Calculating Machine.

In 1833 Mr. Whitworth commenced business in Manchester, under the firm of "Joseph Whitworth and Co.," as a maker of engineers' tools; which firm has gained a world-wide reputation, and is still in full activity. Here he devoted his abilities to the attainment of greater accuracy in mechanical work. In 1840 he read his paper "On the Preparation of Plane Metallic Surfaces," at the meeting of the British Association at Glasgow. The universal adoption of the system described and advocated in this paper inaugurated a new era in the history of mechanical science, and rendered possible a degree of accuracy that had hardly been imagined before. In the meantime, Mr. Whitworth's attention had been drawn to the desirability of introducing an uniform system of screw-threads. At that time almost every engineer had his own form of thread and pitch; the nuts made by one maker would not fit the screws made by another; and the annoyance and loss entailed by this want of system can only be fully realized by those who were actively engaged in the business before Mr. Whitworth's system was fully introduced. The orderly and methodical mind of Mr. Whitworth could not tolerate such a state of disorder. After considerable research and many experiments, he arranged an uniform system of pitches, and determined the best form and angle for the thread. The results of these labours were embodied in a paper "On an Uniform System of Screw-Threads," communicated to the Institution of Civil Engineers in 1841. The system is now universally adopted, and is inseparably connected with Mr. Whitworth's name.

Other questions of public interest did not fail to attract Mr. Whitworth's notice, and to engage his attention. About this time he brought out his "Street-Sweeping Machine," which was speedily introduced in Manchester, London, Birmingham, and other large towns, where it proved of great efficiency, and contributed to popularize the inventor's fame. Meanwhile, Mr. Whitworth's attention was continually directed to the devising of new mechanical tools, or to the perfecting of those already designed. His efforts were directed, first, to accuracy of workmanship, then to economy of labour. The duplex lathe, and the reversing tool of the planing machine, and the standard gauges of size were the fruits of his labours between 1840 and 1850. At the Great Exhibition of 1851, Mr. Whitworth exhibited such a collection of engineers' tools, remarkable for the excellent workmanship and admirable design, as had never before been brought together; and demonstrated the great advances that had been made during the past ten years. He then exhibited the Measuring Machine designed by him, in which the sense of touch was employed, instead of that of sight; and by which he was enabled to detect differences of only one-millionth of an inch in extent. For this machine Mr. Whitworth received the Council Medal. In 1853, he was appointed one of the Royal Commissioners to the New York Exhibition; and on his return, in 1854, he drew up a special report on American manufactures, which was presented to Parliament by command of Her Majesty. In 1854, Lord Hardinge, the then Commander-in-Chief, requested Mr. Whitworth to undertake the requisite researches in order to determine the proper construction of rifle-barrels. This Mr. Whitworth undertook to do, on condition that a covered experimental gallery, 500 yards in length, should be erected in his grounds near Manchester. This was done in 1855, and Mr. Whitworth at once entered upon a series of experiments, from which he deduced the proper form of the projectile, and the principles which govern the construction of a perfect rifle-barrel. Subsequent experience has proved the accuracy of these deductions; and the data laid down by Mr. Whitworth have been gradually approximated to in almost all rifles and systems of artillery. In 1857, the first complete rifle made by Mr. Whitworth was fired at Hythe, in the presence of Lord Panmure, the Secretary of State for War, and his staff. The "figure of merit" (or the average distance of each shot, from the centre of the group) obtained at 500 yds. range, was 4½ in. No "figure of merit" under 27 in. had previously been obtained with any rifle. So complete had been Mr. Whitworth's investigations, and so accurate his calculations, that this figure of merit of 4½ in. has never yet been surpassed with any rifle fired in the open air.

The principles thus discovered and enunciated by Mr. Whitworth for the manufacture of rifle-barrels were equally applicable to all calibres of artillery. This fact was at once recognized by Lord Hardinge, who was so much impressed by the shooting that he witnessed in the gallery, on the occasion of his visit to Manchester, that he requested Mr. Whitworth to rifle some brass field-guns on his polygonal principle. This Mr. Whitworth undertook to do ; some 6, 9, and 12-pounder brass field-guns were accordingly rifled, and the results obtained with them fully confirmed the expectations awakened by the success with the rifle. In 1858, Mr. Whitworth also rifled a 68-pounder cast-iron gun-block on his plan, and with this gun he fired a 68 lb. solid shot through a 4 in. armour-plate, fixed on the side of H.M.S. *Alfred.* This was the first instance in which iron armour-plates were completely penetrated. In 1862, Mr. Whitworth inaugurated another great advance in artillery by firing his patent flat-fronted steel shell through a target representing the side of the *Warrior.* This shell weighed 131 lbs.; it contained a bursting charge of 3½ lbs. of powder, and pierced a 4½ in. iron plate, backed up with 18 in. of teak. Mr. Whitworth was the first to demonstrate the advantages of the flat-fronted projectile, both for penetrating armour-plates at varying angles, and also for penetrating under water. He also first demonstrated the possibility of exploding armour-shells without the use of any kind of fuse.

Of late years Mr. Whitworth's energies have been directed to the perfecting of a material suitable for the construction of rifled guns of the largest calibre; and it is to his successful efforts in this direction that the results lately obtained with one of his 9-inch guns at Shoeburyness, are in great measure due. This gun has been found capable of firing a projectile weighing 310 lbs., with a charge of 50 lbs. of powder; and with it the unparalleled range of 11,243 yards has been obtained.

Mr. Whitworth was elected a Fellow of the Royal Society in 1857; next the degree of LL.D. was conferred upon him by the Senate of Trinity College, Dublin; and the same year he was honoured with the degree of D.C.L., by the University of Oxford. At the Exposition Universelle at Paris, in 1867, Mr. Whitworth exhibited a fine collection of engineers' tools, and also of his rifled ordnance and projectiles. This exhibition attracted great attention, and fully maintained the high character of all Mr. Whitworth's productions. In consideration of the eminent services rendered to the cause of industry and science, one of the five "Grand Prix" given to England was conferred upon Mr. Whitworth. In September, 1868, at the visit of the Emperor of the French to the camp of Châlons, one of Mr. Whitworth's field-guns which was being tried there was fired before His Imperial Majesty, who subsequently expressed his sense of the services Mr. Whitworth had rendered on the subject of artillery by conferring upon him the distinction of the Legion of Honour.

We have much pleasure to add that the Council of the Society of Arts have awarded the Albert Gold Medal to Mr. Whitworth "For the invention and manufacture of instruments of measurement and uniform standards by which the production of machinery has been brought to a degree of perfection hitherto unapproached, to the advancement of arts, manufactures, and commerce." This Medal was instituted to reward "distinguished merit in promoting arts, manufactures, or commerce," and among the recipients in former years have been Sir Rowland Hill, Professor Faraday, and Messrs. Cooke and Wheatstone.

The munificence with which Mr. Whitworth has bestowed the sum of £3,000 a year upon the perpetual encouragement of young students of mechanical and engineering science, remains to be recorded. By this plan he proposes to devote this sum to the foundation of thirty scholarships of £100 each, to be held either for two or three years, ten of which scholarships are to be competed for and awarded in May, 1869. This part of the scheme is left subject to any modification deemed fit, after the first batch of Whitworth scholars are elected. The competition will be open to all Her Majesty's subjects, whether of the United Kingdom, India, or the Colonies, not exceeding twenty-six years of age, and of sound bodily constitution. The subjects in which they will be examined are twofold—namely, sciences and handicrafts. The maximum number of marks obtainable by knowledge of the prescribed theoretical subjects will be about equal to the maximum obtainable by the most skilled workmanship; but a practical acquaintance with a few simple tools will be required of *all* the candidates. No candidate will obtain a scholarship who has not learned to use one or more of such tools as the axe, the saw, the plane, the hammer, the chisel, the file, and the forge; while none will succeed who has not shown a satisfactory knowledge of elementary mathematics and mechanics, practical geometry, both plane and perspective, and freehand drawing. The first competition will be in the following theoretical subjects: Mathematics (elementary and higher), mechanics (theoretical and applied), practical plane and descriptive geometry, and mechanical and freehand drawing, physics and chemistry, including metallurgy. And in the following handicrafts: smiths'-work, turning, filing and fitting, pattern-making and moulding. The examinations are to be conducted by such persons, and in such place and manner, as the Committee of Council may appoint; but Mr. Whitworth himself, with the aid and advice of his friends, who are practically conversant with manufacturing industry, will arrange what shall be the examination in the use of tools.

The students in the enjoyment of the full allowance of £100 a year will be enabled, within certain limits, to use their own free discretion in the mode of carrying on their subsequent education. Mr. Whitworth only stipulates that the successful candidates should be required to spend the period of holding the scholarship in the further satisfactory prosecution of the studies and practice of mechanical engineering, and pursue their studies according to the spirit of the endowment, making periodical reports of them. If the student wish to complete his general education instead of continuing his special scientific study, he may be

permitted to do so.   He may go to the universities or colleges
affording scientific or technical instruction, or he may travel
abroad.

The Committee of Council on Education have acceded to the
request made by Mr. Whitworth that the Science and Art
Department may conduct the necessary examinations and correspondence.   Their Lordships will also give every assistance in
their power to secure the success of the scheme which Mr. Whitworth supports with such patriotic munificence.   Mr. Whitworth
also suggests, for the consideration of the Committee of Council
on Education, whether honours in the nature of degrees might
not be conferred by some competent authority on successful
students each year; thus creating a faculty of industry analogous
to the existing faculties of Divinity, Law, and Medicine.   Mr.
Whitworth is of opinion that such honours would be a great in-
centive to exertion, and would tend in a considerable degree to
promote the object he has in view.   The writer expresses a hope
that the Government will provide the necessary funds for endow-
ing a sufficient number of Professors of Mechanics throughout
the United Kingdom.   As the scholarships scheme could only
come into full operation by degrees, Mr. Whitworth proposed
to create at once, from the fund ultimately available for the
scheme, sixty exhibitions or premiums, of the value of £25 each,
tenable until April, 1869, and placed them at the absolute disposal of the governing bodies of several educational institutions
and towns which he named, in order that they might award them
to youths under twenty-two years of age, who are thus aided
to qualify themselves, and have undertaken to compete for the
scholarships of £100 in May, 1869.

CONTENTS.

YEAR-BOOK OF FACTS.

Mechanical and Useful Arts.

THE WHITWORTH GUN AT SHOEBURYNESS.

WITH a new Whitworth Gun, at Shoeburyness, has been obtained the longest range on record. The piece of ordnance in question is a 9-in. steel-rifled 310-pounder gun, weighing 14 tons 8 cwt.; breech preponderance, 6¼ cwt.; length of bore, 1·10·06 in.; over all, 163·80 in.; calibre, major axis, 0·025 in.; minor axis, 8·250 in.; rifling, Whitworth's hexagonal, spiral uniform, one turn in 171 in. Vent through the cascable in prolongation of centre of axis, the hole being covered with a metal tube-catcher for naval service. The gun is constructed on the built-up system, the inner tube being of Firth's steel, the same as the Woolwich guns. This is covered by a second steel tube, over the rear portion of which is a steel jacket. Over this again are two jackets of Whitworth metal, or steel, compressed by hydraulic pressure. This metal can be made of any degree of hardness or ductibility, and Mr. Whitworth states that the tensile strengths of cast-iron, wrought-iron, and his steel-metal, as used for ordnance, are respectively as 30, 100, and 250. Some preliminary trials of this gun were made by the Ordnance Select Committee in the middle of September last, when, after firing seven rounds, it was tested by Mr. Whitworth's machines (which gauge to the ten-thousandth part of an inch) for detection of the slightest enlargement of the bore, or any permanent set, Mr. Whitworth considering that the first sign of the yielding of the metal marks the commencement of the destruction of the gun, and these delicate testings enable the immediate determination of the maximum charge to which the gun could be exposed without injury. The measurements showed a set of one two-thousandth of an inch at the extreme rear end of the chamber, of one seven-thousandth at the front of the chamber, and thence to the muzzle there was absolutely no difference before and after the firing. The exceedingly minute difference shown may be readily accounted for by the wear even of the instrument or of the face of the bore, or by compression of the mass of metal. In fact, there was no real or actual distension of the bore.

Mr. Whitworth's projectiles are entirely of iron, hexagonal in form, and made spiral to follow the rifling of the gun, having a windage over the major axis of 0·065 in., and over the minor axis of 0·070 in.

In the preliminary trials the projectiles were of three kinds, viz.:—Common shells, having parallel rears, and weighing 200 lb.

empty; length 31·6 in. ; diameter, major axis, 8·96 in. ; minor axis, 8·18 in.; capacity for bursting powder, 18 lb. Common shell with taper rears, 285 lb. empty: 31·6 in. long ; in diameter, 8·96 in. by 8·18 in.; bursting charge capacity, 18 lb. And hollow shot with taper rears, 219 lb. weight ; 21·7 in. long; in diameter, 8·96 in. by 8·18 in. Mr. Whitworth's cartridges are specially arranged so that the powder may be ignited well to the front first. A thin copper tube, perforated with a number of small holes for half its length, is passed through the centre of the charge. Into this at one end is inserted a small funnel-shaped primer cartridge, containing 120 grains of powder, the object of which is to ignite the charge of the gun rapidly, and to begin to move the projectile before the great explosion takes place. Disc papier mâché wads, weighing 18 oz., and fitting the bore, are used. The experiments on the occasion referred to were instituted with the view of ascertaining the suitable charge for the gun, and the difference of range between shells of 310 lb. weight, having parallel *versus* taper ends, and hollow shot of 249 lb., having taper ends. The elevation was 10 deg. through-out, and the following results were reported in the *Standard* at the time:—

Nature of Projectile.	Round.	Powder Charge.	Time of Flight to Graze.	Range to First Graze.	Deflection Right.
		lb.	secs.	yds.	yds.
Common Shell, parallel rear	1	45	13·4	4607	25·0
Ditto	2	45	13·6	1088	31·0
Ditto	3	45	13·0	4730	36·0
Ditto	4	50	13·7	3895	35·4
Ditto	5	50	13·6	4850	27·0
Ditto	6	50	13·5	5485	27·4
Ditto	7	55	13·0 {	About 5880 }	Fell in water.
Ditto	8	55	14·2	5074	11·0
Ditto	9	55	14·0	6005	18·9
Ditto	10	55	14·3	6480	26·4
Hollow Shot, taper rear	11	55	14·4	5229	39·0
Ditto	12	55	14·3	5184	27·0
Ditto	13	55	14·5	6283	31·0
Common Shell, taper rear	14	55	13·0	1881	13·6
Ditto	15	55	13·7	1910	11·0
Ditto	16	55	Not observed.	1881	18·9

The experiments which took place on Friday and Saturday last, November 19 and 20, gave some of the most extraordinary results for range ever known. On Friday, the range was at least 10,300 yards, with a 250 lb. shot, a 50 lb. powder-charge, and a maximum elevation of 33 deg. On Saturday, this gun beat even its previous performance, and with 33 deg. 5 min. elevation, and a 50 lb. charge, throw a 310 lb. shell 11,127 yds. to the first graze, being about 1,000 yds. further than any projectile was ever hurled by any other gun. In these results we may

congratulate Mr. Whitworth upon having obtained ranges which
we believe to be unapproached by any other gun in the world.—
Mechanics' Magazine.

PROJECTILES FOR PENETRATING UNDER WATER.

A PAPER has been read to the British Association, " On the
Proper Form of Projectiles for Penetration under Water," by Mr.
Joseph Whitworth. The author exhibited a photograph showing
the actual effect produced on an iron plate in an experiment made
by him with three descriptions of projectiles. The iron plate
shown in the photograph is 30 in. long, 13 wide, and 1·2 in. thick,
and was immersed in water 39 in. deep. The gun used was the
1-pounder, from which all the former experiments were made
previous to the first penetration of 4-in. armour plates from a 70-
pounder rifled gun in October, 1858. The angle of depression of
the gun was 7° 7'; the distance which the projectile passed
through the water from the point of entering it to the bull's-eye
is 80 inches. No. 1 projectile is Whitworth steel, and of the flat-
headed form always advocated by the author for use at sea. The
photograph showed that it was not deflected by passing through
water. No. 2 shot, with hemispherical form of head, was deflected,
and struck 9½ in. above the bull's-eye. No. 3 projectile is of white
cast-iron, commonly called the Palliser, or chilled shot, and it
struck 19 in. above the bull's-eye, its conical form of head causing
it to rise quickly out of the water. The advantages of No. 1 projectile are, first, its power of penetration when fired even at extreme angles against armour plates; secondly, its large internal
capacity as a shell; thirdly, the capability of passing through
water and of penetrating armour below the water-line. The No.
3 projectile is advocated by Major Palliser on account of its cheap-
ness and its power of penetration, which latter quality, however,
depends upon its being fired at a near approach to right angles
against armour plates. Its adoption is also supported by the
Director of Ordnance (at the War Office) and the President of the
Ordnance Select Committee. The author regretted that he had
for so many years been so frequently obliged to differ in opinion
on mechanical subjects with these gentlemen. His objections to
this projectile are, first, that when it is fired at any considerable
angle against an armour plate its form induces it to glance off,
and the brittleness of the metal causes it to break up; and it is
to be observed that in naval actions oblique fire is the rule, and
direct fire is the rare exception. Second, that the brittleness and
consequent weakness of the metal necessitate a greater thickness
of the sides, and reduce its internal capacity as a shell. And,
third, that its form renders it useless for penetration under water.
If the First Lord of the Admiralty would have a few rounds fired
at sea from the Whitworth 7-in. gun and the Woolwich 7-in. gun,
with each kind of projectile, at a range of, say 500 yards, and at

various angles, against an armour plate fixed on the side of an old ship, the result would show which description of projectile was the best adapted for the service. This power of penetration under water with the flat-fronted projectile, was first brought by Mr. Whitworth under the notice of the War Office and the Admiralty in 1857, simultaneously with the introduction of armour plating: and, by desire of the late Lord Hardinge, who was then Commander-in-Chief, a 24-pounder howitzer was rifled and was sent, with some projectiles, for trial to Portsmouth. The experiment was perfectly successful. Captain Hewlett, of the ship *Excellent*, says, in his Report to the Admiralty, January 25th, 1858, " The penetration into wood at this depth under water has, I believe, never before been obtained." In 1854 the matter was brought before the Armstrong and Whitworth Committee, and an experiment was made from the *Stork* gunboat with a Whitworth 70-pounder gun. On that occasion the projectile penetrated at 3·75 ft. under water the side of the *Alfred*, which was of oak 24 in. thick

NEW BLACKFRIARS BRIDGE.

This magnificent Bridge is expected to be finished to the utmost completion of its most ornamental details by May next. The granite buttresses are highly spoken of ; more magnificent have never been erected for any bridge in England. Viewed from any point their colossal proportions are in perfect harmony with their bulk and huge solidity, and it is not overpraising them to say that of their kind, and for their purpose, it would be difficult to have designed anything more suitable. In their vastness and strength they more resemble the great Egyptian works in stone than anything else we have yet seen in London. The blocks which compose these massive towers are all of dressed granite, and though weighing from 12 to 15 tons each, they are fitted together with the accuracy of cabinet-work, and, in fact, are able to bear comparison with the granite work of the Thames Embankment itself, and more than this it would be impossible to say in praise. Each arch of new Blackfriars is built of nine massive wrought-iron ribs set at a distance of 9 ft. 6 in. asunder. This is a very much greater distance apart than that at which those of Westminster are set, but, on the other hand, those at Blackfriars are very much more than twice as strong, being 3 ft. 10 in. deep at the crown of the arch, and 4 ft. 7 in. at the springing, or, to speak more plainly, at the point where they rise from the masonry of the piers. The cross braces between these ribs are of proportionate strength and depth, and are placed at intervals at 17 ft. apart. Above these, again, come what are termed bearers, and bolted over these again will be powerful buckle plates for the roadway. These buckle plates will be thickly coated with asphalte, then a layer of hard stone rubble, and over all, the

usual granite paving, such as that on London Bridge. There
will be no test need to prove the strength of the bridge before it
is opened, simply because the sectional area of the wrought iron
under each part is ten times in excess of the strain it would have
to bear under the most trying exigencies of metropolitan traffic.
The gradient of the whole bridge will be only 1 in 40,— a great
relief to traffic, when it is remembered that the rise in the old
bridge was 1 in 22. The centre arch will have a span of no less
than 185 ft., the two immediately adjoining this on either side a
span of 175 ft. each, while the two smaller arches at the shore
ends joining the abutments will have a span each of 155 ft. The
height of the centre arch from the water is to be 25 ft.; of the
two next side-arches 21 ft. 6 in., and of the two shore-arches
13 ft. 3 in. All the arches will be of a flat elliptical shape, and
will give together a water-way more than one-third greater than
that of the ruinous obstacle which was for so long an impediment
to the navigation of the Thames. The total length of the new
bridge will be from end to end 960 ft., or about 60 ft. longer than
Westminster. Its breadth will be 80 ft., or almost exactly double
that of the old bridge. The roadway alone is to be 45 ft. wide,
or about 2 ft. wider than the whole of the old bridge from outside to outside, and there are besides these to be two footways
of 17 ft. width each. These dimensions would almost sufficiently
show of themselves what a spacious and noble structure the new
bridge is to be; but the best idea of the real magnificence of its
proportions is to be got on its west side, where the two Surrey
arches are completed, The grace of its lines and the whole style
of its design can hardly be overpraised.—*Times, abridged.*

THE THAMES EMBANKMENT.

As far as it has gone, the Thames Embankment is a very
beautiful piece of work, and as a great link in the promised
chain, it is deserving of all praise. The designs for the Thames
Embankment, as now constructed, both on the north and south
sides of the river, are entirely original, having been prepared for
the Board by their engineer, Mr. J. W. Bazalgette, and approved
and adopted by them. Those for the north side were completed
and contracts let and the works commenced in February, 1864.
The works for the south side were commenced in September,
1865.

Only those who inspect it from an engineering and architectural
point of view can appreciate the difficulties that had to be overcome in its construction, and the exquisite finish with which the
works have been perfected. In the opinion of engineers, both
English and foreign, there has seldom been so colossal a work in
granite put together with the same completeness. It literally
fits with the neatness of cabinet-work, and some of the landing-
stages and piers will remain as standards of what such works

should be. Some idea may be formed of the magnitude and importance of the undertaking when we say that a river-wall in granite eight feet in thickness has been built so as to dam out nearly 300 acres of the river; that this wall is nearly 7,000 ft. long; that it averages more than 40 ft. high, and its foundations go from 16 to 30 ft. below the bed of the river. In the formation of this wall and the auxiliary works of drainage, subways, and filling in with earth behind it, there have been used nearly 700,000 cubic feet of granite, about 30,000,000 bricks, over 300,000 bushels of cement, nearly 1,000,000 cubic feet of concrete, 125,000 cubic yards of earth have had to be dug out, and no less than 1,200,000 cubic yards of earth filled in. Such stupendous quantities of material expended over so short a space of ground have never been heard of till now, and would, if so employed, have been equal to building half-a-dozen structures like the Great Pyramid.

The Northern Embankment, which extends between Westminster and Blackfriars Bridges, is let in three contracts; the aggregate length being 6,040 feet, and the cost of the works as tendered for £675,000. The Southern Embankment extends from Westminster Bridge up the river towards Vauxhall Bridge, and a portion of the works consists in widening and a part in narrowing the river. The total cost of this contract is £309,000, the length of the new roadway from Westminster Bridge to Vauxhall Bridge being 5,000 ft., and its width 60 ft. The footway from Westminster to Lambeth Bridge, in front of St. Thomas's Hospital, was opened to the public in March last, and has been very much used by them since that date. The paved footway next the river, from Westminster Bridge to the Temple, on the Northern Embankment, together with the Westminster steamboat-pier have likewise been opened. The approaches to the footway are from Villiers Street, Wellington Street, and Essex Street, Strand, and at a subsequent period, which is not stated, the road will be continued from the Embankment along the new street to the Mansion House; altogether 37 acres of land being reclaimed from the mud and slime of the river side by the Embankment. These are to be laid out in approaches, ornamental grounds, gardens, and houses.

The end of the Embankment next to Westminster Bridge, and for a long way past Whitehall, is finished, with the exception of the roadway. As a steamboat-pier for arrival and departure, it is now open to the public. A noble flight of stone steps, 40 ft. wide, will give entrance from Westminster Bridge to this portion. As far as it has yet been constructed, there are six piers along the face of the Embankment,—one at Westminster, for steamboats; one at York Gate, for the landing of small boats; one at Hungerford, extending on each side of the piers of the present bridge, for steamers; one at the Adelphi, for small boats; one at Waterloo, for steamers; and one at Temple Gardens, also for

steamers. Small boats will be at liberty to use these landing-
places, but York Gate and the Adelphi are built especially for
their accommodation. York Gate will be one of the *prettiest*
stations on the bank, but the landing-place at Temple Gardens
will be of its kind unsurpassed. . The great frontage of this
pier—nearly 600 ft.—the width of its stone stairways, the
solidity and height of its abutments or terminals, and, above all,
the carved granite arch which will give access to it from the
land, will make this station one of the most conspicuous ornaments of the river. The arch which leads to it is a triumph of
granite work. All the piers and landing-places are of different
designs, though they mostly all keep the same type of massive
and enduring architecture, as befits a great work designed to
last for centuries to come.—*Ibid.*

THE HOLBORN VIADUCT.

THIS grand work is nearly completed. Its leading features are
as follows :—The Holborn Viaduct itself will be 1,400 ft. long
from end to end, and a little over 80 ft. wide. Of this space
50 ft. is given to a roadway throughout, and 15 ft. on each side
for footways. The Viaduct forms a gentle curve from the western
end of Newgate Street, and then is continued in a straight line
to the western side of Farringdon Street, occupying nearly the
whole of the space which recently formed Skinner Street and a
small portion of the churchyard of St. Sepulchre. From Farringdon Street westward it is carried by a gentle curve to the
end of Hatton Garden, occupying the sites of the houses which
formerly stood on the south side of Holborn Hill, the greater
portion of the old roadway, and a large part of the churchyard
of St. Andrew's, Holborn. For all purposes of traffic the road
may be called a level, only sufficient inclination being given to
insure the surface drainage running off. The Viaduct is built
on a kind of double system of arches. Those which support the
roadway are plain solid double archways of 24 ft. span, and built
of the same strength as ordinary railway arches. The footways,
however, are supported by a system of, so to speak, cellular
arches. These are 10 ft. diameter, and rise from one tier to three
tiers. At the commencement of the incline, where the dip down
is slight, there is only one tier or ground-floor of these footway
arches ; but as the descent goes on increasing with the slope of
the hill, it becomes necessary to add another tier of arches above
the first in order to keep the surface of the Viaduct at its proper
level. Thus, at one, the deepest, part at the foot of the hill there
are three tiers of these cellular arches, one above another. All
these arches are lofty, clean, and well ventilated, and will be
used as cellars to the warehouses which will be built up by the
side of the Viaduct. In front of the cellars, and between them
and beneath the main road on either side, runs a subway along

the whole length of the Viaduct. This subway is 11½ ft. high
and 7 ft. wide. It has three rows of cast-iron brackets along its
sides—one for gas-pipes, one for water, and one for telegraph
wires. Thus these at any moment can be reached and repaired
without in any way interfering with the footway above. The
sewage is provided for along the centre roadway in a similar
manner. All the brickwork of these portions of the structure is
most massive. In some parts the rings of the arches are as
much as eight bricks thick, an ample guarantee of their strength
when it is remembered that the Board of Trade only exact five
rings of brickwork for a railway arch. In every case the foundations for the masonry have been taken down to the London clay,
and bedded in 4 ft. of solid concrete. In some cases the clay
was easily reached; in others more than 30 ft. had to be excavated
before it was got at.

THE METROPOLITAN MAIN DRAINAGE.

THE Abbey Mills Pumping-station, at West Ham, near Strat-
ford-at-the-Bow, has been opened. The pumping-station at
Abbey Mills is a most important portion of the scheme for the
main drainage of London. It will be remembered that one prominent feature of the design is the attempt which has been made,
as far as possible, to remove the sewage by gravitation, and thus
to reduce the pumping to a minimum. It is, however, impossible for sewage to fall by gravitation for a distance of ten or
twelve miles from districts which are lower than or near the level
of the river, and yet at their outfall to be delivered at the level
of high water without the aid of pumping. Thus it happens that
all the sewage on the south side of the Thames, and the sewage
of a portion of the north side, has to be lifted, and for this purpose there are four pumping stations, two on each side of the
river. Of those on the south side, one is situate at Deptford
Creek, of 500 nominal horse-power, and the other at the Cross-
ness Outfall, which is also of 500 nominal horse-power. Of the
two on the north side, the largest and most important is that
of the Abbey Mills, which is 1140 nominal horse power. The
fourth will be the smallest station, of 240 nominal horse-power
only, and situated at Pimlico. The Abbey Mills pumps will lift
the sewage of Acton, Hammersmith, Fulham, Shepherd's Bush,
Kensington, Brompton, Pimlico, Westminster, the City, White-
chapel, Stepney, Mile End, Wapping, Limehouse, Bow, and
Poplar, being an area of twenty-five square miles, and a
height of thirty-six feet from the low-level to the high-level
sewers.

The whole of the very extensive works contained in the Abbey
Mills Pumping-station have been constructed after the designs of
Mr. J. W. Bazalgette, C.E., Engineer-in-chief to the Metropolitan
Board of Works, and under the personal superintendence of Mr.

Edmund Cooper, the resident Engineer. The buildings have been erected by Mr. W. Webster, contractor, of London. The engines, boilers, and pumping machinery have been made and erected by Messrs. Rothwell and Co., of the Union Foundry, Bolton, Lancashire, one of the oldest engineering establishments in the country.

The station covers an area of seven acres, divided into two portions by the northern outfall sewer, which passes diagonally across it on an embankment raised about 17 ft. above the surface. On the south-west side of the embankment stands the engine and boiler houses and chimney-shafts, together with the coal-stores and wharf for landing coals and other materials from Abbey Creek. On the north-east side of the embankment are the cottages for the workmen employed on the works, and a reservoir for storage of water to supply the boilers and condensing water for the engines. The engine and boiler houses form one building, the engine-house being arranged on a plan in the shape of a cross, and the boiler-houses forming two wings extending north-west and south-east of the north-eastern arm of the cross. The extreme dimensions of the building, taken across two of the arms, is 142 ft., 6 in.; the width of each arm being 47 ft. 6 in. Each of the two boiler-houses measures 100 ft. in length by 62 ft. in width; and there is a workshop situated between the two, measuring 49 ft. 6 in. by 33 ft. The engine-house consists of four stories in height, two of which are below and two above the surface of the ground, the height of the two lower stories being 36 ft., and that of the two above ground, measured from the engine-room floor to the apex of the roof, being 62 ft. At the intersection of the four arms of the cross the building is covered by a cupola of an ornamental character, rising to a height of 110 ft. from the engine-room floor, and at each of the internal angles of the cross rises a turret in which is formed a circular staircase giving access to the several floors of the building. The boiler-houses are of one story above the finished ground level, the boilers and stoking floor being below that level. The total height from stokehole floor to apex of roof is 33 ft. The style of building adopted is mixed, and the decoration consists of coloured bricks, encaustic tiles, and stone dressings, carved work being introduced at the caps of piers and columns. The chimney-shafts, of which there are two, one on each side of the engine-house, are 209 ft. in height from the finished surface, and 8 ft. internal diameter throughout. They are externally octagonal in plan, rising from a square battered base. They correspond in style with the main building, and are similarly enriched with coloured bricks and stone dressings, and are capped at the top by an ornamental cast-iron roof, pierced with openings for the egress of the smoke. The foundations of brickwork and concrete extend to a depth of 35 ft. below the finished surface.

The engines, which are about 1200-horse power, are eight in

number, of the class known as single-cylinder condensing beam-engines, each having a cylinder 5½ in. in diameter, with a stroke of 9 ft., working two double-acting pumps, 4 ft. diameter, with a stroke of 4½ ft. direct from a strong cast-iron beam 40 ft. long by 6ft. deep in the middle, placed over-head upon very richly ornamented cast-iron entablatures and columns. To ease the working of the pumps, there is placed in the centre of the engine-house, below the floor, a large cast-iron air-vessel, 13 ft. diameter and about 20 ft. high, through which the sewage is pumped into a cast-iron tube or culvert, 10½ ft. diameter. There is also a fly-wheel, 28 ft. diameter, weighing about 40 tons, attached to each engine; and to supply them with steam there are sixteen boilers, 30 ft. long by 8 ft. diameter. Any one of the engines, when in working order, is capable of pumping 1,000,000 gallons of sewage per hour.

The sewage is brought into the pump-well, which forms the lowest story of the building, from the low-level sewer, but, before admission, is strained of any extraneous matters which may be brought down with it, and which would either not pass or be detrimental to the pump-valves, by means of cages of wrought-iron bars, which are placed in chambers in front of the engine-house, and which are capable of being lifted and emptied when full. The building containing the machinery and appliances for this purpose stands in front of the centre of the engine-house, and from the chambers beneath it are three sewers conveying the sewage, after being strained, to the pump-wells in three of the arms of the engine-house. From the sewage-well the water is lifted through rectangular cast-iron pipes, situate at the sides of the building, into the sewage-pumps, and it is from them forced through cast-iron cylinders 6 ft. in diameter, running along the centres of three of the arms of the building, and below the engine-room floor into the large cast-iron air-vessel in the centre of the building. From this vessel the sewage is lifted by the power of six engines, and forced, through the huge iron culvert above mentioned, into the outfall sewer, arrangements being made at its junction therewith for regulating the discharge. Provision is made for disconnecting any of the pumps from the discharge culverts. The boilers, of which there are sixteen, are arranged side by side in the length of the building, eight in each house, the stokeholes extending from end to end at the rear of the building. The boilers are Cornish cylindrical, 8 ft. in diameter, 30 ft. long, with two tubes, each 3 ft. 3 in. diameter, discharging into the main flue running at the back of the boilers, and extending on each side to the chimney shafts.—*Illustrated London News.*

THE ST. PANCRAS TERMINUS OF THE MIDLAND RAILWAY.

THIS magnificent Terminus has been completed. At its entrance the rail level is 17½ ft. above that of the road and the adjoining streets. At the frontage the height is 12 ft. above the Euston Road. The opportunity thus opened has been taken advantage of, and turned to good account in cellarage. In that under the terminus cast iron columns, about 12 in. in diameter, being used instead of brick piers; these columns being put upon brick piers, capped with stone. There are in all about 700 columns, ranged in rectangular lines, and the large space over which these columns extend, is about 700 ft. by 240 ft. On the top of these columns are placed wrought-iron girders, about 2 ft. deep, the main girders running across the building, and having cross girders between them, the flooring being made of Mallet's buckled plates, which connect all the main and cross girders together. The main and cross girders take their bearing upon the columns, and are consequently all of a uniform span, thus rendering the manufacture of so large a quantity of similar work simple and easy to contractors. It was at once seen by the engineer that these girders would form a most excellent tie to a large arched roof; and accordingly, a roof was designed which is *the largest single span in the world.* Those most nearly approaching to it, though of different construction, are the roof of the Riding School at Moscow; and that of the Birmingham Station roof. The tie in this roof being completely tied, there are no tie-rods, ring-posts, or other fittings, which detract considerably from the appearance of a structure of this kind. The clear span is 240 ft., springing from platform level, and the rib is a four-centred arch, Gothic in character. The height from platform level is 96 ft., and the total depth of roof 600 ft. The principal ribs are composed of plate flanges with diagonal lattice bars connecting them, the total length of each main rib being 6 ft., and are placed at a distance of 29 ft. 4 in. apart, centre to centre. The ends of these ribs are connected to the transverse girders, as before stated. It should be noticed that provision has been made in the transverse girders not only to take the thrust of the roof, but strength enough has been given to them to carry the weight of the platform, rolling load, and ballast. Between the main principals are placed three intermediate ribs to support the planking, &c., at equal distances of 7 ft. 4 in., and the whole is braced together by trussed purlins, which occur at equal distances of 17 ft. 6 in., and correspond to the vertical joints of the arched main ribs. The arched main and intermediate ribs carry the gutters, which form a springing for each frames, and about 70 ft. of each side from centre of the roof is glazed, and the remainder being slated on boarding. For about 30 ft. from the springing at each side the main principals are made of plate and angle iron-work instead of lattice-work, and is fastened to a plate box which rests on the walls, and is well secured by holding down bolts to anchor plates built in the foundations of the walls. We notice that provision has been made in the two end main ribs for taking the horizontal thrust in the gable end, occasioned by the wind. These two principals are placed 14 ft. 8 in. apart, and are braced together horizontally.

Although the space of this roof is large, yet there are several elements of economy in its construction as compared with other roofs of large span. In the first place, the tie is provided for it in the girders which support the flooring. Secondly, there is no wrought work in forging, turning, and screwing, and the fitting of gibs, cottars, and fastening requiring expensive workmanship, but the whole is formed of riveted plate ironwork, such as is employed in ordinary bridges. Thirdly, the side walls are not required to support the weight of the roof, and are therefore made much lighter than would be necessary were the great weight of the roof borne upon the walls. And, fourthly, no provision is required for expansion by temperature, the effect of expansion being a slight elevation in the crown of the roof. Thus the complications of roller frames and adjusting bearings are avoided in this construction.—*Abridged from the Mechanics' Magazine.*

NEW METROPOLITAN MEAT AND POULTRY MARKET, IN SMITHFIELD.

THIS much-needed Market has been completed and opened for business. In 1860 the Corporation obtained an Act for erecting market buildings on the site of Smithfield, and in the following year they procured one giving them power to abolish Newgate Market. A design was prepared by Mr. Horace Jones, the City Architect, and the building has been erected by Messrs. Browne and Robinson. Perhaps the most important element in the whole scheme is that the basement story of the market is a "through" railway station, from which there will be communication, not only with all parts of the country, but with all the suburban lines. The first step taken to prepare the site was in the shape of extensive excavations, by certain of the railway companies. About 3,500,000 cubic feet of earth, weighing about 172,000 tons, had to be removed. When this was done, 21 main girders were carried across the entire width of the excavation, 240 ft., on wrought iron stanchions. On the main girders were laid cross-girders, 2 ft. 6 in. deep and 7 ft. 6 in. apart. Between the latter brick arches were turned; and concrete and asphalte were set in above to form a roof for the railway and a bedding for the wood pavement of the market. In these foundations there are five miles of iron girding carried on no fewer than 180 wrought-iron stanchions, and retaining walls of considerable thickness rise all round.

The market is a parallelogram, 631 ft. in length by 246 ft. in width, and covers 3½ acres. The style would probably be called

Italian, though it resembles more the Renaissance architecture of France. The prevailing feature is a series of arcaded recesses between Doric pilasters, fluted on the upper two-thirds, and elevated on pedestals. The general height of the external wall is 32 ft. Between the pilasters, which are of Portland stone, the recesses are built in redbrickwork. The semicircular heads of the arches are filled in with rich iron scrolls. The keystones of the arches are richly carved. Under the iron openings are windows with stone sills and trusses, architraves, and cornices. At the angles of the building are four towers of Portland stone. The lower story, or vertical compartment, of each tower is square, with double pilasters at the corners, and a carved pediment on each face. Above this height the towers are octagonal. The junction between the square and the octagonal portions is effected by stone griffins, the supporters of the City arms, in a conchant position. There are windows with carved friezes on each face of the octagon. The dome crowning each tower is pierced on four sides by dormer windows, and above is a lantern, surrounded by an ornamental railing.

The best points of architectural effect are the two facades of the public roadway, which runs across the market, and divides it into two equal parts. This roadway is 50 ft. wide between the double piers, which carry a richly-moulded elliptical arch and pediment of cast iron. Over each double pier is an emblematical figure, sculptured in Portland stone, representing one of the four principal cities of the United Kingdom. Those on the south front represent London and Edinburgh; those on the north Dublin and Liverpool. The sides of the roadway are shot off from the market by an elaborate screen of open ironwork 14 ft. high; and at its intersection by the central avenue, which runs east and west, the market is closed by gates of ornamented iron-work, having enriched iron spandrils and semicircular heads, similar to those on the arcades outside. Towards the north end of the roadway is a gate giving access by a double staircase to the railway department below. In the northern quarter of the building are also the post and telegraph offices and rooms for the market officers. The gates at the east and west end entrances—the main entrances to the buying and selling parts of the market—are 25 feet high and 19 ft. wide, and each pair weighs 15 tons. The ironwork is in very elaborate scrolls, and strikingly handsome. The central avenue inside is 27 ft. wide, and there are six side avenues, each of a width of 18 ft. The shops are arranged on each side of these passages, which, crossing from north to south, intersect the central avenue. One bay at the eastern end of the market is reserved for dealers in poultry and game, but no fish or vegetables will be sold in the market. There are 162 shops in the market. With the view of securing the best possible light and ventilation, the Mansard form of roof has been used throughout.

The lower part of this covering is filled in with broad glass louvres, so placed that, while air is admitted freely, the direct rays of the sun are kept out. The effect was tested by thermometers during the extraordinary heat of last summer, and it was found that the interior of the building was generally 10 degrees cooler than the temperature in the shade outside. Some might object to the quantity of wood in the roof as being dangerous in case of fire below. There are, however, 12 hydrants on the floor level, which will always be at high pressure, and which will be kept in almost constant use washing the avenues and shops.

When the meat arriving in by rail reaches the depôt underneath the market, it is raised to the level of the floor-way by powerful hydraulic lifts. The Metropolitan, the Midland, the London Chatham and Dover, and the Great Western Railways have direct communications with the depôt. The passenger trains of the Metropolitan run through it every two minutes, and the Great Western Company have an extensive receiving-store for goods there.—*Abridged from the Times.*

IRRIGATION OF UPPER LOMBARDY.

A PAPER has been read to the British Association "On the Irrigation of Upper Lombardy by New Canals to be derived from the Lakes Lugano and Maggiore," by Mr. P. Le Neve Foster, Jun. The author, after referring to the high pitch of perfection to which irrigation in Lombardy has been carried, pointed out that, although the lower part of Lombardy is well watered by existing canals, the whole tract of country to the north of Milan, extending to the foot of the hills of Varese and the Brianza, is too high to be watered by them, and is almost unirrigated. He then described the technical details of a scheme undertaken by Signori Villoresi and Meraviglia, by which it is proposed to irrigate the higher lands by canals from the Lake Lugano, and the lower lands by canals from Lago Maggiore, this lake being situate at too low a level to water the whole region, while the supply from Lugano is not more than sufficient to water half of the whole district. Permission has been obtained from the Swiss Government to store up the flood-waters of Lugano, and regulate their discharge for use in droughts. The same system will be adopted in reference to Lago Maggiore, and works for this purpose, consisting of dams, gates, and locks, will be erected in connection with both lakes. The waters will be distributed by principal canals—five in number, secondary canals, communal canals, and private canals. The total number of the works to be constructed, such as locks (of which there will be forty-seven), bridges, aqueducts, syphons, &c., will be about 260, and the canals are estimated to supply 8,000-horse power for mills, &c., and to irrigate and thus improve the agriculture of 400 com-

munes. The cost is estimated at two millions and a quarter
sterling. One of the most remarkable features of the scheme
is the manner in which it is proposed to raise the capital. Con-
sorzii, or companies of consumers of water, are to be promoted
by the local authorities. In this manner, the provinces, com-
munes, and other corporate bodies bind themselves to take a certain quantity of water, either by payment of a fixed sum down,
or by annual payments, which payments they are able to guarantee from the receipts they will derive from the sale of the
water to the various consumers, or, if necessary, from their other
sources of revenue, and the capital is raised on bonds issued on
this basis. The concession is granted for ninety years, after
which the works become the property of the State. The works
were commenced in autumn.

TUBE-WELLS.

This new arrangement for obtaining Water was described in
The Year-Book of Facts, 1868. Mr. J. S. Norton, of La Belle
Sauvage Yard, Ludgate Hill, the patentee of the Tube-Wells,
has addressed a letter to the *Times*, in which he embodies the
details of his own patent, as well as those of another inventor.
The communication is as follows, abridged :—

"Your New South Wales correspondent's letter, which appeared in your columns of the 4th inst., contains a description of
a Tube-well exhibited at a local agricultural show. In this description he draws some comparisons with my patent, designating
it 'the now notorious Abyssinian tube-well,' but which comparisons, coupled with his remarks, are calculated, if left uncon-
tradicted, to be very prejudicial to the reputation of my wells.

"He states that my wells cannot be made available beyond
28 feet. He is correct so far, that with an ordinary suction-
pump this would be the case, but he is wrong in inferring that
I have no other appliance than a common suction-pump.

"I have patented various arrangements for raising water any
reasonable depth from my patent tube-wells. I accomplish this
by simply converting a short piece of the well-tube itself into
the working barrel of a pump, by lining it with copper, and
fixing therein a bucket and lower valve, so constructed that they
can at any time when needed be brought to the surface for
inspection or repairs.

"This working barrel I place about 25 ft. from the lower
extremity of the tube, and drive the same in the usual way,
adding lengths of tube until water is reached ; the necessary
length of rods to connect the bucket with the handle at the surface is all that is required to bring the water up.

"Your correspondent further observes that the invention to
which he refers supersedes mine, from the fact that it will pene-

trate rock. Allow me to state that in October, 1867, and February, 1868, I patented a plan for penetrating my patent tube-wells through rock, which is in some respects similar to the one described by your correspondent, and which latter appears only recently to have been brought out. My plan is, however, more expeditious, as I dispense entirely with the tedious process of bringing the tool to the surface in order to extract the particles of rock."

PUDDLING IRON.

A PAPER has been read to the British Association "On Puddling Iron," by Mr. C. W. Siemens. The author began by observing that, though cast-steel had recently been introduced to a great extent for structural purposes, the production of wrought iron and steel by the puddling process still ranked among the most important branches of British manufacture, the quantity turned out per year exceeding one and a half million tons, and representing a money value of about nine millions sterling. Notwithstanding its importance and the interesting chemical problems it involved, the puddling process has received less scientific attention than other processes of more recent origin and of inferior importance. The analyses made by Messrs. Calvert and Johnson, of Manchester, of the contents of a puddling-furnace during the different stages of the process had, however, supplied most valuable information, which was confirmed by the investigations of Messrs. Price and Nicholson and M. Lan, and from which Dr. Percy had drawn several important general conclusions that had only to be followed up and supplemented by some additional chemical facts and observations, in order to render the puddling process perfectly intelligible, and to bring into relief the defective manner in which it was at present carried out. It appeared from the investigations above referred to, that the molten pig-iron in the puddling-furnace was mixed intimately in the first place with fused oxides of iron or "cinder," derived partly from the lining or "fettling" of the cast-iron puddling-chamber, and partly from iron oxides of different kinds thrown in with the charge. On forming, by means of the rabble, an intimate mechanical mixture between the fluid cast metal and the cinder, the silicon contained in the iron was first attacked, and was rapidly removed, passing into the cinder in the form of tribasic silicate of protoxide ($3 Fe O, Si O_3$). After the removal of the silicon, the cinder next acted on the carbon of the pig-iron. The mass foamed or "boiled" violently, giving off carbonic oxide, which, in rising in innumerable bubbles to the surface of the bath, might be seen to burn in an ordinary puddling-furnace with the blue flame peculiar to that gas. It was popularly believed that the oxygen acting on the carbon and silicon of the metal was

derived directly from the flame, which should on that account
be made to contain an excess of oxygen; but the very appearance of the process proved that the combination between the
carbon and oxygen took place throughout the body of the fluid
mass, and was to be attributed to the reaction of the carbon
upon the fluid cinder in separating from it metallic iron. It
had been argued, however, that the oxygen might yet be derived
indirectly from the flame, which might oxidize the iron on the surface, forming a fluid cinder that might become transferred to the
carbon at the bottom, in consequence of the general agitation
of the mass. This view the author was in a position to dis-
prove from his recent experience in melting cast steel upon the
open flame bed of a furnace, having invariably found that no
oxidation of the unprotected *fluid* metal took place so long as it
contained carbon in however slight a proportion; and supported by this observation, he felt convinced that the removal of the carbon and silicon from pig-iron, in the
ordinary puddling or "boiling" process was effected entirely by the iron oxides added to the charge or melted
down from the "fettling" of the sides of the puddling-
chamber, and that the waste of iron by the oxidizing action of
the flame commenced only after malleable iron had been already
formed. It followed that if the spongy malleable iron were not
wasted by the flame after it had "come to nature," the weight
of puddled iron produced should greatly exceed that of the pig-
iron employed. The fluid cinder might be regarded as approaching the composition of magnetic oxide of iron ($Fe_3 O_4$); and in
taking the usual equivalents, $Fe=28$ and $Si=22.5$, it was evident
that for every $4 \times 22.5 = 90$ grains of silicon abstracted from the
metal, $9 \times 28 = 252$ grains of metallic iron were liberated from
the cinder; and, similarly, for every $6 \times 4 = 24$ grains of carbon
removed from the metal, $28 \times 3 = 84$ grains of iron were liberated
and also added to the charge. Assuming ordinary forge pig
to contain about 3 per cent. of carbon and 3 per cent. of silicon, it followed that, in removing this silicon and carbon by the
puddling process, 18.9 per cent. of metallic iron was added to
the charge, making a total increase of $18.9 - 6 = 12.9$ per cent.;
while the actual yield of puddled iron was in practice about 12
per cent. *less* than the weight of the pig-iron employed. Thus
a charge of 420 lb. of grey pig-iron ought to produce 474 lb. of
puddled bar; while the yield in practice seldom exceeded 370 lb.
In order to realize the theoretical result, it was, of course, necessary that a sufficient amount of iron oxide should be added to
the charge to effect the oxidation of the carbon and silicon, and
to form tribasic silicate with the silica produced. This amount
calculated as magnetic oxide was about 139 lb. per charge—a proportion of "fettling" that was generally exceeded in present
practice. The author referred next to the elimination of sulphur
and phosphorus, and suggested that the crystals of metallic iron

which formed throughout the boiling mass as it " came to nature,"
were probably chemically pure, and that the freedom of the resulting metal from impurities depended mainly upon the temperature, which should be high, in order to ensure the perfect
fluidity and complete separation of the cinder; and he summed
up with the conclusion that the process of puddling as at present
practised was extremely wasteful in iron and fuel, immensely
laborious, and yielding a metal only imperfectly separated from
its impurities. How nearly it might be possible to approach the
results indicated by the chemical reasoning that had been
brought forward he was not prepared to say; but that much
might be done had been proved by the results of eighteen months'
working of a puddling-furnace which had been erected to his
designs at the works of the Bolton Iron and Steel Company, in
Lancashire. This furnace consisted of a puddling-chamber of
the ordinary form, heated by means of a regenerative gas-furnace,—an arrangement of which the principle was now well
known. The chief advantages of this method of heating were,
that the heat could be raised to an almost unlimited degree; that
the flame could be made at will oxidizing, neutral, or reducing,
without affecting the temperature; that indraughts of air were
avoided; and that the gaseous fuel was free from the impurities
which were carried over into the puddling-chamber from an
ordinary fire-grate. The results of the practical working of the
furnace had been most satisfactory. The yield of puddled bar
was always within one or two per cent. of the weight of pig-iron
employed, and not unfrequently exceeded it, while the cost of
fettling was no greater than in the other puddling-furnaces in
the works; and, owing to its facility of management and unlimited command of heat, the gas-furnace was found to be capable of turning out easily eighteen heats in twenty-four hours,
while in working the same quality of iron twelve heats were the
limit of production in the ordinary furnaces. The iron puddled
in the gas-furnace had proved, as might be expected, much better, more uniform, and more free from cinder than that from the
common puddling-furnaces. The consumption of fuel could not
be directly measured, as the furnace drew its gas from a flue which
supplied at the same time several reheating furnaces; but there
could be no doubt, from extensive experience in the working of
similar furnaces for reheating iron, for glass-making and for other
purposes, that it would be little, if at all, greater in quantity
than one-half the consumption of a puddling-furnace fired
directly in the ordinary way, while a very inferior class of fuel
might be used. No repairs had been necessary since November
last, and the roof was reported to be still in perfectly good condition. In conclusion, the author strongly recommended that in
the working of the gas puddling-furnace, the puddlers should be
employed in three separate shifts of eight hours each per day of
twenty-four hours each, each shift representing as many heats as

are now turned out in twelve hours; and he suggested that the labour of the puddler might be still further reduced with advantage by the introduction of the mechanical rabble. Several of those puddling-furnaces had been erected abroad, and they were also being taken up in England by the Monkbridge Iron Company, Leeds, and a few other enterprising firms.

CAST-STEEL.

(Communicated to the Times.)

NOT long ago one of your Correspondents gave an account of a visit to Krupp's Works, at Essen. He saw the Casting of a 10-ton Ingot of Steel, and I have reason to know, accurately described the process. The metal was melted in crucibles, as in the old and common process of casting steel at Sheffield. The organization was so perfect that hundreds of crucibles were ready to be taken out of the furnaces at the same time, and the molten contents of each were poured in quick succession into a common receptacle. This done, the metal was immediately let out from that receptacle into the ingot mould. At the International Exhibition of 1862, Krupp exhibited ingots of cast-steel so produced, which surprised and delighted every person experienced in the casting of steel. A single ingot weighed upwards of 20 tons, and subsequently, at the International Exhibition in Paris in 1867, an ingot of much greater weight was shown by Krupp. Those ingots were not only large, but sound, and it was this combination of qualities, enormous dimensions, and entire soundness, which made them so remarkable.

I have great satisfaction in informing you that similar ingots of cast-steel are now produced in our own Sheffield, and precisely in the same manner as at Essen. I have had ocular demonstration of the fact on the occasion of a recent visit to the well-known Steel Works of Messrs. Firth and Sons, in that town, in company with several officers of the Royal Artillery. I subjoin an account of what I there witnessed:—The casting-house was rectangular, and along three sides were 106 furnaces, with two smaller in each, the ingot mould being in the middle. Each crucible contained about 50 lbs. of molten steel. At a given signal the pouring of the metal commenced, and in 16 minutes the ingot mould had received the contents of the 212 crucibles. There was no bustle, no chattering, no confusion. The men—200 in number—knew their drill and did their duty without a hitch. The ingot weighed 86 cwt., and was designed for the steel tube of an Armstrong gun. It will be observed that the receptacle employed by Krupp for receiving the steel from the crucibles in the first instance was dispensed with, but that is only a point of minor detail. The Messrs. Firth have cast 9-ton and 10-ton ingots. A 9-ton ingot is required for the 23-ton Armstrong gun. The steel tubes are bored out of solid ingots, and this is one cause of the costliness of our modern ordnance. If larger ingots of steel should be required, they can be made at the same establishment. The enormous ingots of Krupp are to be regarded as *tours de force*, and were specially intended as such. There will be no necessity henceforth to have recourse to foreign steel makers for large shafts or other large articles of crucible-cast steel; and, indeed, if I am correctly informed, the importation of articles of this kind has of late much declined.

I may add that, in expectation of orders from the Government, the Messrs. Firth expended about £32,000 in the erection of two 25-ton Nasmyth hammers, reheating furnaces on Siemens' principle, and other appliances. The anvil block of each hammer is cast-iron, cast in one piece, and weighs 16½ tons. The length of stroke was 9 ft. 6 in. These facts may possibly be of use to the Ministers responsible to Parliament for our great military and naval manufacturing establishments, and prove to them that our steel-makers, at least, are prepared to meet all the requirements of the nation. So, doubtless, will all our manufacturers be equally prepared. They have done so in the matter of armour-plates, and never did the War Office and the Admiralty act more wisely than in throwing the manufacture of those plates open to public competition.

THE SOLID LEVER BRIDGE.

THIS Bridge, in its essentially original features, is the invention of Messrs. Cottrell, Liscom, and Merrill. These features may be best explained by describing the mode of construction, as we find it in the *American Railway Times*, and which is as follows :—In throwing a span from one abutment to another, the first step of all is to lay down upon each abutment a timber projecting over the edge of the bank a short distance. The length of the portion of the structure projected over the chasm is to that portion upon the bank (which is the balance-power), as ten to three. Upon the first layer of timber is placed another, extending further, and firmly trenailed to the first, each successive course projecting further outward over the chasm, and well secured by bolts and trenails, so that when by successive projections the final layer is brought to reach the middle of the span and unite with the corresponding lever from the opposite side, it presents a well-proportioned lever, nearly as solid as if sawn entire of one piece, its projecting weight balanced by that portion of the lever resting on the abutments, and the two united and strengthened by two or three additional courses laid over all. The two levers thus making the sides of the bridge are connected by girders, or floor timbers, on which the flooring is laid—the shore ends bound together to form a cradle for the reception of ballast. Thus completed, a span of 200 ft. would reach back 30 ft. upon each bank, and a span of 500 ft. 50 ft. back. Over this structure two wooden arcs are placed, of slight and graceful curve, united and securely fastened to the main structure by upright iron rods, and abutting in the shore ends of the bridge.

The advantages of this bridge may be stated thus:—1. Economy and facility of construction. It is claimed that an almost unlimited length of span is attainable, and the building of piers, which in deep and rapid rivers is very costly, and which obstruct the navigation, is avoided, and the cost of scaffolding saved. The plan of construction is such that any carpenters and blacksmiths can furnish the materials and do all the work. 2. Strength. From the peculiarity of the structure, being lightest in the centre, all the material bears its true proportion of strain, and there is no waste, and the distribution of the component parts makes them all act at the greatest mechanical advantage. The sides of the bridge being levers balanced by the shore end, the bridge is relieved of the necessity of sustaining its own weight in the centre where an ordinary truss-work sags the most; and being very light in the centre, there is no unnecessary strain upon the balancing or shore end. 3. The arch adds great compactness and increases the strength of the whole in a peculiar manner. Like any other arch it thrusts laterally against its abutments, but its abutments are the two ends of the bridge into which it is built, and it thus forms with the string-courses of timber a wooden bow with a wooden string. In order to spread this arch, the chord or upper string timber would have to be dragged apart by longitudinal strain. Thus there can be no deflection of the bridge after it is well set together. It is this combination of the arch on the one hand, with the re-enforced lever on the other, which gives this bridge a strength immeasurably superior to that which the same amount of material would give when arranged in any other form.

In the iron bridge, patented by the same inventors, and involving the same principles, the strength attained is practically illimitable. The advantages of a cheap, easily constructed, and strong wooden bridge are manifest to all; but those connected with railways are best aware of the immense importance of cheap and strong iron bridges. The obstruction of the western rivers by piers is so serious an impediment to navigation, that this invention becomes of the utmost importance, and comes just in time to solve the problem of how the railways are to cross the Upper Mississippi and the Ohio rivers, without navigation being destroyed.

———

GLOSS ON SILK.

THE method of giving an artificial Gloss to the woven pieces had been invented three years previously to this; that is, in 1663. The discovery of the method was due to pure hazard. Octavio Mey, a merchant of Lyons, being one day deep in meditation, mechanically put a small bunch of silk threads into his mouth, and began to chew them. On taking them out again into his hand he was struck by the peculiar lustre they had acquired, and was not a little astonished to find that this lustre continued to adhere to the threads even after they had dried. He at once bethought him that there was a secret worth unravelling in this fact, and being a man of wits, he set himself to study the question. The result of his experiments was the *procédé delustrage,* or "glossing method." The manner of imparting the artificial gloss has, like all the other details of the weaving art, undergone certain changes in the course of years. At present it is done in this wise:—Two rollers, revolving on their axis, are set up at a few feet from the ground, and at about ten yards, in a straight line, from each other. Round the first of these rollers is wound the piece of silk of twenty, forty, or a hundred metres' length, as the case may be. Ten yards of the silk are then unwound and fixed by means of a brass rod in a groove on the second roller, care being taken to stretch the silk between the two cylinders as tightly as possible. A workman with a thin blade of metal in his hand, daintily covers the uppermost side of the silk (that which will form the *inside* of the piece) with a coating of gum. On the floor under the outstretched silk is a small tramway, upon which runs a sort of tender filled with glowing coals. As fast as one man covers the silk with gum, another works the tender up and down so as to dry the mucilage before it has had time to permeate the texture. This is a very delicate operation ; for if, on the one hand, the gum is allowed to run through the silk, or if, on the other, the coals are kept too long under one place, the piece is spoiled. In the first instance it would be stained beyond power of cleansing, and in the second it would be burned. None but trusty workmen are confided with this task ; and even with the most proved hands there is sometimes damage. When ten yards of the piece have been gummed and dried, they are rolled round the second cylinder, and ten more are unwound. This is repeated till the end. But the silk, with its coating of dry gum, is then stiff to the touch, and crackles like cream-laid note-paper when folded. To make it soft and pliant again, it is rolled anew some six or seven times under two different cylinders, one of which has been warmed by the introduction of hot coals inside, and this is sufficient to give it that bright, new look which we all so much admire in fresh silk.—*Cornhill Magazine.*

PRACTICAL SUGGESTIONS CONCERNING THE SELECTION OF A SEWING-MACHINE FOR FAMILY USE.

1.—*It should be easy to learn,* because ladies will not generally take the necessary time to master a difficult one. However useful a machine may be in the hands of a skilled operator, if an apprenticeship be necessary to learn to work it well—to do good work with it—it will be unused and unprofitable, or, at the best, it will require some person to be specially qualified to do the family sewing.

2.—*It should be easy to work.* Every member of the family should be able to use it, including children and invalids. A good easy-running sewing machine affords pleasant occupation and recreation to children, and assists in teaching them the useful arts of cutting and putting together garments. It can hardly be a family machine if children cannot use it, and take pleasure in it. Ladies in delicate health cannot have more agreeable and healthful exercise than the occasional use of a proper family sewing-machine.

3.—*It should be easy to change from one kind of work to another.* The "family use" of a sewing-machine consists of all sorts of work, one after another; and generally a change in the work requires some change of the machine. Changes of needle, cotton, or both, are frequent. In some machines these changes are easily made; but in others they require time and the careful attention of even a skilful operator. Before buying, a lady should always set the needle herself, notice whether there is any liability of setting it wrong, whether there is any means of knowing whether it is set wrong or not, and the consequences of setting it wrong; she should remove the cotton, and thread up the machine herself; and remember that these changes and many more occur just when she has little time to lose, as she "must get her work out of the way." She should use the hemmers also; see whether they turn the hems and fells *under*, as in sewing by hand, or *up* on the right side, causing a garment to appear wrong-side out.

4.—*It should be easy to keep in order.* It should require only cleaning and oiling, and should not require skill to do even that. It should never require to be taken to pieces for any purpose whatever; it should be so contrived that no part need be removed to get at any other part. If any derangement occur, as will occasionally be the case in all machinery, both the difficulty and the remedy should be so obvious that no serious consequences can follow. The machine should right itself, when the obstacle is removed.

5.—*It should be simple in its mechanism.* A skilful saleswoman can display the attractive features of any machine, and hide its defects. A lady should try to learn to use a few of the best machines before buying. She will not then need the caution against a complicated one, having learned a timely lesson from its continual derangements.

6.—*It should be noiseless.* The old, heavy, cumbersome, noisy style of machinery, is very unwelcome in the home circle. A truly noiseless machine does not interrupt reading or conversation.

7.—*It should be so well made as to require no repairs.* The

best made machines, when worked by steam-power for manufacturing purposes, will wear out in a few years, but in family
use, they will last a generation, accidents excepted.

8.—*It should do all kinds of work well, and make a strong,
secure, and beautiful seam, that will stand washing, ironing,
and wear.* Some machines do light work well, others heavy
work; rarely one does both equally well. In regard to the kind
of stitch, one should not act upon the opinion of any other person.
She should subject the different stitches to actual test, and judge
for herself.

9.—*It should make the best stitch.* There are four kinds
of stitch in common use; the chain stitch, the lock stitch, the
Grover and Baker stitch, and the Willcox and Gibbs stitch.

The chain stitch is made only by the "cheap machines," and
it is believed to be quite worthless; at least, the machines are,
for they drop stitches, and are otherwise defective; it may therefore be left out of the comparison.

The lock stitch is made by two threads, one lying nearly straight
on the under surface of the work, (except on thick cloth, when it
is nearly alike on both sides), the other passing through the
material and crossing the under thread. This stitch is preferred
on leather, and is much used on heavy cloth, but on thinner
materials, the seam is less satisfactory on account of its in-
elasticity, and the difficulty of equalizing the tensions of the
two threads.

The Grover and Baker stitch is made by two threads looped
together, one of which passes through the cloth and with the
other forms a ridge upon the under side; the under thread is
much finer than the upper, in order to make the ridge as light
as possible. This is an excellent embroidery stitch, the most
showy of all. It is also in use for general purposes, having
an advantage over the lock stitch in being secure, and, at the
same time, easily taken out when desired. The mechanism by
which both this and the lock stitch are made, is, however, very
complicated, and much skill is requisite to use it successfully.

The Willcox and Gibbs stitch is made by one thread direct
from the reel by means of a very simple mechanism, which does
its work with greater certainty than any other, and almost entirely
without noise. Each loop grasps the preceding one and is firmly
twisted around it. The seam is fastened off by taking two
stitches beyond the cloth. This ties a firm knot. When the
fastening is required before coming to the end, it is done by
simply drawing the thread backward upon the seam. It is the
strongest, most secure, and the most beautiful stitch known. If
three tucks be made side-by-side with the three stitches respectively, and cut across at short distances, and the cloth pulled, the
lock-stitch gives way first *by the drawing out of the stitch*, the
Grover and Baker next *by the breaking of the under thread*, the
Willcox and Gibbs last.—*Mechanics' Magazine.*

COAST DEFENCES.

COLONEL JERVOIS, C.B.R.E., has delivered, at the United Service Institution, a lecture upon "Coast Defences, and the Applications of Iron to Fortification." Colonel Jervois commenced by pointing out the necessity and uses of permanent fortifications, which do not, as is sometimes supposed, involve the shutting up of bodies of troops to render them unavailable for employment in the field, but rather create a great opportunity for economy, because a position of great importance can be held by a comparatively small body of men. Behind the ramparts of permanent forts our reserve forces could become first-rate soldiers, while holding the lines of works that defend vital spots against everything but the slow operations of a regular siege. A beaten army can rally under the walls of a fortress. An attacking enemy must either delay his advance and besiege the place, or weaken his army by leaving strong detachments to watch the works with their garrison. From fortresses sallies can be made and the enemy's line of communication endangered. Furthermore, such places as Portsmouth must be guarded somehow or other, if not by fortifications then by a large army in the field. Armies are set free by land defences, fleets by sea forts. Colonel Jervois then pointed out that the strength required for land works is much less than that for coast defences; because neither twelve ton, eighteen ton, nor twenty-three ton guns can be dragged over the open country by an invading force without great risk and delay, though they may be carried in ships. Therefore, against ships heavier guns and stronger protection are required than against attack from the land. To apply the same reasoning to land and sea defences would be manifestly absurd. Since something exceptionally strong and carrying very heavy guns must be provided for harbour defence, the question arose, should such structures be fixed or floating? Colonel Jervois said there was no *versus* in the question. Both forts and floating batteries must be used, and and the lecturer claimed to have suggested the necessity of taking the subject into consideration ten years ago. The committee of which Colonel Jervois was a member, in conjunction with a naval and an artillery officer, met and actually recommended the employment of vessels strongly resembling the American Monitors of later times. Obstructions, too, are required to hold ships back from advancing to places where their guns would be dangerous, and torpedoes of various kinds to attack the vessel from the surface of the water or from beneath. The labours of a committee lately appointed have been crowned with perfect success, and both naval officers and engineers are passing through courses of torpedo instruction.

Mr. Abel, chemist to the War Department, deserved praise for his scientific and practical labours in this field. Yet torpedoes without forts would soon be removed by the enemy, or rendered innocuous

by the death or captivity of the men placed on shore in order to explode them. Forts, then, there must be. Barbette batteries have
the disadvantage of exposing men and guns to fire from the enemy;
shrapnel soon drives exposed men from their pieces. But if men
must be protected—how? Earthworks with iron shields have
been provided in some situations; granite casemates with iron
shields, and sometimes with turrets on the roof; and, lastly, iron
forts, where there is no room for earth and something stronger
than granite is wanted. Colonel Jervois puts so much confidence
in the superiority of the fire from forts over that from ships that
he would be willing to hold by the American system, and retain
granite casemates with many guns, as being cheaper than iron
forts. He pointed out that our casemates are stronger than
those of foreign nations, and asserted that no fort had ever yet
been breached by the fire of ships. All such effects during the
American war came from shore batteries. Mortar fire should be
provided to attack the vulnerable decks of ironclads, and if Moncrieff's system of mounting guns were applied extensively, as
Colonel Jervois explained had been recommended by him, the
chances of success for a naval attack against earthworks and
granite forts would be small indeed.

The question of iron structures was then touched upon. Many
plans were shown in diagram, but all the foreign ones appeared to be either weak or very expensive. No nation has
worked out the subject as English officers and scientific men have,
and it is not wonderful that we should have discovered more facts
bearing on efficiency and economy than could possibly occur to
those who have no experimental knowledge of the subject. Colonel
Jervois showed a drawing of a Russian shield which cost some
£2200 per gun, and was weaker than the famous Gibraltar shield,
which cost less than £1000. Several forms proposed by various
individuals were shown and explained. The lecturer seemed to
be somewhat in favour of a backing of concrete or masonry enclosed in iron casing, but said that the subject would soon undergo further investigation. The Plymouth Breakwater fort lately
fired at on the practice-ground at Shoeburyness was described,
and the lecturer spoke rather warmly of the nature of the trials.
In those trials "The power of the guns in the forts was not considered," though that is at least half the battle, and "the science
of submarine mining was ignored." If the whole system of defence is strong enough to prevent any enemy from attacking,
what more can be wanted? All increase in the power of artillery
is in favour of the defence, and enables gunners on shore to keep
ships at greater distances, or to protect a wider circle of torpedoes. Speaking in detail upon the late experiments, Colonel
Jervois pointed out that the system of plate upon plate is cheap
and strong, giving nearly the same amount of protection as solid
armour of the same thickness, and at a much smaller cost. The
two 15 in. plates were dashed to pieces at Shoeburyness, and the

smashing of the whole thickness of defence would be more fatal than the detachment here and there of part of a plate or plank in the compound structure. Finally, Colonel Jervois asserted that when people talked of defences not being strong enough, they forgot that nothing could be easier than to make forts impenetrable to guns, if only the money were forthcoming. But cost must always influence the quantity of any pleasant thing procurable, and the Department of Works claimed the credit of having done the best it could with its resources, and, further, to have produced iron defences equal or superior to those of any other nation; strong enough, if supported by all proper scientific means—torpedoes and the like—to defend the harbours of England against all the navies in the world.—*Abridged from the Times.*

THE MILLWALL SHIELD.

THE Millwall Shield at Shoeburyness, designed and constructed by Mr. John Hughes, of the Millwall Iron Works, has undergone another test, from which it came out signally victorious. It had already withstood fourteen rounds from the 9 in. and 10 in. guns without any serious damage to the backing; and on the present occasion it was fired at by the 12 in. gun at seventy yards, with battering charges of 74 lb. of powder. The Gibraltar shield was destroyed with ten rounds from the 9 in. gun. The Millwall shield, however, with Mr. Hughes' hollow stringer backing, stopped a 600 lb. Palliser shot, which imbedded itself 22 inches in the iron and backing of the structure, which it drove bodily back 6 ft., notwithstanding the heavy timber struts by which it was secured. Four rounds were fired, the first shot striking the upper proper left corner of the port jamb, on the 6 in. face plate portion of the shield, gouging out a circular notch and glancing through the porthole. The second shot struck the lower proper right corner of the porthole on the head of a 3½ in. bolt upon the 9 in. plated portion, where the hollow stringers are filled in with teak. The shot penetrated 15 in. into the shield, and broke up into fragments, which were scattered in front of the target. The stringers stood well and the skin remained intact, some rivet-heads only being sheared off. A piece of the front plate, however, was broken off, the shot taking effect between two previous shot-holes. In the third round the shot struck the upper 6 in. plate on a level with the top of the porthole. The shot-hole measured 12 in. in diameter and was 32 in. deep, and consequently 7 in. short of total penetration. Two previous shot-holes were in close proximity, and three fissures, having a gape of about 1½ in., extended from them across the plate. Behind the shield the whole body of the rear girders was thrown back at top about 4 in., breaking away the inner flange of the proper left strut, and casting the fragment, about 4 ft. 6 in. long, with the wedge piece, and a large length of a 3½ in. bolt

attached, nine paces to the rear. In this portion the hollow
stringers have no infilling. Another 3¼ in. bolt was stretched
4½ in. by the bulge, without breaking. The fourth, and last shot
fired, struck the lower proper right end of the 9 in. plate, behind
which the backing is filled in with teak. The shot penetrated
the target 22 in., leaving 3 in. of its rear projecting from the face
of the plate. The projectile split into four nearly equal parts,
and the general result was—as we have already mentioned—that
the right wing of the shield was knocked back 6 ft. from its previous line, and the left wing slewed forward 1 ft. The wood
piling and iron blocks, by which the shield was stayed, were
carried away, and the foot-plate of the shield was torn away by
the shearing of the rivets. The covering strip was thrown away
from the side of the shield 8 ft., the ends of the hollow stringer
backing being exposed. No damage of the slightest importance
occurred to the rear of the shield, which has certainly withstood,
in a manner totally unprecedented, the tremendous pounding to
which it has been subjected. Whether we look at Mr. Hughes'
hollow stringer principle, the Millwall iron used in the construction of the shield, or Mr. Parsons' bolts, which held the parts so
splendidly together, we can only find good cause for congratu-
lation. The results of this trial will certainly constitute an era
in the practical history of guns and targets.—*Mechanics' Magazine.*

THE NEW COLONIAL DEFENCE SHIP.

THE first Monitor constructed under the auspices of our
Admiralty has been built in Messrs. Palmer's yard, at Jarrow-
on-Tyne. This vessel, the *Cerberus* is intended for the defence of the city of Melbourne. The colonial authorities have
agreed to pay one-fifth of the first cost of the vessel, to furnish
her armament, and to get her navigated to Melbourne at their
own expense. The vessel in question is an example of the breastwork principle, which, we believe, was first publicly propounded by the Chief Constructor of the Navy in 1867, at the
Institution of Civil Engineers. The reason for adopting this
plan is to protect all the principal openings in the decks to a
height of 10 or 12 ft. above the water, thus adding greatly to
the security of these low-decked vessels. The dimensions of the
Cerberus are as follows:—Length, 225 ft.; breadth, 45 ft.;
draught of water, 15 ft. 6 in.; and burden, in tons, 2,107 (builders' measurement). She carries two turrets, one at each end of
the breastwork, each turret furnished with a couple of 18-ton
guns. These turrets stand about 5 ft. 6 in. above the breastwork, and can be turned either by manual or steam-power. They
are protected with 10-inch solid armour in wake of the ports,
and with 9-inch elsewhere. Next to the armour comes the usual
teak backing, and inside all are two thicknesses of half-inch

plating. The tops of the turrets are covered with half-inch
plating worked upon beams. The breastwork stands upon the
midship portion of the upper deck, and is 112 ft. long, 34 ft. wide,
and 6 ft. deep. Its ends, which are circular, are protected with
nine-inch armour, and the remainder of it with eight-inch, the
top being covered with two thicknesses of half-inch plating.
Within the breastwork are the funnel, air-shaft, turret machinery,
steering wheels, cooking ranges, and the hatchways leading
directly to the ammunition store and the lower deck. The upper
deck, outside the breastwork, is covered with two thicknesses of
three-quarter-inch plating, and is pierced for three skylights.
Each of these skylights is surrounded with armour, and is provided with an armour-plated cover for use in action. The
amount of free-board given to the *Cerberus* is 3 ft., the side
being covered with armour from stem to stern to about 4 ft.
below the water-line. This armour is in two strakes, the upper
one being 8 in. thick, and the lower one 6 in., both tapering
towards the extremities of the vessel. Behind the armour comes
the usual teak backing and skin plating. The armour and backing are worked upon a recessed armour-shelf, so as to make the
line of the side a continuous curve, as in broadside vessels, and
contrary to the practice adopted in the American monitors and
the *Glatton*. Above the turrets is a hurricane-deck, extending
the whole length of the breastwork, with the interior of which
it communicates by water-tight iron trunks. All openings in
this and the other decks are protected by iron gratings from vertical fire. A pilot-house is erected on the top of the breastwork,
and extends 4 ft. above the hurricane-deck ; its sides are protected with 9 in. armour, and its ends with 8 in. The frames of
the vessel are of the combined longitudinal and bracket description now usually adopted in the Royal Navy. She has also a
complete double bottom, and numerous transverse water-tight
bulkheads.

The *Cerberus* will be propelled by twin screws, driven by
engines of 250-horse power (nominal), constructed by Messrs.
Maudslay. She is expected to attain a speed of about eight
knots, high speed not being considered essential in a vessel
intended for such duties as she will have to perform. The turret guns will have an uninterrupted arc of fire, since the vessel
is entirely devoid of rigging of any description, and even the
boat davits can be lowered out of the way. The arrangements
for ventilation are very perfect. Fresh air is admitted by means
of the air-shaft and of the openings in the deck over the stoke-
hole. There is a fan at the bottom of the air-shaft, and another
at the fore-end of the stoke-hole, both worked by steam. These
will drive the air through main pipes running fore and aft, from
which branches are led wherever required. One peculiarity of
this vessel now-a-days is that she is not a ram, properly speaking,
but still, as she has a perfectly upright stem, she could be used

for ramming. The absence of the projecting prow ought to be
an advantage, so far as speed is concerned, since there can be no
doubt that the enormous wave which it raises somewhat impedes
a vessel's progress. We must not forget to add that the *Cerberus* is fitted with that valuable invention the balanced rudder.
—*Pall Mall Gazette.*

OUR IRON-CLAD SHIPS.

THERE has been read to the Society of Arts a paper on "The
Turret and Tripod Systems of Captain Coles, as exhibited by
Admiral Halsted in Paris." The gallant Admiral explained at
very great length the principles of the system, and advocated a
system by which the turrets could be used on a vessel having an
upper deck, so that the sailing power of our navy might still be
kept up. The class of ships he expressed a desire to see constructed was exhibited in model—a vessel of great length and
fine lines, and with an upper deck for sailing power. He stated
that his views had been concurred in by most eminent men in
naval architecture, and by men who well understood what was
wanted for naval warfare under existing circumstances. Mr.
Scott Russell said the great difficulty constructors, artillerists,
and engineers had in founding a good vessel was in finding a
sailor to say what he did want. Admiral Halsted, however,
had said pretty distinctly what he wanted, and had demanded
for the navy the best possible turret-ship. In his (Mr. Scott
Russell's) opinion, the group of ships designed, and the plans of
which had been laid before the meeting, was excellent; but he
thought the broadside guns might be more protected by the
vessel being made broader. He urged upon the consideration
of the meeting whether it was not well to build the turret class
of vessels small, leaving "bigness" to the broadside class; but,
whichever way it might be, he hoped our sailors would find some
means of testing the vessels and the guns without having a war
for the purpose. Mr. E. J. Reed said naval officers generally
expressed an opinion that the large vessels of the class advocated
by Admiral Halsted were undesirable; and, in entering upon a
discussion of the details of construction, he said the estimates
and proportions made out by unofficial and irresponsible persons were always open to correction. The turret and broadside
vessel proposed by the Admiral, he said, was too long and too
unwieldy, and to show how undesirable it was to adopt any plan
or system of ship as final, he mentioned that he had had before
him the plans of the "Monitor" class of vessels proposed by a
Russian admiral, and which were so designed that they could
come out of the Baltic and sink such vessels as were here proposed. Before we went too far, we should see what the *Captain* and *Monarch* turned out to be. Sir F. Grey urged that
we should be careful, and not go too fast in the construction

of any class of ships. Mr. Lamport said we wanted two classes
of vessels—one for the defence of our shores, the other for service in aggression. Admiral Sir Edward Belcher said that ships
of great length were unmanageable under canvas, and he did
not admire the "scaffolding" appearance of the deck over
the turrets. Then such a ship as the *Sovereign* could not be
sent on a long voyage, as we should desire to send a ship-of-
war in time of war. As to submarine artillery, he advocated
instead a Nasmyth hammer being below the water-line of a ship,
and worked with steam, so that it could "punch" the antagonist
when in close quarters. The Earl of Hardwicke said the first
considerations were to float these turret-ships, and give secure
platforms from which to work the guns. As, he said, it had
been allowed that iron could be penetrated, what was the use
of the iron? That was a question yet to be settled; and whatever
other class of vessels was built, we must have the line-of-battle
ships to act in masses; for, after all, the fighting would be in
masses, as it ever had been, whatever sort of ship was used.
Then, as to the gun, the view he had was that the moveable platform in the centre of the ship was the best mode of fighting;
and he added, amid cheers, our men would want no turrets, for
they could fight better, as they had ever done, in the open. It
was frightful work firing in a confined space, as all who had had
experience in the old men-of-war knew, and firing in these turrets would deafen the men. What was wanted was a class of
vessels which could cruise in safety, have uniformity of speed,
or nearly so, and with ready means of working the guns from a
centre platform.—*Times*.

Mr. E. J. Reed, in a lecture on Iron-clad Ships, has ably illustrated the subject in a popular manner at the Sailors' Home.
Mr. Reed commenced by describing the floating batteries which
were need against the Russian forts during the Crimean war, and
then explained the various steps which had since been taken by
the Admiralty for increasing the naval defences of this country.
He minutely described the peculiarities of each successive ship
which had been devised and launched under his own direction,
and entered especially into the question of their vulnerability or
otherwise. Comparisons were also instituted between French
and English vessels of war, which resulted in a demonstration of
the superiority of the latter over the former. The various kinds
of backing, and of inner skins employed in the construction of
iron-clad ships, and of experimental targets, for testing the
penetrative powers of the Whitworth and other guns, were
explained, and then the lecturer advanced to a consideration of
the fleet of the future. Although carefully guarding himself
from divulging the plans of the Government, it was quite apparent that Mr. Reed's inclination was towards the introduction of
shorter ships, thicker armour-plates, increased steam-power, twin-
screws, and greater coal stowage.

The Chief Constructor of the Navy has likewise read a paper before the Royal Society "On the Relation of Form and Dimensions to Weight of Material in the Construction of Iron-Clad Ships." It is an important question, for experience enough has already been gained to prove that the proportion of length to breadth in a ship, and the form of her water-lines, should in a great degree depend on the weight of the building material. The proportions and form of an armour-plated ship should be very different from those of a ship without armour; and any increase in the thickness and extent of the armour should be followed by a diminution of the length of the ship, and an increase in the breadth and of fulness in the water-lines. Influenced by these considerations, Mr. Reed designed the *Bellerophon*, a short, handy ship in comparison with some other iron-clads, which, like the *Affondatore* at the Battle of Lissa, would have to run out of action in order to turn round. A ship that will "come about" in less than two minutes will be a much more trustworthy ship in these days of quick movements than one that requires seven minutes and extra hands on the steering-tackle for the operation. A long ship weighs 12,570 tons, reckoning only the hull; a short ship, 7,576 tons. But the latter has been reduced in the *Minotaur* to 7,100 tons, and in the *Bellerophon* to 4,460 tons; and on this there hangs a conclusion which will perhaps make an impression on the British taxpayers, if not on the functionaries who shape the estimates—namely, that taking the cost per ton at £55, which is the average cost for the hulls of armour-clad ships, the saving made by adopting the new ship of the *Bellerophon* type would amount to £283,250!

<hr />

CIRCULAR IRONCLADS.

Mr. JOHN ELDER, has read, at the United Service Institution, a paper in which he discussed his original theory of constructing circular iron ships with immersed propellers. Having adverted to the comparison of the respective merits of the turret and broadside systems as at present adopted in the Navy, the lecturer said that in the opinion of almost all who were competent to judge, the former was the preferable one, inasmuch as the vessels built on its principles were less vulnerable than broadside ships, and were also capable of carrying heavier guns. He proposed to develop that system more fully by making the turret carry itself—that is to say, by extending its diameter, and by arranging its bottom so that it would represent a small section of a large sphere. To make the subject intelligible to non-professional readers, we may observe that the hull of a ship built on Mr. Elder's plan, as illustrated by his models, would be somewhat similar in shape to a saucer with a flat covering, or to a small section of an orange the rind of which would represent the skin of the vessel. Mr. Elder stated that a ship of this shape would draw only about half the water which would constitute the draught of an ordinary-shaped vessel of equal displacement, though the midship section would, of course, be much greater. At first sight it might appear impracticable to drive a vessel so constructed through the water at any considerable speed, but the lecturer's own experience, he said, had afforded ample evidence that such a belief was erroneous. He had made two models—one of an ironclad of the most modern design, and another of a vessel built according to the plan he was advocating—and he had found, after repeated experiments in smooth and rough water, that the circular model required no more power to propel it than the other. It was proposed, for the purpose of propulsion, to employ hydraulic machinery in vessels built on what we may now call the circular system, similar to that used in Her Majesty's ship *Waterwitch* — the suction-pipe and water-jet being in a line with each other, and it was estimated that there would be no difficulty in obtaining a speed of twelve knots an hour, if, indeed, the circular vessels would not attain to a speed commensurate with that of our fastest ironclads.

The machinery for manœuvring the vessel is also very ingenious. On each side of the suction-pipe and of the delivery pipe or water-jet, two other pipes were placed, curved at their outer ends in opposite directions, and through these the water might be taken in and given out instead of being received and delivered through the straight pipes referred to. By this means the vessel might be made to revolve in any direction, and the several guns, which were placed at frequent intervals round the vessel, could each in its turn be brought to bear on the same spot. Supposing the case of a ship represented by the mode used for illustrating the lecturer's theory, carrying twenty-six guns, each throwing 600 lb. solid shot, and revolving once in a minute, the entire fire of the twenty-six guns could be directed against any particular object, or series of objects, within the time occupied by one revolution of the vessel. The method of steering—and this was the particular in which the greatest difficulty suggested in the subject under discussion—was somewhat complicated, though readily intelligible to those who inspected the models exhibited. A centrifugal pump or turbine was made to revolve by means of a rack and pinion—the shaft carrying the pinion having at the other end a similar arrangement driving another rack, which again was connected with a revolving pilot-house. Whenever the pilot-house turned, the turbine also moved, and the water-jet was consequently similarly influenced in its direction. A curious and valuable result was thus obtained. By having a "look-out," or line of sight, in the pilot-house, corresponding with the suction-pipe of the turbine, the person in the pilot-house, while steering the ship, would have his back to the water-jet, and would "look-out" in a line with the suction-pipe. The ship would thus be caused always to travel in the corresponding direction, or, in other words, the steersman would only have to continue looking at any particular place in order to direct his vessel towards it as a destination. The lecturer then proceeded to indicate the other peculiarities of the ships constructed under his plan, and dwelt particularly on their remarkable stability, enabling them to carry the heaviest guns on a high tower, if required, to fire into forts or over high projections. From the great angle of the sides, they could only be hit by ricochet shot. The great capacity of the ships for stowage was also illustrated by the fact that they could carry sufficient coals for twenty-five days' consumption. It was also suggested that as there were now now several mounted mortars for the carriage of which there were no ships suited at the present time, they might be advantageously employed in vessels constructed on Mr. Elder's theory, which would, it was asserted, be admirably adapted for the purpose.—*Times.*

NEW DUTCH ARMOUR-CLAD RAM.

MESSRS. R. NAPIER & SONS have launched from their shipbuilding yard at Govan, near Glasgow, an Armour-clad Twin Screw turret Ram, named the *De Buffel*, built for the Dutch Government. This new war-ship is of about 1,473 tons, builders' measurement. She is 205 ft. in length by 40 ft. beam, and 24 ft. deep. Her sides are plated with armour 6 in. thick, backed with 10 in. of teak on an inner skin, right forward and aft, extending 3 ft. below, and 2 ft. above the water-line, and thus protecting the most vulnerable part of the vessel. The main deck consists of 6 in. of teak on a 1 in. plate. The wall on the main-deck round the base of the turret, which is constructed on Captain Coles's principle, is composed of " eight armour backed by twelve " teak on an inch inner skin. The armour and backing on the turret are similar to those on the wall. The *De Buffel* is to carry two 300-pounder 12½-ton Armstrong guns in the turret, and four smaller broadside guns on the main-deck. Her guns train right round the circle, with the exception of a few degrees on each side the keel aft, the funnel preventing the circle from being completed. The turret may be worked by steam, under the care of one man. Accommodation is provided on the main-deck for the officers and crew, whose comfort has been considered in every respect in the construction of the vessel. It is expected that the *De Buffel* will attain to a speed of about 13½ knots per hour. Her engines, also constructed by the Messrs. Napier, are of 400 nominal horse-power, fitted with surface condensers, super-heaters, &c. The *De Buffel* is the second ironclad for the Dutch Government, which has issued from the works of Messrs. Napier, within three weeks, the *D Tijger*, a monitor, having been launched within that period.—*Mechanics' Magazine.*

"WARRIOR," "MINOTAUR," AND "BELLEROPHON."

THE Controller of the Navy reports the result of a trial of the
speed of these three ships at the measured mile, and for six con-
secutive hours' steaming outside the Isle of Wight, in fine weather.
The speed of the *Minotaur* at the mile was 14·411 knots per
hour, but at the six hours' trial only 14·165; of the *Warrior*
14·079 and 13·936; of the *Bellerophon* 13·874 and 14·053.
The indicated horse-power in the *Minotaur* declined from
6,702 at the mile to 6,193 in the six hours' trial; in the *Warrior*
from 5,267 to 5,092; but the *Bellerophon* increased the horse-
power from 6,002 at the mile to 6,199 at the six hours' trial,
showing that if the expected horse-power were realized, the calculated speed of the ship would be assured. The *Bellerophon*
had the disadvantage of having been twenty-one days out of dock
when tried for six hours, or twice as long in the water as the
other two ships, at a time of year when the growth of weeds is
very rapid. The Controller considers that these experiments prove
that, with good coal and good stoking, there is but little difference between the result of a trial at the measured mile, and one
lasting for six hours on the open sea ; that the bad performances
of the *Warrior, Minotaur,* and *Bellerophon*, during their previous
cruise, arose from causes over which the designers of these ships
had no control, and that any judgment on their qualities dependent on such records would have been illusory and misleading.

THE MONCRIEFF CARRIAGE.

THE experiments made at Shoeburyness with this invention
have an importance far above any that attaches to the question
of superiority between one iron shield and another. If Captain
Moncrieff's invention is adopted—as it must be, sooner or later
—iron shields for ports and embrasures become superfluous, because embrasures and ports will cease to exist. The weak spots
in our forts will be removed, and our gunners will work their
guns behind parapets or walls impervious to the heaviest projectiles ; while, at the same time, they will reap great advantages in
facility of laying their guns, and in largely increased lateral
range. Nor is this the only benefit to be derived from this remarkable invention. Ships, as well as forts, may be improved,
and the advantages of a turret-ship be obtained without a turret.
And on shore, guns may be mounted where they can sweep the
land or the sea with their fire, and yet afford no mark for the
enemy's shot, though no parapet has been raised for their protection.

Such statements as these may seem paradoxical, but they contain no semblance of exaggeration. An entirely new principle
has been introduced, and it brings with it entirely new conditions
of warfare. The labours of thirteen years, which Captain Mon-

crieff has expended on maturing his system, are now meeting their reward in the completely successful development of an invention that will add tenfold to the strength of England's national defence.

As far back as the time of the Crimean campaign—in which Captain Moncrieff, an Edinburgh artillery militiaman, took part as a volunteer, in order to see the actual conditions under which guns are worked in a siege,—he began to design lifts for guns, with a view to obtain a system of firing over a solid parapet, while preserving free lateral range, and neither exposing the gun and detachment, nor involving the labour of raising and lowering the piece. In the course of his work it occurred to him that in the recoil of the gun—that hitherto untamable and destructive enemy—he had an agent suited for his purpose. He has tamed that fierce and violent force; he has brought it to do his bidding and to be a useful servant, supplying the place of many men.

It is difficult to explain the action of such a machine as Capt. Moncrieff's invention without a drawing. In the natural or firing position, the gun pointing over the parapet is supported on a "carriage," consisting of two triangular brackets, attached in front to two "elevators," and having in rear two rollers, resting on the front of the slides of the platform, which are inclined downwards to the rear. The gun is kept in this position by a counter-weight, sufficient to balance the weight of the gun, placed between and attached to the elevators. When the gun is fired, the carriage runs backwards down the slides of the platform, pulling the elevators. These roll backwards on the bottom of the platform, causing the gun to descend in a curve, while at the same time the counter-weight rises (at first with increasing velocity). The periphery of the elevators is circular at the part which is first rolled back on the platform, and the centre of gravity of both the gun and counter-weight together is also the centre of the circular part of the periphery. Hence, the common centre of gravity of the gun and elevators travels backwards in a horizontal plane. Soon, however, the elevators pass off the circular arc on to the greater curve, and the leverage in favour of the counter-weight goes on in an increasing progression, until it becomes sufficient to meet the utmost force of the recoil; and thus the recoil is absorbed. As the elevators roll back, a rack of peculiar form attached to them turns a pinion, giving motion to a large toothed wheel outside the elevators. When the gun has recoiled as far as it will go, it is held in position by a self-acting pawl catching this toothed wheel. It is now under cover, entirely concealed by the parapet. In this position the gun is loaded—the detachment being entirely protected by the parapet. When loaded it is only necessary to raise the pawl, and the energy of the recoil, stored up in the counter-weight, raises the gun up into the firing position. If left to itself, it would run up with some violence; so its rate of movement is regulated with the greatest ease by a friction-band, worked by one gunner. In the experiments of August, the break-wheel became damaged, and preventer ropes applied to the carriage were found to answer the purpose equally well. This had been repaired before last week, and the friction-band answered its purpose admirably.

When the gun has run up into the firing position, the necessary elevation can be given, without exposing a man, by means of a trunnion-pointer and a segmental scale on the cheek of the carriage, and the gun can be laid by a reflecting sight, also invented by Captain Moncrieff, and consisting of a small mirror fixed to the gun in front of the left trunnion, in which the fore-sight and object aimed at can be seen at the same time from below. A line through zero on the fore-sight, and parallel to the axis of the gun, passes through the intersection of two lines on the mirror. The elevating screw can be worked from the front by the man looking in the reflecting sight, and the traversing can be performed by one man, also below; so that a moving object can be followed and fired at while the gun is actually in motion —not a man of the detachment being visible to the enemy. If the object is stationary, it becomes unnecessary to use any sight after the gun has once been laid; for the platform does not move when the gun is fired, and it is only necessary to make a pointer upon its foot correspond with a mark made at the first round on the circular racer round which the whole gun traverses. Excellent practice at 1,500 yards was made at Shoeburyness last week, when the gun was laid in this manner, and worked by only three men. The Government called upon Captain Moncrieff to apply his principle, in the first instance, to a 7-inch 7-ton rifled gun. He has done so with perfect success. His theory and principle are unerring; their application to greater weights is a mere matter of detail. We are virtually in possession of a system of mounting guns so as to entirely conceal gun and detachment from direct fire. Let us see in what manner it can be applied.

In the first place, as regards earthworks. Take Southsea Castle for example: the parapet is pierced with embrasures, each offering a large opening on the outside, converging like a funnel to the neck of the embrasure, and guiding the enemy's shot, while the parapet is very weak where it is cut away on each side. Thus there is great danger for the detachment, and it is considered necessary to close the embrasures by iron shields. The unsatisfactory shields for the Gibraltar works cost £1,000 a-piece; moreover, the great horizontal strain with ordinary carriages and platforms renders very strong and expensive foundations necessary, while the ordinary iron carriage and platform for a 7-inch gun cost about £500. Now take the Moncrieff carriage. It costs about £800 or £900; part of the difference is saved in foundations, for there is no horizontal strain, no friction being used to check the recoil. The embrasure is filled up with earth and disappears, the whole parapet becoming one uniform shot-proof thickness, and the costly shield is not wanted. Perfect immunity from direct fire is obtained by the gunners, and the gun which *en embrasure* had an extreme lateral range of sixty degrees, and that only at the risk of blowing the sides of the embrasure to pieces, has the range of the whole circle, or practically, over the parapet in front, of 150 degrees. In confined situations, where earthworks cannot be used, as for example, Plymouth Breakwater fort, the difficulty lies in the shields. The body of the fort is easily made shot-proof, but the shields for the ports are limited in thickness by the necessity of allowing the gun to traverse. With Captain Moncrieff's system the fort may have walls of one uniform thickness, over which the guns appear for an instant to fire, disappearing as they deliver their shot. Captain Moncrieff proposes to introduce gun-pits sunk in the natural ground, which would offer no mark of any kind to an enemy—supporting and connecting them by trenches for infantry. "If," he says, in a paper published in the Proceedings of the Royal Artillery Institution, "a few powerful guns were judiciously placed in Moncrieff batteries, connected and supported by trenches for infantry, could anything more embarrassing be imagined for ships than to receive a deadly fire from the most peaceful-looking hillocks, and when they looked for their enemy, to see no mark of his position except a cloud of white smoke passing gently to leeward, until their attention was distracted by the same phenomenon in some unexpected quarter?"—We quote this excellent paper from the *Athenæum*.

THE DARDANELLES CANNON.

ONE of the giant cannons of the Dardanelles has, we understand, with a quantity of its shot, been received at Woolwich Arsenal. This curiosity in the way of ordnance has been presented to our Government by that of Turkey, and will be placed in the museum of artillery in the repository grounds at Woolwich. The gun is of bronze, and in two parts; the hind part or powder chamber, screwing into the fore part, or shot chamber. The screw in the hind piece is a curiosity in itself. It is 1½ in. long, and has five threads about 3 in. wide and deep; the diameter of the screw is 2½ in. The hind piece itself very much resembles in form a windlass or capstan, having hand or lever holes all round at both ends for the purpose of screwing it to the front piece. The gun is without any trunnions, being intended to be laid on sleepers on the ground, as other similar guns are now placed in the batteries on the Dardanelles. It is evidently of great age, and similar to that described by Gibbon in his *Decline and Fall*, as employed by Mahomet II. at the siege of Adrianople in 1453. The powder chamber in the breech part is 6¼ ft. long and 10 in. in diameter. The bore, or shot chamber, is 25 in. in diameter. The shot for this monster are round and of white granite; they weigh 650 lb. each. The following are about the weight and dimensions of this interesting piece of artillery:—Breech part, 9 tons 18 cwt.; length, 7 ft.; front part, 8 tons 17 cwt.; length, 10 ft.—*Mechanics' Magazine.*

OLD GUNS.

THE Report of the Ordnance Select Committee upon Palliser's system of lining old cast-iron guns with coiled tubes, laid before the House of Commons, is understood to be favourable to the system, which is likely to be adopted to a considerable extent. The many acres of old cast-iron guns which encumber the ground in the Arsenal at Woolwich will be utilized, either by converting them into rifled 64-pounders and 7-inch guns, or by breaking them up and passing them through the operation of puddling. Wrought-iron has never before been made at Woolwich. Bronze, or gun-metal, field-pieces formed in old days the staple manufacture. During the reign of Colonel Wilmot over the gun factories before the Crimean War, a foundry was erected for casting iron guns. Experiments were set on foot which might have resulted in obtaining cast-iron ordnance of as good material as that possessed by the Americans. But the Armstrong wrought-iron construction pushed cast metal, both iron and bronze, out of the service as far as guns are concerned. It is still a moot point to what extent converted ordnance can be used with battering charges if the calibre is large. In the meantime, it is certain that a valuable addition can be made to the armament of the country by the use of Palliser guns up to seven or eight inches calibre.

COATING CAST IRON.

IN Germany objects of cast-iron are coated with copper by the following process:—The surfaces, after having been well cleaned with a brush and with diluted muriatic acid, are steeped in water slightly acidulated. The articles are then placed in a bath composed of 25 grammes of oxide of copper, 176 grammes of muriatic acid, half a litre of alcohol, and quarter of a litre of water. The copper is equably deposited over the surfaces, the alcohol reducing the rapidity of deposition, and thus giving greater density to the copper film. These coppered objects may be zinced by placing them in a bath composed of 10 grammes of chloride of iron and one litre and a half of alcohol, and in contact with pieces of metallic zinc. A coating of antimony may be given to the coppered objects by mixing chloride of antimony with alcohol and adding muriatic acid until the mixture becomes clear. In this bath the objects may be left for three-quarters of an hour. For silvering glass or vases four solutions are prepared: the first composed of 10 grammes of nitrate of silver in 100 grammes of water; the second, an aqueous solution of ammonia of 0·98¾ density; the third of 20 grammes of caustic soda in 500 grammes of water, and the fourth of 25 grammes of sugar in 200 grammes of water, to which is added a cubic centimetre of nitric acid and 50 centimetres of alcohol. Mix the three first solutions and then add the last, when the deposition of silver will take place.— *Illustrated London News.*

PAINTING ZINC.

IT is a difficult matter to get a coat of paint to adhere well to zinc, which rapidly oxidizes when exposed to air and moisture; and, as most engineers know, galvanized iron goes very quickly when once the covering of zinc has decayed. Many means have been tried to obtain the firm and close adherence of paint to zinc. The last we have met with is due to Dr. Bottger, who professes to have completely succeeded. He makes a solution of one part of chloride of copper, one part of nitrate of copper, and one part of chloride of ammonium, in sixty-four parts of water and one part of commercial hydrochloric acid. This solution acts as a sort of mordant. It is paid with a wide brush over the zinc, which immediately becomes of a deep black colour, forming, according to the doctor, a basic chloride of zinc, and what he calls an amorphous brass. The black colour changes in the course of twelve or twenty-four hours to a grey, and upon this grey surface any oil-paint will dry and give a firmly adhering coat. Summer heat and winter rain will have no effect in disturbing this covering, which affords complete protection to the zinc.—*Mechanics' Magazine.*

WIRE ROPE.

ACCORDING to the Pottsville *Miners' Journal,* the largest, longest, and heaviest wire-rope ever made in one piece, for an inclined plane, has been completed at the works of John A. Roebling, of Trenton, N. J. It is intended for Plane No. 2 of the Lehigh Coal and Navigation Company, located at the north slope of the Wilkesbarre Mountain, in Hanover township, Luzerne County, for the purpose of lifting coal out of the Wyoming Valley. The load hauled up each trip consists of ten coal-cars, weighing thirty tons, laden with fifty-five tons of coal, making a total weight of eighty-five tons. Speed of rope, nine miles per hour; inclination of plane, 14 ft. 8 in. in 100 ft. The length of this wire rope is 3,700 ft.; its diameter over 2¼ in., and weight twenty tons net. It is stated that ropes of double the above size and length can be made by the powerful machinery used in laying up this one.

STOVES AND OVENS.

BLACKLEAD is a great institution in this country, and probably few but cooks and housemaids would care to see its use diminished.

It certainly has its recommendations, but it can hardly be said to be ornamental while it entails an immense amount of labour on our servants. In Germany, where a stove and sort of kitchen-range is continually to be found in the common sitting-room of a respectable family, the unsightliness seems to have been felt, and a suggestion has been made to do away with the blacklead, and paint the stoves and ovens. Oil-paint, of course, cannot be em-ployed, but water-glass (silicate of potash) coloured with pig-ment to match the paint of the apartment, is the material recommended. Before this is applied it must be thoroughly cleansed from grease, and all rust-spots must be rubbed off with a scratch-brush. Two or three coats of the paint may then be put on and allowed to dry, after which the fire may be lighted without fear of injury to the colour, which may, indeed, be heated to redness. Grease or milk spilt over the paint has no effect upon it, and it may be kept clean by washing with soap and water. Dutch ovens and like utensils may also be coated with the same materials, and the labour spent in polishing be saved. A good coating of the paint, the author says, will last a year or two.—*Mechanics' Magazine.*

NEW BREECH-LOADING RIFLE.

AT Wormholt Scrubs, the rifle practice-ground of several Lon-don Volunteer regiments, a new breech-loading rifle, especially designed and adapted for military use, has been exhibited and tested before many practical riflemen—not for the first time—and with perfect success. The rifle is that which, under the name of the *Walker-Money* has attained a wide renown in a very few months, although only a few specimen weapons have been manufactured. The patentees are Mr. Mowbray Walker and Lieutenant-Colonel Money, and the principle for a breech-loading weapon put forward by these gentlemen has been accepted by practical riflemen of all ranks of life as being one of the simplest in construction, perfectly safe in action, economical, and one of the most rapid in firing. The appearance of the weapon is, in barrel, stock, and sighting, similar to the Enfield, only that this weapon has neither nipple nor hammer, and when the breech is closed, the *Walker-Money* might, if the look-sight were turned from the spectator, be taken for an Enfield with the nipple off and the lock out. With regard to the breech principle itself, there is this most remarkable difference between it and other contrivances—that the whole of its work is done without the aid of a single spring of any kind—and so quickly done, too, that 20 rounds in a minute can be fired with aiming, and 26 without aim-ing, and this with but a few hours' drilling. The breech is opened by the raising of what is called "a balance breech-block"—a block about 6 in. in length, one end of which enters the breech, while the other rests in the stock, the block itself working in its

D

centre upon a stout "pin." The raising of the end of the block depresses what is called the "extractor," a small bent arm shaped like a finger slightly bent, which works upon a "pin," and when the one end is pressed down by the breech-block, the other end thrusts out the exploded cartridge case, leaving the passage perfectly clear for the insertion of a new cartridge. The firing without the fall of a cock is another feature, and this is done by pressing a trigger to release a needle, which penetrates the cartridge and causes the explosion.

Some of the experiments which have been carried out with this rifle in all weathers, and in the presence of riflemen at different ranges, are worth relating.

Mr. Walker has taken the barrel and breeching out of the stock, and having cut and damaged a cartridge so as to render it defective and open to an escape of gas, he has closed the breech and fired the weapon without the stock, and dirt and sand having been thrown into the breech, the weapon has still been fired without any difficulty; and in the case of cut cartridges without any breech escape of gas. It is perfectly obvious that the breech cannot be thrust back by the explosion—a charge, and a most serious one, which has in several instances been made against other systems. Nor can a breech escape of gas occur, as has been reported to have occurred with the Snider. The cartridge used is that called the "central fire," of Mr. Daw, Threadneedle-street, which gained the £100 prize given by the War Department; and this cartridge, it is said, is made at a much cheaper rate than the *Boxer*, and shoots better. The advantage of the Daw over the "improved," the *No. 5 Boxer*, was apparent to every one who saw the two yesterday. The *Daw*, being perfectly metallic, cannot be affected by the weather, while the *Boxer*, being covered with brown paper, is not unlikely to "stick" on a wet day in the breech, or to clog in one place through the paper scraping off. Though the rifle was given into the hands of men who had not handled it before, the centre was kept at 200, 300, 400, and 500 yards, notwithstanding that there was a strong wind, what riflemen, who class winds between "No. 1, a gentle wind," and "No. 6, a hurricane," would class as a No. 4. Ensign Hyam, of the 36th Middlesex, and Sergeant Ault, fired with the weapon, and made quite as good shooting as they thought it was possible to have made in that light and wind, as with their well-known Enfields. It may be added that the breech principle is applicable to any rifling and to any bore, but the Enfield three-grooves and Enfield five-grooves have generally been used. These advantages of the gun are among others presented by the patentees. With regard to construction and economy, they say, "No portion of the breech mechanism comes apart or is separated from the gun. The area of motion, and consequently the friction, is the smallest possible, being simply the surface of the cartridge base. This rifle can be manufactured at a less cost in material, time, and

labour than any other—namely, in large quantities, London-made,
and London-viewed, at 52s. per arm." With regard to safety
they urge :—" The detonation of the cartridge cannot by any possibility occur unless and until the breech is entirely closed. The
very act of firing closes the breech if left open by accident. Accidental discharge is provided against by a very simple contrivance
rendering the trigger immovable. The gun can also be carried in
perfect safety, loaded, and uncocked, and can be re-cocked without moving the cartridge. The danger of a defective cartridge
bursting is entirely avoided by the fact that the breech-block,
being held firm on all sides, is absolutely immovable at the
moment of discharge." Also, that " the breech-block may be removed in a few seconds, and the whole gun can be taken to pieces
at once by the removal of the one single screw in the arm. So
simple is the gun, both in construction and action, that it is next
to impossible for any recruit, either by clumsiness or by wilful
intention, to put it out of order."—*Times.*

STEAM ROLLING-ENGINE.

A MAGNIFICENT upright Steam Rolling-engine has been constructed at the South Brooklyn Engine and Boiler Works,
United States, for the Trenton Iron Company. An immense
cast-iron open pedestal sustains the cylinder steam-chests, and
connections, the connecting-rod and crank-work inside the column near the bottom. The fly-wheel and spur-wheel are secured
to the shaft by three massive feathers forged on the shaft, the
intervals between which and lugs cast in the interior of the hubs
are filled with hard-wood wedges, intended to receive and diminish the jar and concussion to which an engine employed for
driving rolls must be subjected. The fly-wheel is 22 ft. diameter, and weighs nearly 25 tons. The cylinder is 46 in. diameter,
and the stroke 40 in. The total weight of the machine is 67 tons
14 cwts. The engine is calculated to make 75 revolutions per
minute, at a steam pressure of 80 lb., and is, although extremely
compact, of 1,200 estimated horse-power.—*Mechanics' Magazine.*

NEW STEAM-ENGINE.

AN extremely simple Steam-engine, in which piston, crank,
steam-chest, &c., are dispensed with, has, it is said, been invented
by Mr. Benjamin Franklin, of Westmoreland, Penn., U.S. It
depends entirely upon centrifugal force; friction is almost
entirely overcome, and will produce 1,500 revolutions per minute,
with one-fourth the steam usually required, although the same
amount of horse-power is developed. This centrifugal steam-
engine condenses almost all its steam (which in itself is a great
saving), whilst from its simplicity it can be constructed at one-
fourth the ordinary cost, and is not liable to get out of order.

AMMONIA MACHINE.

A MACHINE has been invented by M. Fromont to be driven by a mixture of steam and ammoniacal gas. Strong liquid ammonia is used in the boiler, and the vapour generated is said to be a mixture of at least 80 parts of ammoniacal gas and 20 parts of steam, so it may be fairly called an Ammoniacal Engine. The principal recommendations of ammonia, when applied as a motive-power, consist in the small amount of fuel required and the short time it takes to get up the steam, so to speak. The economy in fuel is very considerable, being about one-fourth of that required to generate steam alone. As regards the boiler, it may be of either of the ordinary forms, the only complete novelty being the apparatus for condensing the steam and ammonia. The gas disengaged (about six atmospheres at 110 deg. centigrade with an ordinary solution of ammonia) does its work in the cylinder and then escapes into the tubes of a condenser, where the steam is condensed and the gas is cooled. The gas then meets with water from an injector, which dissolves it, and the solution is carried on into a vessel called the "dissolver," from which it is pumped back into the boiler to do its work over again. The water for the injector is taken from the boiler, and is cooled before meeting with the ammoniacal gas by passing through a worm surrounded with cold water. These arrangements are necessarily a little complicated, and could not be fully understood without drawings. It is, however, satisfactory to see that an ammonia engine is a possibility, and thus power is obtainable where fuel and water are both scarce.—*Mechanics' Magazine.*

WATER VELOCIPEDE.

AN ingenious application of the principle of the velocipede to water-locomotion may now be seen on the Lake of Enguien, near Paris. The form of this new species of naval construction, says *Galignani*, is that of the twin-ship tried some years back on the Thames, the motive-power being placed in the middle instead of on each side, as in ordinary paddle-steamers. A pair of hollow water-tight pontoons, about 12 ft. long, 10 in. wide in the thickest part, and tapered to a point at each end, are fastened together about 20 in. apart by transverse bars near the extremities. In the centre is placed the seat, rising about 2 ft. above the water, and supported by iron rods. In front is the paddle-wheel, about 3 ft. in diameter and 8 in. broad, provided with sixteen floats, the axle turning on stout iron uprights, and the rotary motion being obtained from cranks worked by the feet. This little vessel is steered by rudders at each of the sterns, and moved by lines. The pontoons being made of thin mahogany planks, the whole construction is very light, and glides along with astonishing rapidity. This water velocipede, having been built as a first ex-

periment, is no doubt susceptible of improvement in some of its
details, but the principle may be already pronounced a complete
success. The inventor is M. Thierry, an architect, of Paris.

ERICSSON'S SOLAR-ENGINE.

THE important announcement reaches us from America that
Ericsson, the eminent Swedish engineer, who has long been resident in that country, and who is well known as the inventor of
turret-ships, and the co-inventor of the screw-propeller, has
lately matured what he calls a solar-engine, the motive heat of
which is derived directly from the sun. Such a device was one
of the dreams of the early mechanists. It is well known that
Archimedes, at the siege of Syracuse, is said to have set fire to
the Roman fleet by concentrating upon it the sun's rays collected
by reflecting mirrors; and in 1615, Solomon de Caus, in a work
published in Heidelberg, propounded the scheme of a solar foun-
tain, in which the sun's rays were employed to elevate water.
But no permanent result was derived from these ancient projects
—and it has been reserved for Ericsson practically to utilize this
important idea. Ericsson states that he finds that the heating
power of the sun's rays falling on ten feet square of surface will
evaporate 489 cubic inches of water in the hour, which is equivalent, if skilfully expended, to somewhat over 1-horse power;
and he reckons that the solar radiation received by a square mile
in the latitude of Sweden would suffice to keep in action 64,800
engines, each of 100-horse power. The details of the invention
are not yet published; but the contrivance preferred for utilizing
the heat appears to resemble that of the caloric engine, in which
motive power is produced by the expansion of air by heat, only
that in the solar-engine the heat is produced by the concentration of the sun's rays instead of by a fire. In the case of the
steam-engine, the power developed no doubt also comes from the
sun, the coal being merely an intermediary, in which, as in a
coiled spring, power is stored up. But the sun, by its action on
vegetation, builds up a structure of carbon from carbonic acid, in
which act power is absorbed as in winding a spring, which power
is afterwards given out when the carbon is burned. As, however, coal or carbon is now becoming very scarce, the necessity
of finding some new source of power has long been manifest, and
tide-mills have been suggested as one important expedient. But
tide-mills are applicable only in exceptional situations; they are
both cumbrous and intermittent, and the sun appears to be
greatly preferable, not merely as a source of power, but as a
source of heat, for metallurgic and other purposes where a high
temperature is required without the oxydizing or soiling action
of a fire. No doubt the solar heat is available only when the
sun shines; but it is available at every spot on the surface of the
globe; and Ericsson proposes to generate and store sufficient

power, either in compressed air reservoirs or otherwise, during daylight, to keep his engine going during the night wherever that is important. Without this adjunct, however, there are a vast number of cases in which the solar-engine might be at once applied, as in pumping water or sewage, irrigating and tilling land, and the like. In such a country as India, too, it would be most valuable in keeping the punkahs in every house going, without any expenditure of fuel; and it would be easy during the day to pump sufficient water into a cistern on the roof of each house to keep the punkahs in action during the night. It cannot be doubted that if ten feet square of receiving surface is adequate to generate a horse power, the solar engine is destined to obtain a very wide introduction in every part of the world, especially in those parts where fuel is scarce and dear; and even in those parts where fuel is cheap, it must come into use for certain purposes, as no fuel can be so cheap as that of the sun, which costs nothing at all.—*Scientific Results : Illustrated London News.*

NEW CALORIC ENGINE.

A NEW species of caloric engine has been proposed by Mr. Wenham, in which the fuel is first turned into carbonic oxide, which, being exploded in a cylinder with a proper quantity of air, generates the power. This will be a more economical kind of engine than Lenoir's or Hugon's gas-engine, as the carbonic oxide gas may be produced at a cheap rate.

STABILITY OF MONITORS UNDER CANVAS.

MR. E. J. REED, the Chief Constructor of Her Majesty's Navy, has read to the Institute of Naval Architects a paper, namely, "On the Stability of Monitors under Canvas." Mr. Reed said that proposals were often made to mast monitors and send them to sea as first-class ships, and the purpose of his paper was to exhibit the dangers of such a course. He then proceeded, by means of diagrams, to show that, using the word "stability" in the sense, as it is used by naval architects, of meaning the efforts made by a ship to return to an upright position, it was impossible for vessels with a low freeboard, as the monitor class have, to roll under canvas without capsizing. He said the case of what barges did was not to be taken as a fitting illustration, for though barges were low down, yet their cargoes and ballast were so placed as to give a low centre of gravity, whereas, in monitors there was an upper weight by the armour, the turret, and guns, and the tendency was to increase this weight, which caused a high centre of gravity. He acknowledged that monitors might be perfectly safe under steam, but he said they would be quite unsafe under canvas. As to his opinion of the low freeboard, he said that a vessel of this class had been built under him for a harbour defence for Melbourne; but before it started he had a freeboard added, although the vessel was going to steam across to its destination. He demonstrated that whatever use should be made of the monitor class of vessels, they could not be used, as at present designed, for first-class men-of-war, and resumed his seat amid warm applause. After remarks by Mr. Hunwood, Mr. Scott Russell said he considered the statements of Mr. Reed to be very lucid, and confirmatory of the views of naval officers as to the dangers of a low freeboard, for it would be remembered the old frigates with a low freeboard were called "coffins." The experience of these ships with low freeboard was unsatisfactory, and proved conclusively that such ships had a tendency to turn upside down. Whatever had been advocated of these vessels with a heavy deck-weight, there was no question that many of them had gone to the bottom under circumstances when they ought, according to their designers, to have floated. There ought to be in all vessels a great deal of spare buoyancy, for "quite sufficient" was not enough in a vessel liable to be wounded. He had himself a great love of the cupola vessel. As a labour of love he had constructed Captain Coles's first cupolas, and he had entered heartily into the question of the construction of this class of vessel, but he did not intend to lay claim to have invented anything at all in the system, and he warned all constructors to be very careful how they made any such claims, for he assured his hearers that before anything was heard of the American Monitors, the committee appointed by the House of Commons, sitting in 1859, had every class of vessel before them, and a gentleman in Mr. Reed's position was the repository of so many official secrets as to views and plans of construction, that other constructors would do well not to question where particular plans came from, or lay a claim to any particular ideas. The speaker went on to urge that though the cupola ship would not do for a "first-rate," yet 100 of them now would be an admirable defence to our harbours. The views expressed in Mr. Reed's paper were further supported by other practical men, and the discussion was continued by Professor Rawson, Admiral Sir Edward Belcher, and Captain Selwyn. Mr. Reed, in reply, referred to the remarks of Mr. Scott Russell as to the usefulness of 100 cupola ships in our harbours, and said that, though it was a matter upon which he had no authority to speak, it must be remembered that the cost of such vessels being constructed could not be well borne by the country at the same time as the heavy fortifications were being constructed at the "tops of the hills."—*Times.*

PROPULSION OF SHIPS.

PROFESSOR MACQUORN RANKINE has read to the British Association a paper, "On a Probable Connection between the Resist-

ance of Ships and their Mean Depth of Immersion." The author,
after referring to previous researches of his own and of Mr. Scott
Russell in relation to waves, stated that the object of his paper
was to call the attention of the British Association, and especially the committee on steam-ship performance, to the probable
existence of an element in the resistance of ships hitherto
neglected, viz., that every ship is probably accompanied by
waves whose natural speed depends on the virtual depth to which
she disturbs the water, and that, consequently, when the speed
of a ship exceeds that natural speed there is probably an additional term in the resistance depending on such excess. The
author suggests that suitable observations and calculations
should be made in order to discover its amount and its laws.
Amongst observations which would be serviceable for that purpose might be mentioned, the measurement of the angles of
divergence of the wave-ridges raised by various vessels at given
speeds, and the determination of the figure of those ridges, which
were well known to be curved; and amongst the results of calculation the *mean depth of immersion*, as found by dividing the
volume of displacement by the area of the plane of flotation;
and that, not only for the whole ship, but for her fore and after
bodies separately, it being probable that the virtual depth of
uniform disturbance, if not equal to the mean depth of immersion is connected with it by some definite relation.   In
an appendix, the author gave the results of three observations
he had been able to make; and few as they had been, he thought
they were sufficient to prove the existence of waves whose speed
of advance depended on the depth to which the vessel disturbs
the water.   The connection between these waves and the resistance remains for future investigation.

Mr. Merrifield has read to the British Association a paper
"On the Necessity for further Experimental Knowledge respecting
the Propulsion of Ships." He began with a short review of
what was already known on the subject of the law of the resistance to which a ship was subjected by its having to force its way
through the water. He showed that, although there was a general consent that the resistance varied, with a certain degree of
approximation, according to the law of the square of the velocity,
yet there was abundant proof that that law was inexact, and that
the nature and causes of this discrepancy, although much discussed, were still in need of experimental determination. He
considered that the first requisite was to have the direction and
velocity of the currents of water which accompany a ship's motion
determined by actual observation. For this purpose he submitted
to the Section a rough scheme of experiments, which, however,
he wanted to get corrected by the experience of a Committee of
the Association. He suggested that a vessel of the corvette class
should (at separate times) be towed, and also driven by her own
screw, instruments being used to measure both the power em-

ployed, the speed of the vessel, and the velocity and direction of the accompanying water, at various rates of speed. He pointed out serious difficulties in ascertaining the direction of the currents of water, and was unable to suggest for this purpose anything better than direct vision. He exhibited certain instruments for assisting the eye in looking through the disturbed surface,—one of them being a common water-glass, a simple trumpet with a sheet of plate glass at the bottom, which was dipped below the water; the other being Arago's scopeloscope. He also described an electrical log, patented by M. Anfonso, of Mende. But he thought all these things required further consideration. He proposed to apply for a Committee of the Association to discuss the subject, with a view of considering what experiments might best be made; and if the Committee were of opinion that satisfactory results might be expected from such experiments, then to memorialize the Admiralty to detail vessels and officers for the purpose of carrying them out in the course of the summer.

HER MAJESTY'S SHIP "HERCULES."

THE screw (armoured iron) frigate *Hercules* Captain Lord Gilford, the latest ironclad that we have sent afloat. She is paid for in full, but she stands on the national books for " maintenance," and now that she is ready to enter upon her duties as a fighting machine in the first line of the nation's defence, it may be as well to criticize impartially her apparently good or bad qualities, and ascertain, as far as may be possible, what the Chief Constructor of the Navy has given the nation for its money.

First, then, if we put aside for the present one objectionable feature in the *Hercules* construction, or rather design, as a broadside gun-ship, it must be freely admitted that her strength both for offensive and defensive purposes excels greatly that of any other ship yet afloat and belonging to the British navy. It has been said that the *Hercules* is in reality no stronger than the *Bellerophon* which, although the strongest ship yet doing duty with the "Channel Squadron," is nevertheless incapable of resisting the shot and shell from her own 12-ton guns. If, however, we compare briefly the armour-plating of the two ships and its packing, it will be found that the *Hercules* is much the stronger ship of the two, if, as is presumed, by that term is meant the ship's strength for defence. Firstly, the *Hercules* has a belt of 9 in. armour at the water-line, a couple of inches in excess of the *Bellerophon*; secondly, all the armour at and below the water-line throughout nearly the entire length is backed by huge logs of teak, which are again backed by a second iron skin and sets of iron frames, none of which exist in the *Bellerophon*; thirdly, the 6 in. armour-belt above the water-line belt, and at the top of the graduated thicknesses of the side armour, is very much deeper on the *Hercules'* sides than on the *Bellerophon's*;

and, fourthly, the *Hercules* has an armoured stern battery for
a 12-ton gun, which the *Bellerophon* is without. All these
are obviously very substantial additions to the fighting strength
of the *Hercules* over that of the *Bellerophon*, and they also
place the *Hercules* in this respect above all comparison with
other previous or present broadside gun-ships, with the single exception of the huge Prussian ship *King William*. This ship,
which was also designed by our Chief Constructor, and was built
in an English private ship-building yard on the Thames, is more
extensively plated with 8 in. armour than the *Hercules*, but
she has no 9 in. armour, and is without the additional internal
defensive strength of teak logs and an iron skin below the water-
line. A close comparison of the *King William* and the *Hercules*
in the disposition of their armour shows the former to be the
stronger plated above the water-line, but the *Hercules* at and
below the water-line; and in a contest at sea between the two
ships it appears probable that the latter would have the longer
life afloat, the offensive powers of each being taken as equal. In
the *Hercules*, in fact, the defensive strength of armour and
backing is concentrated at the water-line; and to illustrate how
far we have gone in this concentration we may compare the
water-line defence of the *Hercules* with that of our first ironclad,
the *Warrior*, the latter consisting of 4½ in. armour backed by
18 in. of teak, and ¼ in. inner iron skin; while the *Hercules* line
is composed of 9 in. armour, 10 in. of teak, and iron skins 2½ in.
thick. Of the entire disposition of the armour-plating of the
Hercules it may be briefly described as consisting of a band of
plating entirely round the ship from 5 ft. below the water-line
to 9 ft. above it, the thickness of the band graduating, but reaching its *maximum* thickness of 9 in. in the row of plates at the
water-line. From this band rises the great central battery, and
here the plating is disposed in eight tiers, which are thus arranged:—At the water-line one width of 9 in. plates, and next
above this one width of 8 in. Above the 8 in. are five rows of
6 in., and then another width of 8 in. The battery is closed at
each end by 6 in. plated bulkheads. The weight of the ship's
side and its armour at the water-line is, approximately, 795 lb.
per square foot, and at the level of the gun-deck 582 lb. per
square foot.

Thus much for the defensive powers of the ship. Her powers
of offence next claim notice. Her armament is of a character
such as no other ship possesses, exemplifying most fully the
present tendency to substitute a small number of very large guns
with a wide range of training for a large number of less powerful
guns, with a comparatively limited range; yet it cannot, we
think, be disputed that the means taken to obtain this increased
range of training constitutes the weak point in the ship's defence, and is, in fact, the objectionable feature in the ship's design
to which reference has been made, and to which we shall refer

again presently. In this central battery the *Hercules* carries
eight 18-ton guns, which throw 400 lb. shot. Four of these
monster guns can be fought through ordinary side-ports, or four
of them can be fought out of indented ports, one at each corner
of the battery, where they can be fired at an angle of 15 deg.
from the line of the ship's keel. The guns are mounted on massive double-sided iron carriages, designed by Captain R. Scott,
R.N., the slides and rack-racers carrying them being also on the
plans of the same officer. The port-sills of this tremendous
battery of guns are about 11 ft. from the water. It is certain
that 12-ton guns are now worked with ease and security on the
broadside at sea, and Captain Scott anticipates an equally satisfactory result with the 18-ton guns of the *Hercules*. This,
however, can only be determined by actual experience at sea.
As yet these guns have been loaded and fired from their present
position. The remainder of the ship's armament on the main
deck consists of two 12-ton guns throwing 250 lb. shot, and
both fought behind armour-plated ports and firing in the line
with the ship's keel. One of these guns is mounted under the
forecastle, and looks out under the bowsprit through a port in
the ship's stem. The other is mounted in the captain's cabin,
and looks out in like manner directly over the ship's stern. The
armament of the upper deck consists of four 6½-ton guns, throwing 115 lb. shot, and fought through unprotected ports. Now,
in this arrangement of guns and gunports lies the objectionable
feature in the ship's design, and the weak point in her defensive
strength—that is, in the four indented or recessed ports at the
four corners of the central battery. Each of these indentations
is about 26 ft. in length, funnel-shaped openings, and admirable
guides for an enemy's shot or shell to enfilade the *Hercules* own
battery. Mr. Reed has, however, discarded this fatal principle in
a ship's fortification in his later-designed ships, having in them
followed the plan he pursued with the *King William*, where he
gives the guns at each end of the main battery the required angle of a line of fire from the ship's keel by mounting them in
projecting bastions or sponsons, forming the section of a circle
beyond the ship's side.

The important feature of "ramming" has not been forgotten
in the construction of the *Hercules*, the bow being specially designed and built of extraordinary strength to fit it for this purpose, while the improved balanced rudder which has been fitted
to the ship will doubtless render her very handy and quick in
answering her helm, and therefore add materially to her efficiency
as a ram. The naval action of Lissa sufficiently proved that a
ship's quickness and powers of turning in obedience to her helm
are essential to success, both in regard to ramming the ship of
an enemy and avoiding being rammed by her. The *Hercules'*
rudder is an improvement upon that of the *Bellerophon*, which
is also on the balanced principle, and is expected to yield even

more satisfactory results than were obtained in the trials made
with the latter ship, especially under sail. It was the expressed
opinion of commanding officers of the *Bellerophon* that the
large area of her rudder had been on occasions the cause of the
ship missing stays when she was put about, owing to its large
area suddenly checking the ship's way. To avoid this, the rudder
of the *Hercules* is jointed at the axis, so that it can be used
as a plain balanced rudder when under steam, and as an ordinary rudder under sail. This jointing of the enormous rudder,
like the original adoption of the balanced rudder, was a bold
mechanical expedient, but so far as it has yet been tried, at the
Nore and between the Nore and Spithead, it has proved successful in working.

With respect to the structural arrangements of the *Hercules*, she embodies all the improved methods of construction
introduced by Mr. Reed in iron shipbuilding, and which he
claims to embody the maximum provision for the strength and
safety of the ship combined with remarkable lightness of material and cost of production.

We have compared the water-line defence of the *Hercules*
with that of the *Warrior*; but in order to do full justice to
the *Hercules* as a broadside gun-ship of the latest type, it is
necessary to compare her more fully with the improved *Warrior*, the "*Achilles*, which may be taken as the early representative broadside ironclad of the British navy, as the *Hercules*,
has been taken as the latest. The *Achilles* is 380 ft. long,
58 ft. 3½ in. broad, of about 9,500 tons displacement, and of 6,120
tons tonnage. She carries a total weight of armour of 1,200
tons, the greatest thickness being 4½ in.; her armament weighs
about 300 tons, and she carries 620 tons of coals.

The *Hercules* is 325 ft. long, 59 ft. broad, of about 8,600 tons
displacement, and of 5,226 tons tonnage. She carries a total
weight of armour of 1,480 tons, the thicknesses being 0·8 in. and
6 in.; her armament weighs about 510 tons, and she will carry
600 tons of coal. Thus the *Hercules* is 55 ft. shorter, of 900
less displacement, and 894 tons tonnage less than the *Achilles*,
yet she carries 280 tons more armour and 210 tons more of
armament than the latter ship, and this with greatly increased
defensive strength. The *Achilles* carries 20 tons more coals
than the *Hercules*; but the latter has engines of the modern
type, with super-heaters and surface condensers, specially designed to economize fuel, and weighing nearly 300 tons more
than ordinary engines of the like nominal power; so that the
coal she carries would enable her to steam much farther than the
Achilles could, both starting with full bunkers.

The machinery of the ship consists of a pair of Messrs. John
Penn and Sons' trunk engines, of 1,200 nominal horse-power,
estimated to indicate six times their nominal power on the official measured-mile trial. The diameter of the cylinders is 127 in.

and the diameter of the trunk 47 in.; diameter of cylinder, effective, 118 in.; length of stroke, 54 in. Each cylinder weighs
32 tons 17 cwt. The cylinders are jacketed all over, the covers
being cast hollow for the reception of steam. The main slides
are on Messrs. Penn's usual principle. The cut-off is effected by
gridiron expansion valves, travelling on faces on the upper sides
of the slide-valve boxes. The condensers are vertical cast-iron
cylinders standing at the side of the crank-shaft farthest from
the steam cylinders. They are 11 ft. 4 in. in diameter, and the
length of their copper tubes measures in the aggregate 12 miles.
The condensing water is driven through the tubes by two Appold
centrifugal pumps, drawing water either from the bilge or sea,
each capable of discharging 60 tons of water per minute, and
worked by a pair of auxiliary engines of 40 horse-power. There
are two boiler rooms, each containing four boilers with their
stokeholes amidships. There are 40 furnaces in all, the size of
the fire-grates being 2 ft. 10 in. by 8 ft.; number of tubes, 3,600.
Length of tubes, 7 ft., and their inner diameter, 2 in. The screw
is a two-bladed Griffith, cast in metal, and weighs 23 tons 10
cwt.; the crank-shaft weighs 34 tons 16 cwt., and the screw-
shaft 24 tons. The total weight of machinery, boilers filled with
water, and spare gear is estimated at 1,090 tons, or rather less
than 3 cwt. per estimated indicated horse-power. It is almost
superfluous to say that the engines of the *Hercules* are magnificent specimens of work in metal, or that with them in motion
under full steam-power the vast apartment in which they stand
becomes the most impressively interesting part of the whole ship.

All the internal arrangements of the *Hercules* deserve the
highest praise. Officers nor crew were never better or so well
berthed on board a ship-of-war, and the arrangement of baths
and lavatories for the accommodation of all on board, from the
smallest boy 'tween decks to the captain aft, is really superb, nor
are the dispensary and sick bays less deserving praise.

In point of workmanship the *Hercules* is a marvel inside
and outside; but in this respect a most lavish and utterly need-
less expense has been incurred. The ship is none the better for
such an extravagant outlay upon her, both in the form of labour
and material, while at the same time her cost has been thereby
largely increased and the national purse so much the worse for
the process. Altogether, the *Hercules* may be summed up
briefly as being the best ship the present Chief Constructor has
yet added to our navy, but at the same time possessing the great
weakness in her defensive strength alluded to in the four indented
ports of her central battery. In general appearance the ship is
exceedingly noble in all her proportions, and even singularly
handsome for an iron ship. Consideration being given to the
proved handiness of the ship under steam (due to her comparative shortness and her balanced rudder), the thickness of her
armour, and the power of her guns, it may be safely asserted

that, notwithstanding the four objectionable ports, the *Hercules* is capable of performing any service that was formerly performed by our unarmoured wooden ships, and that she need not shrink from an engagement with any ironclad broadside ship at present afloat.—*Times.*

A NOVEL GUNBOAT.

A GUNBOAT named the *Staunch*, built for the Admiralty upon the proposition and plans of Mr. George Rendel, of the firm of Sir William Armstrong and Company, has been completed, and tried at sea off the mouth of the Tyne, with the Admiralty inspectors and a numerous party of officers on board, including some members of the Ordnance Select Committee. A correspondent of the *Pall Mall Gazette* gives the follow account of the boat :— This vessel, though wholly insignificant in appearance and cost, represents some very novel principles. She is only 79 ft. long and 25 ft. beam ; her draught of water when loaded 6 ft., and her displacement 150 tons. She has twin-screws driven by two pairs of condensing engines of 25-horse power (nominal) combined, giving her a mean speed of seven and a half knots. She carries as heavy a rifled gun as any in the navy, and to all appearance carries it most efficiently. The gun, a 12½-ton 9-inch Armstrong, is mounted in the fore part of the boat in a line with the keel, and fires through a bulwark or screen over the bow, which is cut down and plated something like that of a Monitor. Thus placed, it is easily worked in a rolling sea, and its change of position by recoil does not appreciably affect the trim of the vessel. At the same time, to provide for heavy weather, it is made capable of being lowered into the hold, so as to relieve the little vessel of its deck-load, and enable it to carry the weight as cargo. Machinery is also employed for the purpose of working the gun, by which means more than half of the ordinary gun's crew can be dispensed with. The operation of lifting and lowering is performed by simple but powerful machinery.

During the trials, the gun, with its carriage and slide, and the platform carrying them—weighing in all 22 tons—was raised and lowered in a rough sea, with the boat rolling 11 deg. each way, in from eight to six minutes. When the gun is lowered the gun-well is closed and the deck left perfectly clear, but in a few minutes the gun can be again brought up ready for action. For working the gun small capstan-heads on deck are used. These are turned by machinery from below, and instead of the gun-tackles being hauled by a large gun detachment, one man on each side has merely to take a turn with his rope round the nearest of the revolving capstan-heads. The capstan, upon his tightening his end of the rope, draws the rope for him, and on his slackening his end frees it. Thus the gun is run in or out, or trained right or left, with great ease and precision. In the same way shells are run up out of the shell-room, and other analogous manual services performed. This simple method of economizing labour has been already applied in many ways, such as for drawing trucks and moving heavy weights in railway goods stations, in conjunction with hydraulic machinery. With such assistance, during the trials of the *Staunch* the 12½-ton gun was easily handled by six men, instead of sixteen, and fired with extra charges of 56¼ lb. of powder, and 285 lb. shot. It must be observed that very little, if any, training is requisite with the gun of the *Staunch*. The vessel is so small as to be a sort of floating gun-carriage. Her twin-screws enable her to turn rapidly in her own length. Her helmsman is placed just behind the gun. The gun, therefore, can be laid by rudder right or left with far more ease and speed than any gun of similar weight otherwise mounted. During the recent trials with the engines driving reverse ways, the vessel made the full circle in her own length in 2¾ min. With both engines going full ahead, she made by the helm a complete circle of 75 yards diameter in 2¼ min. The *Staunch* is wholly unarmoured. Her strength and security lie in her great gun and her diminutiveness. And she must be considered as one of a flotilla of similar vessels. Sixty such could be built at the price of a single armour-clad frigate, and ten of them, acting different points, doubling in their own length, escaping into shallows, sheltering under forts, would drive off, or render a good account of any hostile vessel venturing to attack our harbours. Primarily they are intended for harbour defence; but the power of lowering the gun and carrying it as cargo would afford great security for these vessels at sea, and enable them to be sent from harbour to harbour with safety. The *Staunch* is now to be sent round to Portsmouth, where she is to be attached as experimental gunboat to the gunnery ship *Excellent*.

THE STEAM-MAN.

The Steam Man in New York, is a new locomotive for common roads, invented at Newark, New Jersey, and is intended to walk or run about as a man would, and draw carriages after it. The steam-man is person of commanding presence, 7 ft. 9 in. high, weighing 500 lb., measuring 200 in. round the waist, and decidedly stout in general appearance. His legs are made up of iron cranks, screws, springs, and other intricate machinery, and have a motion similar to the human extremities; his stomach is a furnace, his chest a boiler, and the smoke passes up through his head and towering hat. He bears a good-humoured countenance, with a handsome moustache, while in his mouth is fixed a steam-whistle, and a gauge and a safety-valve ornament the back of his head. He is the figure-head, as it were, of a phaeton, capable of

accommodating four persons, together with a tank to carry water
and a box for coals. The driving machinery is at his back, and
within easy grasp of the persons on the phaeton, who can start,
stop, curve, go fast or go slow at their pleasure. The inventor
claims that 20 lb. of steam will set him in motion, and 20 cents.
worth of coal work him a day. He also claims that he can accomplish a mile in two minutes on a level course, and can step
over all obstructions not higher than a foot. His engine is 4-
horse power, and the man takes 30 in. at each stride. This
steam-man, however, has not yet exhibited himself in public,
though he is promised a race down Broadway when the weather
is fit, and meanwhile his owner offers to manufacture steam-men
at short notice for 300 dollars a-piece. Whether they prove of
any practical good or not, the one at New York is unquestionably
a great curiosity, and when we have said this we think we have
said all that can be said.—*Times.*

STEAM ROAD-ROLLER.

THE Town Council of Sheffield have purchased a Steam Roller
from Messrs. Aveling & Porter, of London and Rochester, at a
cost of £900. There was formerly a reservoir on the site of part
of this street, and in comparison with other streets in the town,
the ground was soft and shifty. It was covered to the depth of
about 10 in. with the loose stony material used in road-making,
and before the ponderous machine was driven upon it was as
rough and untraversable as it well could be. The steam being
up, it was not long before the huge machine began to cranch along
the new highway. The weight of the roller is 26 tons. When
the machine had gone over the street two or three times it had
transformed it from a rough, impassable thoroughfare, to one
almost as level and satisfactory—of course not so smooth—as
an ordinary asphalted footpath.

The roller consists of four wheels or rollers, the two front
ones being 3 ft. 6 in. apart, and the hinder ones running close
together, 6 ft. diameter and 2 ft. broad. The hinder wheels over-
lap the front ones 3 in. The total width covered by the roller is
7 ft. 6 in.

The whole machine weighs 25 tons, equally distributed over
the rollers. The boiler is horizontal, and the working parts are
on the top of the engine out of the way of the dirt of the road.
The power is communicated to the rollers by an improved endless
chain of great strength.

The two hinder rollers are fitted in a turn-table, and become
the steerage of the machine, which is perfect in its action. A
boy twelve years old can steer the machine, and it can be completely turned in a road 30 ft. wide with great ease. We may
add that the roller can be worked backwards or forwards. It
is therefore seldom necessary to turn it.—*Builder.*

NEW METHOD OF RAISING SUNK VESSELS.

THE Glasgow and Belfast mail-steamer *Wolf* has been raised out of 42 ft. of water in Belfast Lough, where she was sunk by a collision with the *Prince Arthur*, in a fog, last autumn. Her mails and a valuable part of her cargo had been recovered by divers. As she lay directly in the way of other vessels, it was necessary to remove her; and the owners, Messrs. G. and J. Burns, with the Glasgow underwriters, had to devise the means of doing so. All attempts to lift iron vessels out of deep water had hitherto failed. To blow her up would have been very costly, and the vessel would have been destroyed. It was therefore resolved, upon the advice of Mr. John Wield, the able and experienced salvage agent, to entrust Messrs. Harland and Wolff, shipbuilders and engineers, of Belfast, with the task of lifting her by a newly-invented process. It was found by the divers and by soundings to ascertain the injury done to the hull of the vessel, that the stem of the *Prince Arthur* had made a considerable breach into her fore hold, and that she had sunk about 8 ft. into the mud, so that there was no possibility of passing chains under her in the usual way. With a rise of tide in that locality barely equal to the depth, she had sunk into the mud. These difficulties necessitated an entirely new means of attaching the lifting chains to her, and of bringing some powerful yet simple purchase to bear on them, so as to lift the vessel by them, independently of and in addition to the very limited rise of the tide. The very bold and ingenious plan, which is, we believe, quite novel, was therefore adopted of attaching strong hooks to the ends of fifty stout chains, each capable of carrying twenty-five tons, and of simply hooking each of these into an equal number of side light-holes that were in the vessel, and then passing these chains up over pulleys on a floating raft of tanks, there to be attached to a powerful screw, about 6 ft. in length, with nut, to be worked by a 6 ft. screw-key and five hands.

The *Wolf* was estimated to weigh, under water, about 800 tons, requiring a lifting power, of course, somewhat in excess of this weight. There was some encouragement from the perfect success which had attended Messrs. Harland and Wolff in the lifting of a similar vessel off the shore in that neighbourhood last summer, by means of large iron air-tanks constructed for the purpose. The same tanks, with two others still larger, built especially for this occasion, having in all a flotation equal to 852 tons, were now secured together in two distinct rafts, by stout double cross logs, the one raft to lift the fore body and the other the after body of the vessel; and on these cross logs the lifting-screws were fixed, with stages ranged on each side for the workmen to stand on whilst plying the screw keys. During the spring these ponderous rafts of tanks, with all their gearing,

E

were prepared; and whenever the weather and sea permitted, the divers were at work attaching the hooks into the side-lights of the sunken vessel, and ranging the lifting-chains attached to them on her deck ready to be sent up to the raft when the time came.—*Illustrated London News*, which see for engraving.

IMPROVED PARALLEL RULER.

AN ingenious improvement has been made in the ordinary parallel ruler by Messrs. Reeves and Sons, of 113, Cheapside, and which so enhances the utility of the instrument that it will eventually supersede the ruler hitherto used. The improvement consists in the engraving of scales on the sides of the ruler, and of an alteration in the shape of the metal connecting-links. By these simple means, parallel lines of any desired distance can be at once ruled without the necessity of being previously marked off with any other instrument. The right angle, and the angles most generally required, can also be laid down without the aid of a set square or protractor.

It will be found most useful to all whose profession or manufactures embrace the art of drawing. The ruler is the invention of Mr. Cecil Harrison, of the London University College School.

DRIVING IRON COLUMNS.

A PAPER has been read to the Institution of Civil Engineers "On the Supporting Power of Piles; and on the Pneumatic Process for Driving Iron Columns, as practised in America," by Mr. W. J. McAlpine. The first part of this communication related principally to the experience gained in driving 6,539 piles, an average depth of 32 ft., for the foundation of the Government graving-dock at Brooklyn, N.Y., when the support was mainly derived from the adhesion of the material into which the piles were driven, and slightly from their sectional area. The piles were in rows 2½ ft. apart, and at transverse distances of 3 ft., all from centre to centre; intermediate piles of tough second-growth oak being frequently employed. The main piles were chiefly round spruce spars, very straight, from 25 ft. to 45 ft. long, and not less than 7 in. in diameter at the smaller end, and on an average 14 in. in diameter at the larger end. From a record kept during the progress of the work, it was ascertained that it took two and one-third blows to drive each foot of pile, and that the distance moved uniformly diminished from the first to the last blow, ranging from 8 in. at the beginning to no movement at the end—the average distance moved by the last five blows being 1 in. A considerable number of the piles were driven by a Nasmyth steam piling-machine, with a ram of three tons, and a stroke, or fall, of 3 ft., and making from sixty to eighty strokes per minute. The other machines were generally operated by steam-power, giving an average of a blow per minute; but occasionally the hammers were hoisted by manual and horse power. The rams in the latter machines were of cast-iron, swelled out at the bottom to concentrate the weight at that point, and weighed about 2,200 lbs. each, though some were used of 1,500 lbs.; the fall being 30 ft. It was observed that the heaviest ram, when striking blows of the same effect as lighter ones, did the least injury either to the head of the pile or to the protecting iron ring, and this injury was still less with the Nasmyth hammer. It was also found that no advantage was gained by the fall of the ram being more than 40 ft., as the friction on the ways then prevented any increased velocity to the ram when falling from a greater height. With the Nasmyth hammer, piles were driven 35 ft. in seven minutes, while with the other machines similar piles required one hour, or more, to drive them the same distance.

Experiments were made at different times to ascertain the weight which the piles would sustain. For this purpose a long lever of oak timber was employed, with which a number of the foundation and coffer-dam piles of nearly the same size, and driven under exactly similar conditions, were withdrawn. It was thus ascertained that a weight of 125 tons was required to move a pile, driven 33 ft. into the earth, to the point of ultimate resistance, with a ram weighing 1 ton, and falling 30 feet at the last blow. These trial piles averaged 12 in. in diameter in the middle. From a number of other experiments, it was believed that the extreme supporting power of the pile, due to its frictional surface, was 100 tons, or 1 ton per superficial foot of the area of its circumference. From an analysis of the experiments, the following general laws seemed to have prevailed in these cases:—1st. That the effect of lengthening the fall of the ram was to increase the sustaining power of the pile in the ratio of the square root of the fall. 2nd. That by adding to the weight of the ram, the sustaining power of the pile was increased by 0·7 to 0·9 of the amount, due to the ratio of the augmented weight of the ram. 3rd. That a pile driven by a ram weighing 1 ton, and falling 30 ft., would sustain an extreme weight of 100 tons. The formula based upon these data, as applicable to rams weighing from 1000 lb. to 3000 lb., falling from 20 ft. to 30 ft., was

$$X = 80 \left(W + 0.228 \sqrt{F} - 1 \right),$$ in which X was the supporting power of a pile driven by the ram W, falling a distance F; X and W being in tons, and F in feet. The author was of opinion that, under the most favourable circumstances, the pile should not be loaded with more than one-third of the result given by this formula; and when there was any danger of a future disturbance of the material around the pile, or when there was any vibration in the structure which might be communicated to the piles, the load imposed should not exceed one-tenth.

The bearing support due to the sectional area of the pile had not been considered in the preceding inquiry; but numerous experiments had been made, which gave results of from 5 tons to 10 tons per square foot.

DEAD-STROKE POWER-HAMMER.

THIS hammer, of American invention, has been found to give satisfaction to most of the firms who have tried it. The inventors intend it to supersede the trip hammer. The first working model, with a 60 lb. head, has been in use for some time past in Philadelphia—principally in drawing down the ends of railway spring blades. The hammer consists of a framework of cast iron about 6 ft. by 6 ft. 8 in. by 10 ft. 8 in. high. The weight of the hammer is 1000 lb., or nearly half a ton, suspended from the spring which is attached to the end of a beam worked by the crank shaft, which is driven by a belt pulley. There is a wheel for tightening the belt, so as to regulate the speed at which the hammer works. The inventors say that it must not be supposed that the power obtained from this hammer is due to the momentum of the ram merely; the principal agent is the elastic force of the steel bow-spring, a force that is proved by bracing its extreme points so as to prevent any flexibility, when it will be found that, with all the velocity given to the ram, no power can be obtained from it other than that of an accelerated drop or dead fall. Absolute tests prove that a weight of 2500 lb. must be exerted to compress the spring usually attached to a 1000 lb. hammer, to enable it to reach the maximum point of deflection when driven at its regular speed. By this it will be seen that the increase of force over the mere weight of the ram itself is greatly in excess of that of any hammer known, while the power used is economized in about the same proportion. The machine gives a dead blow, and works with a considerable saving of power as compared with ordinary drop hammers.

LIQUID FUEL.

A GOOD deal of attention has been attracted to the subject of "Liquid Fuel," by Captain Selwyn, in a paper read to the Institution of Naval Architects, in which it is asserted that coal-tar burned in the furnace of a steam-boiler will generate nearly as much steam as three times the weight of coal. Experiments confirmatory of this result were recited by Captain Selwyn, whose statements were corroborated by those of other persons present; and, as coal-tar is at present a refuse material, and as any kind of concentrated fuel, even if more expensive, would be very valuable in enabling a ship of war to keep the sea, Captain Selwyn insisted upon the importance of substituting tar for coal fuel in the vessels of the Navy, the more especially as tar, unlike petro-

leum, is not dangerously inflammable. The tar is introduced into
the furnace through a bent pipe, about ¼ in. in diameter, which
points through a notch in the furnace-door; and beneath the tar-
pipe there is another similar pipe which conveys steam. The bars
of the grate are covered with ashes, which require to be renewed
every twenty-four hours, and the tar and steam form an inflammable gas which, when ignited, keeps up the heat. The evaporative efficacy asserted by Captain Selwyn, though no doubt stated
in good faith, is quite incredible, and there is no reason to conclude that the evaporative power of tar is much, if at all greater,
than that of an equal weight of good coal. It is well known that
coal with much bitumen or tar in it is not more effective than
coal with little or no bitumen in it, and the efficacy of patent fuel
compounded of coal-dust stuck into bricks with coal-tar is not
greater than that of coal. The most probable supposition is, that
in Captain Selwyn's boiler, much of the water supposed to have
been evaporated, was carried over the steam by priming, without
being raised into steam at all, and if this action occurred a great
apparent evaporative efficacy would be obtained, though the real
evaporative power of the fuel was only the same as that of coal.

Nor is there any novelty in Captain Selwyn's plan. In 1834
Mr. Bourne introduced into the steamer, "City of Londonderry,"
apparatus resembling that of Captain Selwyn, for burning coal-
tar in the furnaces. The tar was contained in tanks, upon each
side of the steam-chest, and the surface of the tar was pressed by
the steam to force it through bent pipes, entering the furnaces
through notches in the furnace-doors, precisely as in Captain
Selwyn's arrangement. Instead, however, of a steam jet, a jet
of boiling water taken out of the boiler was employed, and the
two jets were so set that they converged to a point within the
furnace, where they broke up and were vaporized into inflammable gas. Mr. Lamb, now superintending engineer of the Peninsular and Oriental Steam Company, was engineer of the steamer,
City of Londonderry at the time this experiment was made,
and it is not to be supposed that he would have allowed the benefits of the system to have remained so many years in abeyance,
if those benefits were so momentous as Captain Selwyn alleges.
Mr. Bourne made numerous experiments on this subject in 1835
and 1836, and in some cases he raised the tar into vapour as well
as the water, and conducted it into the furnace through hollow
furnace-bars, perforated at the sides and covered with a layer of
broken fire-brick, maintained in a state of incandescence. But
the conclusion finally arrived at was that tar had only about the
same evaporative power as coal, and that, although at the moment
it was a refuse material, it would at once cease to be so if it came
into request as a fuel for steam-boats. — *Scientific Results;
Illustrated London News.*

TOOTH'S SYSTEM OF BOILING SUGAR.

THE *Produce Markets Review* says :—"In an ordinary vacuum pan, the top of the liquid alone forms the evaporating
surface, and evaporation would, of course, be more rapid were a
greater portion of the liquid exposed. The most notable feature
of Mr. Tooth's invention consists in pumping the juice down from
the top of the vacuum pan, at the moment of granulation, through
a rose. The juice is thus distributed in small streams through
the air contained in the evaporating vacuum chamber, and the
surface exposed, as compared with the old system, is said to be
as 1000 to 50. The evaporating chamber differs from the old
vacuum pans in shape, being to speak roughly, a long cylinder,
with the ordinary round pan at the top and bottom. The juice,
before reaching the evaporating chamber, is pumped up through
a number of pipes surrounded by steam in a cylinder. The following advantages are stated by the inventor to be secured by
this process :—1. The juice is protected from excessive and long-
continued heat. 2. Long exposure to the injurious influence of
the atmosphere is avoided. 3. Great rapidity in carrying on the
evaporation is secured. 4. The juice is transferred to the vacuum
pan (or evaporating chamber) immediately after defecation and
filtration, avoiding the necessity of open pans. 5. Any extent of
heating and evaporating surface is easily obtained. 6. The cost
of fuel is greatly lessened. 7. Vacuum pans now in use may be
made available for the improved system at a comparatively small
cost. 8. The finest sugar is produced without the expense of
animal charcoal, and the crystallization being perfect, there is no
loss by drainage. 9. There is no formation of molasses beyond that
naturally existing in the juice, as the temperature never need
exceed 140 deg. to 160 deg. Fahr. 10. The system is also useful
in beet-root sugar manufactories. There is an arrangement by
which the clogging of the rose, through which the partly granulated sugar passes, is remedied. The idea of exposing a greater
surface to evaporation seems to us excellent in theory, but it
belongs, of course, to practical men to say if it will work. Mr.
Tooth has another patent to compete with, Mr. Fryer's Boncrotor,
for rapid and cheap evaporation. This consists in passing the
partially granulated juice through a rose, and letting it drop down
through a long cylinder or tower filled with heated air. The
patentee states that the juice reaches the bottom in the shape of
sugar."

NEW MODE OF VENTILATING MINES.

THE Incorporated Association of South Staffordshire Mine
Agents have paid a visit to the Hamer-hill Colliery, near Dudley,
for the purpose of inspecting and testing one of Guibal's new
patent fans for Ventilating Mines. This being the first applica-

tion of mechanical ventilation in South Staffordshire, it was regarded as a matter of great interest and importance. The inspecting party was met at the pit by Mr. J. F. Swindell, the proprietor of the colliery, and his agent, Mr. E. Foley. The members first proceeded to inspect the fan, which is 16 ft. 8 in. diameter, and 5 ft. wide. It is enclosed in a brickwork casing, and is connected with the top of the upcast shaft by a tunnel of 35¼ square foot sectional area. It is driven by a small 10-horse power high-pressure horizontal engine, the steam for which is supplied from the winding-engine boilers. When started the engine and fan require little or no attention for days together. The air is drawn from the mine up the upcast shaft, and driven by the fan up a chimney which is considerably wider at the top than at its base. Near the bottom of the chimney there is fixed a sort of iron venetian shutter, which regulates the quantity of air according to the requirements of the workings. The fan is fixed to the crank shaft of the engine, so that it performs one revolution to every stroke made by the engine. An experiment with the engine making 65 strokes per minute showed that the fan produced 37,500 cubic ft. of air per minute, with a water gauge of only 1·03 in., thereby clearly proving that the large sectional area of thick coal roads reduces the friction to the merest minimum. By increasing the speed to 90 strokes, with a water gauge of 1·75 in., the fan produced 51,700 cubic ft. per minute, being more than three times the quantity ever likely to be required. The total cost of engine, fan, and necessary apparatus has been about £500. The tunnel connecting the fan with the top of the upcast pit, is approached by a covered way with two air-tight doors. Through these the party passed into the main airway, but such was the intense velocity of the current when the fan was put to run from 60 to 70 revolutions per minute that they were only too glad to beat a hasty retreat. When put at its greatest speed it was scarcely possible for a person to stand upon his feet. It was clearly shown that it only took about 20 seconds to increase the ventilation from a state of nearly stagnation to that of 60,000 cubic feet per minute. Coals are raised at both the upcast and downcast shafts; but the former has to be worked with a moveable wood cover, which is raised by the cage when it comes to the top.

The party next descended the pit under the guidance of Mr. Foley, in order to test the efficiency of the ventilation at the most remote parts of the workings. The colliery comprises about 60 acres of coal, of a fiery character, and the most distant part of the workings is about 500 yards from the bottom of the shaft, which is itself 200 yards deep. At the extreme point a side of work of ten pillars was found partly opened. The party took their stand here, and, according to previous arrangements, the speed of the fan was slightly increased. The velocity of the current immediately increased through the stalls and openings

to such a degree that the lights were all extinguished, and the
members nearly smothered with coal-dust. The gateway was
10 ft. wide and 10 ft. high, and so completely was this filled in
every part with air that it was found impracticable to keep a
naked candle lighted when exposed to the current. At the conclusion of the inspection the members were unanimously of
opinion that the success of this new system is most satisfactory.
—*Times.*

VENTILATION OF PUBLIC BUILDINGS.

THE Institution of Mechanical Engineers have published a
Report of their annual meeting, held last summer in the lecture
theatre of the Conservatoire des Arts et Métiers at Paris, when a
paper was read by General Morin "On the Ventilation of Public
Buildings." Although we cannot give details, the importance
of the subject justifies our mentioning it briefly. For good reasons General Morin holds that outlets for the escape of bad air
should be at or near the floor of a room, and the inlets for fresh air
near the ceiling, or at such a height as to prevent the sensation
of a draught. Why should the carbonic acid produced by the
breathing of the people in the room be allowed to vitiate the
entire atmosphere, when it can be at once discharged at its
source? This discharge is best effected by "suction," and to
maintain this suction nothing more is required than an ordinary
fireplace. This being the case, the same system is applicable to
ordinary dwelling-houses as well as to public buildings. The
displacement of foul air by the mechanical forcing in of fresh air
is, as General Morin maintains, far less effectual, and requires
more attention than the suction system, which, besides the building above-named, is in use at the Théâtre Lyrique, and in certain public schools, where its operation is satisfactory. Striking
evidence of the fact might have been obtained by passing under
the seats of the room in which the paper was read; for there, as
the General stated, "he had felt completely stifled by the poisonous atmosphere drawn off from the room." The diagrams
published with the report show clearly the method of operation,
and the direction of the several currents of air.

FILTER-VANS FOR THE ARMY.

A NEW filter-van has been invented by Messrs. H. Bayley and
Co., of Newington Causeway, for the purpose of supplying pure
water to an army on the march or in the field. The contrivance
is extremely simple, and the idea is very ingenious. The apparatus consists of a galvanized iron reservoir, capable of containing 250 gallons, enclosed within a wooden case, mounted on
a frame with four wheels, and carrying underneath two filters,
through which the water passes to five outlet taps at the tail of
the van. The water is pumped into the reservoir by means of an

attached pump, through a suction hose leading to the source of
supply. Beneath the reservoir is a sediment chamber, into which
the coarser suspended impurities may subside, and immediately
above this are the supply pipes of the filters. These filters are
fed from below, so that the water ascends through them, and
overflows into two reservoirs, each capable of holding 25 gallons.
The filters are intended, as a rule, to act independently of each
other; but, if the water should be very foul, it can be passed
through them in succession. The filters can be cleansed at pleasure by forcing compressed air through them, and an air-pump
for this purpose is carried in a chamber or boot of the van, together with hose and a few tools to meet any emergencies that
may arise. The reservoir and sediment chambers are easily
cleaned. The wheels and framework are of great strength, well
adapted for the rough work of a campaign; and the filters are
so placed as to be protected from all ordinary occasions of injury.
The weight of 250 gallons of water is about a ton, and the
weight of the empty van is about 25 cwt., so that the whole
would not be a heavy tax on the draught-horses, while the top of
the van may be so constructed as to carry light baggage, or tired
or wounded men. The two pure-water reservoirs contain, as
stated, a supply of 50 gallons ready for use; and, for ordinary
muddy river-water taken at a trampled ford, the filters would
discharge nearly 2,000 gallons a day, so that one van would be
sufficient for the wants of a large number of men and horses.
Apart from the filtration, such a means of conveying-water seems
to be of great practical value. In the Crimea several horses of
the Light Brigade perished from thirst during a short expedition
inland from Eupatoria, and the scanty water supply then available for the force was carried in skins on the backs of pack-
horses. In India, too, thirst on the march is often very distress-
ing, and the drinking of foul water is a frequent cause of parasitic
and other disease. On the whole, Messrs. Bayley's invention
will add much to the comfort and safety of the soldier and of the
soldier's horse, and we are pleased to hear that the military authorities have reported very favourably upon it, and have recommended its adoption in the service.—*Times.*

NEW METHOD OF BRICKMAKING.

A VERY excellent and scientifically complete mode of Brick-
making has just been established (says the *Aberdeen Free Press*)
at Ruthrieston, by Messrs. Keith, Harrimann, and Watson, of the
Northern Patent Brick Works. The new works are constructed
on the principles of Hoffman and Licht's patent annular
ovens, only that the kiln—if it may be so called—in place of
being built in the form of a circle, has been built in
the form of an oval. Let the reader, then, imagine an oval
175 ft. in length and about 95 ft. wide, as the extreme width.

The circumference is formed of something like a small railway tunnel, arched with firebricks and strongly bottomed with the same. There runs round the whole oval the tunnel, as we have called it, being a chamber about 15 ft. wide and 8 ft. to 9 ft. in height; the outer walls, 10 ft. thick at the base, and in which are twelve openings or doors, dividing the whole into as many sectional chambers. Inside the tunnel is another brick wall, or rather double wall, filled with earth in the middle, with smoke chambers, and a number of flues pointing towards the centre of the oval, from which rises the chimney stalk, 175 ft. high. In the tunnel we have described the bricks are burnt. As we have said, the tunnel is divided into twelve chambers. Each of these is capable of containing 15,000 bricks, representing a day's produce of the ovens. The *modus operandi* is this :—In place of only one apartment or chamber being filled at a time, the whole are filled, save one, which is being emptied of the burnt bricks, and another next to it, which is being simultaneously filled with "green" bricks, the fire being meanwhile kept up in only one chamber. But it is here that the special value of the patent oven is seen, for while the fire is led round from one chamber to another, feeding coal through a series of small metal-covered holes in the crown of the arch (the whole of the chambers being hermetically closed up to the one that is being emptied, which is shut up at the end nearest the fire by an iron door), the hot air and flame, finding no other means of escape, go forward into the chambers in advance of the fire, gradually heating up the wet bricks. Nor is this all, because the cold air that comes in where bricks are being emptied must pass through a series of chambers in which the bricks burnt on four or five days preceding are cooling. It touches, first, a chamber of nearly-cooled bricks; next, one burnt a day later, which is considerably hotter; then another, hotter still, till by the time it reaches the fire it is quite at the point of combustion, carrying forward the flame in waving lines among the interstices of the burnt bricks. There is thus secured the utmost possible economy of heat, and, at the same time, any smoke evolved is thoroughly consumed, the only "visible emanation" from the top of the chimney-stalk being a little steam off the green bricks when they begin to feel the influence of the heat. The economy in fuel which is secured by these means is, of course, obvious, while the scientific principles on which the power of the fire is concentrated enable the cheapest coal dross to be used in firing. All these arrangements will materially cheapen the price of bricks.

VENETIAN GLASS.

As the art of glass-making was introduced into modern Europe by the Venetians, Mr. Herries, Her Majesty's Secretary of Embassy and Legation at Florence, in his report issued, has furnished some statistics relating to the production of Venetian glass. He states that, besides discovering the art of rendering glass colourless by means of manganese, the Venetians also enjoyed the monopoly of mirrors, the silvering of which was a secret long kept from other countries. These mirrors, however, have now lost their reputation, as foreign competitors produce larger plates. Glass beads are still made in considerable quantities for exportation. Venetian enamels have always been famous, and among the peculiar productions of Venice may be reckoned the beautiful composition called aventurine, the secret of which is said to be in the possession of a single manufacturer. The great glass-works are at Murano, one of the islands of the Lagoon. The number of persons employed in glass-making at Murano and Venice is 5,000, of whom one-third are men, and two-thirds women and children. The annual cost of the substances employed in the manufacture is estimated at about 7,000,000 francs. In the East there is a constant demand for beads and other articles known as *conterie*. There are six glass-works in Turin, three in Genoa, five in Milan, thirteen in Florence, eleven in Naples, and twenty in Venice. Those fifty-eight works produce articles of the annual value of 10,276,725 francs.—*Mechanics' Magazine.*

COLONIAL GLASS.

A GLASS manufactory is shortly to be started in Melbourne. The colonial manufacture of glass has hitherto been very small. Many years ago there was a manufactory in Sydney, but it was abandoned. Lately it has been re-established, and although unable to compete in the ordinary forms of glassware with the home manufactories, it was found that there were many articles, such as glass fish-globes, confectioners' glasses, carboys, soda-water bottles, &c., the importation of which is attended with so much expense, which could be profitably made in the colonies. The resources at the command of the manufacturer at the commencement of operations will enable him to melt and convert into glass 350 lb. of "metal," as it is technically termed, twice a week, but if the enterprise meet with support this quantity can be doubled.

THE LIME AND PLATINUM LIGHTS.

A RENEWED endeavour to make the Lime or Drummond light available for use instead of gas has been made. With that view improvements have been suggested. Arrangements are being made for supplying Perth barracks with the lime-light. The jet of hydrogen being lighted, a separate jet of oxygen will be turned on so as to mix with it at the moment of combustion, when the flame impinges on the lime, which then emits the intense light for which it is noted when white-hot. Various towns in

Scotland are said to be adopting the light. Another light of an analogous description has been suggested by a Frenchman, M. Bourbouse, who uses common air instead of oxygen, and common gas instead of hydrogen, for the sake of economy. In this case the air and gas are admitted into one common tube; thence they pass through a sheet of metal, perforated with a great number of holes, in order to be divided into many small jets: these are delivered through a gauze of platinum wire, when they are lighted. The metal, in being heated, soon becomes red, then white, and thus diffuses a dazzling light. If, as seems to be the case, the air and gas on this plan are previously mixed in the proportions proper for combustion, that is a dangerous element in the proposed light, because such a mixture is explosive. We would suggest, therefore, that the air should be supplied to the gas at the point of combustion. Otherwise, perhaps the platinum light would be less unsuitable for ordinary house illumination than the lime or magnesia light. Has lime ever been tried with a light from street gas and common air instead of pure or mere hydrogen and oxygen, or gas and oxygen?—*Mechanics' Magazine.*

WATERPROOF PAPER OR GLAZED FABRICS.

MR. JOHN THORNE, of Manchester, has patented an invention which relates to the manufacture of a novel description of Waterproof or Glazed Paper, which is applicable for numerous purposes where liquid-proof qualities, or paper with a glazed surface, is required; such as, for instance, for enclosing goods or materials packed by hydraulic or other pressure, or for ordinary packing purposes, for lining or covering boots, hats, or articles of wearing apparel, and for other lining or covering purposes, as table-cloths, bandages, sheeting, tents for preventing liquid penetrating, or for preventing evaporation, or for making receptacles for liquids, or for covering walls or partitions, either to exclude moisture, or as a glazed surface on which to print or paint patterns or designs.

The invention consists in applying to the surface of paper, on one or both sides, a mixture of copal or other varnish (ordinarily used to varnish wood) and linseed oil, containing litharge, litharge, or oxide of lead, to produce a superior quality of liquid-proof paper. The inventor prefers paper made from wood-fibre, but the paper first treated may be varied according to the quality to be produced. He takes about equal quantities of copal or ordinary wood varnish and boiled linseed oil, with which has been mixed sufficient litharge, oxide of lead, or other compound to make it dry readily, and the paper is to be coated on one or both surfaces with this mixture, either by hand or by passing it through a trough containing the mixture (either hot or cold), and thence between flannel-covered rollers, to render the surface even and remove the surplus mixture, or by other ordinary mechanical means. After which the coated paper is placed in a stove heated to about 160 deg. or 190 deg., and removed when dry. The proportions given above are capable of considerable variation, as for purposes the varnish may be employed almost alone. After the paper is thus prepared it is liquid-proof and water-proof, and it may be used in such condition, or it may be grained or printed upon in colour for subsequent use as table-covers, or for other ornamental or covering purposes.—*Mechanics' Magazine.*

JAPANESE PAPER.

PAPER appears to be destined to obtain many new applications. In Japan it is made into coats, and even into pocket-handkerchiefs; and last month we noticed its employment in the production of stereotype casts for printing, and of a new kind of straw hat or bonnet. In America a manufactory has been set up for the production of paper pails, pans, and wash-basins. . The paper is so prepared as to be impervious to water or acids; and it is said that the utensils formed of it may be placed in an oven until the water within them boils. The paper pails are lighter than those formed of wood, and will not crack in the sun.

COMPRESSED LEATHER.

THE inventor of the compressed leather, Capt. J. H. Brown, R.N., has devoted his energies for some years past to converting the refuse cuttings of animal hides and skins into useful products. We remember his patent parchment and vellum skins, which, during the excise duty on paper, were brought before the Court of Exchequer as being subject to that duty. The inventor objected to the impost on the grounds that as it was made from animal skins, and not vegetable fibres, it was exempt from the duty, as parchment and vellum skins were. In the course of the trial, sheets of the patent parchment were exhibited—some of them engrossed. The court admitted that had they not seen the specification they would have called it parchment, but being manufactured in a paper-mill, and being reduced into a fibrous pulp and fabricated into sheets in the same manner as paper was, it came within the meaning of the Act of Parliament, and therefore was subject to the duty. This decision had the effect of preventing the progress of manufacture for a time, but since the abolition of the paper-duty the works have been resumed, and in addition the compressed leather is being produced at the Abbey Mills, Romsey, Hants.

The *modus operandi* in the manufacture of the compressed leather is the reduction of cuttings or waste of shoemakers into fine filaments, cleansing them, in the first place, from dirt and foreign matters. In the next place, the cuttings or refuse of ox and similar hides, which are generally unfit to tan, and are sold to the glue-makers, are also reduced to a fibrous mass. These are combined together with water, to which is added one part of sulphuric acid to one hundred parts of water, until it assumes a plastic mass, when it is pressed into moulds of the size and thickness required. When dried in a steam-heated room they are passed through heavy pulp rolls, glazed on one side and rough on the other, to represent the grain and flesh sides of the leather. The addition of the raw fibre with the tanned filaments is in certain proportions, according to the quality of the leather required, from five to twenty per cent. can be safely employed. It gives vitality to the tanned fibres by agglutinating them and imparting the albumen and gelatine which has been destroyed by the tannic acid. To render the compressed leather more supple or flexible, it is necessary occasionally to incorporate about one pint of glycerine to a hundredweight of mass. In the manufacture of shoes, boots, and similar articles, and other purposes for which leather is employed, the compressed leather will become of great importance from the fact—and we speak advisedly, having seen some specimens—that it is less permeable than ordinary sole leather. It is also harder, closer, and more compact, and can be sold to the consumer at 50 per cent. less than the natural hide. It is not suited for machine bands, or harness, but in the manufacture of boots and shoes, especially for nailed soles and heels, as is also for inner soles, it is also superior to much of the materials at present employed. On the whole, we are much pleased with this invention, which proves that in another great department of economic industry the importance of utilizing waste material is fully recognized.—*Mechanics' Magazine.*

THE ADJUSTABLE BELGRAVIA SEWING-MACHINE.

THIS machine, considered the most perfect and complete of its kind, consists primarily of a Wheeler and Wilson's machine, upon which some very important improvements have been engrafted by Mr. John Mabson, of Newcastle-on-Tyne. These improvements are no less than seventeen in number, and include various appliances for doing different classes of work. Among the principal is an adjustable cloth-presser, a cloth-plate, and a belt-tightener, which may be classed amongst the improvements in the machine itself; whilst a self-acting creaser and an embroiderer may be considered, with others, as extra appliances which can be had separately from the machine. It will make the Wheeler and Wilson, or lock-stitch, the Wilcox and Gibbs' stitch, the Grove and Baker stitch, the pearl stitch, the cable stitch, and, finally, the Belgravia embroidery stitch, which is made on the right or upper side of the material—a thing never before accomplished. The Belgravia stitch constitutes a very important improvement in this machine, and is made by a most ingenious mechanical arrangement. There are a pair of claws or thread-guides resting on the presser-foot, and which open at one stitch and close at the next. By this means, the embroidery threads are crossed, and at each crossing are secured to the fabric, to be ornamented by the machine working the ordinary lock-stitch. Various thicknesses of the embroidery material may be used, and the pattern may be carried to any width. The most elaborate and intricate designs can be correctly followed, the pattern being distinctly seen through the crystal presser. The ordinary Wheeler and Wilson machine, of which the patent has expired, obtained a gold medal last year at the Paris Exhibition. As this machine forms the basis of the Belgravia machine, it is a pity that Mr. Mabson was not ready with his improvements so as to have exhibited, for it is but reasonable to infer that the great advance he has made would have secured him the medal. As it is, however, we are glad to learn that at the Liverpool and Manchester Centenary Agricultural Show, held last year, Mr. Mabson was deservedly awarded the silver medal for "valuable improvements." His mechanical skill has enabled him to produce a machine which commands a wide sphere of usefulness, its work ranging from the simplest stitching to the most elaborate and finished embroidery. Viewed simply as a piece of mechanism, it is a singularly perfect and unique production, whilst its general utility cannot fail to make it a favourite with the public.— *Mechanics' Magazine.*

THE COTTON MANUFACTURE IN SWITZERLAND.

ONE of the most important industries in Switzerland is the spinning and weaving of cotton. Spinning is chiefly carried on in the eastern part of Switzerland. The *Society of Arts' Journal* informs us that the total number of spindles in all the establishments amounts to 1,600,000, of which 607,082 spindles are in 78 establishments in the canton of Zurich ; 266,805 in 22 establishments in the canton of Argovie ; 200,000 in 12 establishments in the Canton of Glaris ; 172,136 in 20 establishments in the canton of S. Gallo ; 109,600 in 4 establishments in the canton of Zug ; 50,400 in Svitto ; 42,800 in Turgovie ; 31,600 in the Grisons ; 30,000 in Bern ; 22,768 in Soletta ; 10,000 in the canton of Bale ; 16,120 in Scaffhansen ; and 6,016 in the canton of Lucerne. The total number of workpeople employed in this manufacture is about 15,400, and adding the number of managers, the clerks, and other employés, and the families of the workpeople, about 30,000 persons may be said to depend on this industry for their livelihood. The annual production of the spinning-mills in Switzerland is estimated at 336,630 quintals. In 1857, the exports of cotton yarn was 18,504 quintals, against 4,818 imports ; in 1863, 53,836 quintals exports against 47,475 quintals imports ; and in 1866, 35,738 quintals exports against

16,686 imports. This decrease must not be attributed to a smaller production, for during the years 1864, 1865, and 1866, the imports of raw cotton and the exports of cotton goods had largely increased; it should rather be attributed to the greater development of the cotton-weaving during the last few years in Switzerland, which has tended to augment the consumption of cotton yarn, and for this reason a greater quantity has been imported.

LEAKAGE OF SHIPS.

A CONTRIVANCE has been invented for indicating the existence of leakage into a ship by placing a buoy at the bottom of the hold, which, if water enters, will be floated up and detach a detent, which will ring an alarum bell. In steam-vessels the existence of leakage is at once perceived by the rise of the bilge-water in the engine-room, and in sailing-ships it is a periodical duty of the carpenter to let down a sounding-rod either through one of the pumps or through a tube provided for the purpose.

DUTCH RAILWAY.

On the 1st of November, one of the most important Dutch lines of railway was thrown open to the public. It runs from Utrecht to Waardenburg, and forms a link of the section Utrecht Bois-le-Duc, leading to Brussels and Paris. Bois-le-Duc was on that same day admitted into the Dutch railway system, and put into communication with the railway that now stretches away from Goes (in Zeeland) to Venlo (in Limburg). It is probable that the section Bois-le-Duc-Waardenburg, which now alone remains unfinished in this part of the country, will be ready by the middle or end of next year, so that 1869 will witness an unbroken communication between Amsterdam and Paris. Of all the lines to be built by the State, according to the Act passed in 1860, this one from Utrecht to Bois-le-Duc is the most difficult and costly. Three tremendous bridges had to be constructed over three large rivers, the Mouse, near Hedel, the Waal, near Zalt-Bommel, and the Lek, near Culemborg. The last is now open for traffic, and the others are nearly ready. The bridge near Culemborg, half-way between Utrecht and Waardenburg, is one of the grandest works of engineering skill. It consists of one arch of 492 English feet span, one of 262 feet, and seven arches of 57 foot span each, or 399 feet; total 2,063 English feet. It is constructed on the "fish-shaped girder" system. The one near Zalt-Bommel will measure 2,680 English feet.— *Athenæum.*

AUXILIARY RAILWAYS.

MR. W. THOROLD has read to the British Association a paper "On an Auxiliary Railway for Turnpike-roads and Highways

Passing through Towns." The author stated that he only required a single line of rails, which he proposed should be laid on one side of the road, out of the way of the ordinary traffic. By an arrangement of grooved wheels under the centre of the engine and carriages, so constructed that they will be capable of maintaining their grip upon curves of 20-feet radius, thereby giving the vehicles the power of turning corners with the greatest facility, the inventor thinks his principle peculiarly adapted to locomotion through new countries, and for passing through ravines, or up and down the sides of mountains, up any gradient not exceeding 1 in 12. He proposed to propel the carriages by steam traction-engines, although they might also be drawn by horses or other beasts of burden. The adhesion of the traction-wheels could be regulated to any weight, and by the application of a special apparatus the engine might be made to lift the traction-wheels out of a soft place. The cost of the new railway he estimated at about £500 per mile.

LONDON STREET TRAMWAYS.

Mr. H. Bright, in a paper read to the British Association, said:—"The London omnibuses were notoriously mismanaged; and when it was remembered that there were six hundred of these vehicles in London, each capable of carrying, on an average, twenty-three passengers, the question became an important one. There could be little doubt that a judicious system of street tramways, or horse railways, would supply a great and rapidly growing demand, which could not be met by steam locomotion on the ordinary railway, where the trains could not work like omnibuses, taking up passengers at every moment when required, but must run through from station to station. Street tramways had proved a success wherever they had been judiciously tried, and would doubtless yield an enormous profit if laid down in London and other large towns. They were extensively used in America and Canada, and had been adopted at Copenhagen and the suburbs of Paris; while it was proposed to apply them to Berlin, Brussels, and Vienna. The objection which might be urged against the interference of tramways with the ordinary traffic would be met by taking the many good and available lines afforded by back-streets, taking care to bring the line at certain points into close proximity to the main traffic. The system he proposed to adopt was somewhat similar to that which was at present in use in Manchester and Geneva, the vehicle being kept on the track by means of a wheel, which the driver could at pleasure drop into or lift up from a grooved rail in the centre of the track. The formation of the lines for the carriage-wheels was peculiar, there being a slope with a depression of only one inch for each wheel, which would be so made as to fit the wheel-ways, while the depression will be so slight that

it could not obstruct the progress of any ordinary vehicle. The
vehicle would be enabled to turn the sharpest curves, and would
carry forty-eight passengers, exclusive of the driver and conductor. It was proposed by an efficient system of breaks, with
a carefully-devised scheme of compensation for the horses, to
enable the driver to stop the vehicle at any moment.

AERONAUTICAL SOCIETY'S EXHIBITION.

This first display was held at the Crystal Palace, and was but
preliminary to a great exhibition. In some instances we had
machines very well designed for aërial flight, but which required
the impossible addition of great power to sustain them in their
course. On the other hand, there were excellent engines light
enough perhaps for practical flight, but too light for safety. It is,
however, well that such an exhibition has taken place, as it will
give each an opportunity of examining into the various principles
presented to his notice, and may eventually lead to a practical
result. At present, while giving all credit to our inventors and
improvers, we have really got but little beyond the old Nassau
Balloon of twenty years ago—except of course, in theory. The
balloon, "Le Captif," of M. Delamarne, was the only aërial
vehicle in which the public were invited to mount aloft, and even
this unfortunate sole representative of practical aëronautics was
accidentally burnt. This event is to be regretted, as it was one
of the principal attractions of the exhibition, and was to make
ascents daily to the height of 1,000 ft., M. Delamarne made an
attempt to inflate the balloon by his new process, and the arrangements were inspected by several engineers acting upon the exhibition committee and council in the interest of the public. The
stupendous machine appeared to be fully inflated in twenty minutes, and then commenced to oscillate. Upon the fourth oscillation, the heating apparatus (the whole weight of which rested
upon the car) was overturned, when instantly the balloon caught
fire upon the south side, where it burst, then fell to the ground
blazing, where it lay and smouldered to ashes. The sympathy
which was shown to M. Delamarne by those around was very
gratifying.

Amongst the few exhibits in advance of the rest, was a working model of an aërial steam carriage, by Mr. J. Stringfellow, the
whole including engine, boiler, water, and fuel, weighing about
12 lb.; cylinder 1 3-16th in. diameter, 2 in. stroke, working two
propellers 21 in. diameter, about 600 revolutions per minute,
gets up steam of 100 lb. pressure in five minutes.

Mr. Morney Moy exhibited a working model to illustrate
a mode of flying vertically by direct action on the air, without any screw in the wings.   He also had a working model
to illustrate natural flying, the wings being used to propel and
sustain; the tail to sustain only. This will fly horizontally for

a short distance. Mr. Moy also exhibited a mariner's kite, for use in rough weather, to communicate from one ship to another, or to the shore. It is rectangular, and stretched between two sticks on one horizontal stick, to which the loop is attached; the kite is ballasted and attached to a float.

Professor Ansell sent a model illustrating a proposition to omit ballast in balloon ascents. By this proposition gas would be withdrawn from the balloon by an air-pump, which would compress the gas into a chamber carried in the car when a descent becomes necessary. An ascent will be obtained by opening a tap, and thus allowing the compressed gas to escape from the chamber by a tube into the balloon. The advantages of this would be that the natural balance used by fishes would be applied to balloons, gas being reserved for use, instead of escaping, as now obtains. The Duke of Argyll exhibited a working model, showing progressive motion by flapping action of the wings. By winding up a clock-work arrangement the wings flapped away beautifully, and the whole machine, which was suspended by grooved wheels from a horizontal wire, made about 12 in. of progress. But we are not at all sure that this progress was not due to the vertical vibratory motion of the wire. Indeed, his Grace does not imitate the action of a bird's wing. Mr. Shill sent an ingenious one-horse power turbine injector steam-engine weighing less than 12 lb. with inclined vanes showing its adaptation for aërial purposes, with rudder and gear for working. Steam here injects water against the turbine, and the force of impact is said to do all that can be desired. The inventor told us he had worked the machine, but he could not admit that he had obtained any available power from it. Mr. Spencer showed a flying machine which, being attached to the body, enables a person to take short flights. The exhibitor of this machine says he has, with less perfect apparatus, accomplished flights to the extent of 160 ft., rising from the ground by a preparatory running action. We have abridged these details from the *Mechanics' Magazine:* they are more ingenious than practical.

INSTITUTION OF CIVIL ENGINEERS.

THE Council of the Institution of Civil Engineers have awarded the following Premiums for original communications submitted to the Institution, and read at the ordinary meetings during the session 1867-8. A Telford Medal, and a Telford Premium, in Books, to G. Higgin, for his paper "Irrigation in Spain, chiefly in reference to the Construction of the Henares and the Esla Canals in that country." A Telford Medal, and a Telford Premium, in Books, to C. P. Sandberg, for his paper "On the Manufacture and Wear of Rails." A Telford Medal, and a Telford Premium, in Books, to Lient-Col. P. P. L. O'Connell, R.E., for his paper "On the Relation of the Freshwater Floods of

Rivers to the Areas and Physical Features of their Basins." A Telford Medal, and a Telford Premium, in Books, to W. Wilson, for his "Description of the Victoria Bridge, on the line of the Victoria Station and Pimlico Railway." A Telford Medal, and a Telford Premium, in Books, to C. D. Fox, for his paper "On New Railways at Battersea; with the Widening of the Victoria Bridge and Approaches to the Victoria Station." A Telford Medal, and a Telford Premium, in Books, to J. W. Barry, for his paper "On the City Terminus Extension of the Charing Cross Railway." A Watt Medal to E. Clark, for his paper "On Engineering Philosophy: the Durability of Materials." A Telford Medal to W. J. M'Alpine, for his paper "On the Supporting Power of Piles; and on the Pneumatic Process of Sinking Iron Columns, as practised in America." A Telford Premium, in Books, to T. Login, for his paper, "On the Benefits of Irrigation in India; and on the proper Construction of Irrigating Canals." A Telford Premium, in Books, to A. Wilson, for his paper "On Irrigation in India." A Telford Premium, in Books, to W. Airy, for his paper "On the Experimental Determination of the Strains on the Suspension Ties of a Bowstring Girder;" and the Manby Premium, in Books, to A. C. Howden, for his paper "On Floods in the Nerbudda Valley; with Remarks on Monsoon Floods in India generally."

STEAM-RAM.

MESSRS. NAPIER, of Glasgow, have received orders from the Admiralty to construct the *Hotspur*, a steam-ram, which bears no resemblance to anything in our Navy at present. She is neither a broadside ship nor a monitor. Like the *Belier*, this vessel is intended to fight end-on. The armour-belt at the water-line consists of two strakes of plating, the upper one being 11 in. thick, and the lower one 8 in. She has a formidable ram. On the main-deck is an armour-plated breastwork extending about one-third the length of the ship, similar to that which has been adopted in the new monitors. From the bow aft to the breastwork the main-deck is plated with 8-in. armour; and at the forepart of this breastwork a pear-shaped battery, covered with 8-in. armour, is brought above the upper-deck. This battery is pierced with several ports, and contains a turn-table, carrying an 18-ton gun, the whole being trained, &c., by suitable machinery situated on the main-deck. The only other gun to be carried by the *Hotspur* is a 40-lb. Armstrong: this will be placed aft.

Natural Philosophy.

THE ROYAL SOCIETY.

FROM a courteous report of the Royal Society's *soirée*, March 7, we select the following more noticeable items:—The philosophical and mechanical apparatus and instruments were well worth examination. Mr. John Browning's group of instruments showed improvements in construction, tending to greater utility and wider application; his silvered glass speculum, 12 in. diameter and 6 ft. 3 in. focus, might be said to demonstrate its excellence by the two views of Jupiter taken by its aid, and exhibited along with it; and his meteorspectroscope, fitted with a cylindrical lens, obviates the objection that the field of view was far too narrow for observations of the spectrum of so swift a fire, for the addition of the cylindrical lens widens the field to 60 degrees, and thereby increases the observer's opportunity. In a darkened room, much too small for the occasion, Dr. Tyndall repeated Faraday's marvellous experiment—the magnetization of light: Faraday's third great discovery, as Dr. Tyndall calls it, likening it "to the Weisshorn among mountains—high, beautiful, and alone." In this instance, the ray, passing from the lamp between the poles of a large horse-shoe magnet, showed itself as a spot of light on a screen: contact was made with the battery, the horse-shoe became powerfully magnetic, and immediately by the shifting of the spot of light on the screen, it was seen that the ray had been deflected from its former course. Some of our readers who remember Faraday's first public demonstration of this remarkable phenomenon, at a Royal Institution Friday evening lecture, will remember also the admiration and enthusiasm which it excited. It is an experiment which few could witness coldly— the strong influence of one imponderable upon another. A new induction machine, which might be called, after the inventor, Sir William Thomson's whirligig, showed how readily electricity might be developed without friction. A vertical fan of sheet brass is made to rotate rapidly between fixed segments of similar metal, and absolutely without contact. Let one of the segments be charged by a spark, it remains ever afterwards in a different state of tension from the others. Consequently, when the fan is made to rotate swiftly, electricity is generated, and can be led off by a conductor in a constant stream of sparks. This instrument, which does for statical electricity that which Wheatstone's and Siemens's do for dynamic electricity, is constructed by Elliott Brothers.

The Master of the Mint (Mr. Graham) carrying on his researches in dialysis, exhibited two experiments, the dialytic

separation of hydrogen from coal-gas, and the extraction of occluded hydrogen from palladinm. For example, inside a glass
vacuum-tube place a smaller tube of palladium, through which is
flowing a stream of coal-gas; heat the metal to 200°, and the
pure hydrogen will bo separated, pass through the metal, and can
be collected in a test-tube. Palladium is so greedy for hydrogen
that it will take up 986 times its own bulk of the gas; and this
occluded hydrogen can be separated from the metal by a roversal
of the experiment.

Wier's Pneumatic Signal apparatus, displayed on a large table
in a well-lighted room, attracted much attention. It has long
been known that by pressure at one end of an air-tube a signal
can be produced at the other. Mr. Wier shows that this can be
done through a length not exceeding 250 feet, and that the smaller
the tube, the more effectually is the work performed. Hence,
the captain of a ship having one of the dials on the bridge, can
send signals to the man at the wheel, who reads them on a similar
dial placed before him; or down to the gun-deck, and order the
firing of any of the guns; or, taking a small cylinder fitted with
flexible tubes under his arm, he can climb to the maintop, or the
crosstrees above the smoke, and from that elevation fire the guns
himself by pressing a pin in the cylinder. The superiority of
this method of giving orders to the shouting of them through a
trumpet is obvious.

THE ROYAL SOCIETY'S MEDALS, 1868.

AT the head of the list of awards made by the Council, stands
Sir Charles Wheatstone for the Copley Medal—an award which
cannot fail to command general approval. It is not the first time
that Sir Charles's merits have been recognized by the Society:
they gave him one of their Royal Medals in 1840; and there are
few who, remembering all that he has done for science in the
subsequent twenty-eight years, will question the present bestowal
of the highest medallic distinction in the Society's trust. In the
dry, brief words of the official phrase, the Copley Medal has been
given for "researches in acoustics, optics, electricity, and magnetism;" but how much of scientific labour and achievement of
high quality do these few words represent! Gifted with a remarkably inventive genius, Sir Charles Wheatstone has produced
instrument after instrument in such numbers that a mere list of
them would be a long one. Some are for determining the con-
stants of a voltaic circuit: the rheostat, the chronoscope, the
electro-magnetic clock, the speaking machine, the "Wheatstone's
bridge," or differential resistance measurer, indispensable wherever
the resistance of telegraph wires or cables has to be measured, or
electro-motive forces determined; and all that variety of instruments which, since the first experiments in 1839, have been
devised for the transmission of telegrams, down to dial-telegraph

working without any clock-power, and the "high-speed telegraph," in which, to quote General Sabine's words, "the messages, previously prepared on slips of paper, are, by passing through a very small machine, constructed somewhat on the principle of the Jacquard loom, made to print the messages at the remote station, in the ordinary telegraphic characters, with a rapidity unattainable by the hand of an operator."

One of the Royal Medals was awarded to the Rev. Dr. Salmon, D.D., Regius Professor of Mathematics in Trinity College, Dublin, for his researches in Analytical Geometry and the Theory of Surfaces; and the other to Mr. A. R. Wallace, an eminent traveller and zoologist, in recognition of the value of his many contributions to theoretical and practical Zoology, among which his discussion of the conditions which have determined the distribution of animals in the Malay Archipelago occupies a prominent place. The question, briefly put, runs thus :—The strait separating the island of Bnly and Lombok is fifteen miles wide only, nevertheless, the animal inhabitants of the island are widely different, the Fauna of the western island being substantially Indian, that of the eastern as distinctly Australian. This Mr. Wallace has described and discussed more completely and definitely than any previous observer. Moreover, in a noteworthy paper, published in the *Proceedings of the Linnæan Society*, "On the Tendency of Varieties to Depart from the Original Type," he has set forth the doctrine of natural selection, which, while travelling in the Malay Archipelago, he had developed independently of Mr. Darwin. Apart from which it was the immediate cause of the publication of the "Origin of Species." — *Athenæum Report*. See also the memoir of Sir Charles Wheatstone, in the *Year-Book of Facts*, 1867, for an account of his scientific labours, accompanied by a portrait.

"FARADAY AS A DISCOVERER."

PROFESSOR TYNDALL in his discourse delivered at the Royal Institution, says :—"For several years, although Faraday was acquiring, not producing—storing and strengthening his mind—yet, between 1816 and 1820, he published various scientific notes; and his first paper 'On Two New Compounds of Chlorine and Carbon" was read before the Royal Society on Dec. 21, 1820, and printed in the *Philosophical Transactions*. In 1823 he succeeded in liquefying chlorine gas; and in 1829 he gave his first Dakerian lecture on the manufacture of glass for optical purposes." "In 1831," says Dr. Tyndall, "We have him at the climax of his intellectual strength, forty years of age, stored with knowledge, and full of intellectual power, and commencing his wonderful series of experimental researches in electricity, which include the discovery of magneto-electricity and terrestrial magnetic induction," of which Professor Tyndall gave a brief analysis, but which our limited space prevents us from touching upon. "Thirty years," said he, "have passed since the discovery of magneto-electricity; but, if we except the extra current, until quite recently nothing was added to the subject. Faraday entertained the opinion that the discoverer of a great law or principle had a right to the spoils;" and "his wonderful mind, aided by his wonderful fingers, overran in a single autumn this vast domain, and hardly left behind him the shred of a fact for his successors." In 1833 and 1834 Faraday read papers before the Royal Society on electro-chemical decomposition, containing a series of experiments, which forced upon his mind the conclusion that the amount of this decomposition depends not upon the size of the electrodes, or the intensity of the current, or the strength of the solution, but upon the quantity of electricity which passes through the cells—the quantity of electricity being proportional to the amount of chemical action. Upon this law he based the construction of his celebrated voltameter, or measurer of voltaic electricity. His next researches related to the "contact theory" on the origin of power in the voltaic pile, and culminated in the utter demolition of this theory in a paper communicated to the Royal Society in 1840. The first great paper on Frictional Electricity was sent to the Royal Society on Nov. 30, 1837; and here says Dr. Tyndall, "we find him face to face with an idea which beset his mind throughout his whole subsequent life—the idea of action at a distance"—which drove him to experiments, with him always prolific of important results. Regarding these profound researches, Dr. Tyndall said, "though often obscure, and sometimes inconsistent, a fine vein of philosophic thought runs through these investigations. They contain dark sayings, difficult to be understood. It must, however, be remembered that he worked at the boundary of our knowledge, and that his mind dwelt habitually on the 'boundless contiguity of shade,' by which that knowledge is surnamed." In 1840 Faraday felt the effect of the great mental strain of so many years, and in 1841, he retired, with his wife, to Switzerland; and "to her loving care," said Dr. Tyndall, "we and the world are indebted for the enjoyment of his presence so long."

CRYSTALLIZATION.

A CURIOUS discovery has been made by M. Auguste Bertsch, and turned to practical account by M. Kuhmann, the celebrated chemist. Who is there that has not, during cold winters, stopped to admire the beautifully symmetrical and yet fantastic figures of leaves and flowers depicted on the window-panes of a well-heated room, the air of which is charged with aqueous particles? M. Bertsch has found that Epsom salts (sulphate of magnesia) dissolved in beer, together with a small quantity of dextrine (artificial gum), and in this state applied to a pane of glass with a sponge or brush, will, on crystallising, produce the identical designs above alluded to, hitherto considered peculiar to water; with this improvement, however, that the liquid may receive any colour whatever, at the option of the operator. The ephemeral productions of frost may thus be easily perpetuated; but M. Kuhlmann, on being apprised of the fact, conceived the idea of going a step further, and transferring those fairy-like creations to stuffs and paper. For this purpose he first got the crystallizations on sheets of iron, on which he afterwards laid one of lead. By means of a powerful hydraulic press the minutest details of the figures in question were durably imprinted on the soft metal, and a copy of them in relief was then obtained by galvanoplastics. But here another difficulty arose. In the impression of cotton stuffs the pattern must be continuous; whereas in M. Kuhlmann's plates the lines at one end would clearly not coincide with those at the other, so that disagreeable interruptions would be caused in the printed designs. This obstacle, however, has been overcome in a most ingenious manner by effecting the crystallisation on the cylindrical surface of a roller. A slight rotatory motion imparted to it will prevent the liquid from accumulating at any particular point before it has evaporated—*Galignani*.

CRYSTALLIZATION OF ICE, AND THE FORMATION OF BUBBLES OF AIR IN THE CONGEALED MASS. BY M. A. BARTHELEMY.

DR. TYNDALL's observations on the decrystallization of ice by heat have proved that an apparently amorphous fragment of ice is not formed of agglomerated crystals. On the other hand, the observations of M. Bertin on the double refraction of ice have confirmed the assertion of Sir David Brewster that ice belongs to the Hexagonal System. I may here also refer to a paper of my own which appeared in the *Comptes Rendus*, in 1857, in which I described some large hailstones having the form of hexagonal transparent pyramids. Since 1856, I have made observations which appear to me to confirm this view of the crystallization form, and which at the same time elicit some remarkable facts in connection with the crystallization of water.

A cask, whose upper end had been removed, was filled with water and exposed, in a northernly situation, to a temperature of 7° or 8° below zero. The ice was first formed in the upper part, the staves of the cask yielding to the expansive force. When the layer of ice was about a decimètre thick, a shock almost perpendicular to the surface produced a rupture into six rays, which formed three diagonal planes of a six-faced prism. At another point on which the shock was exerted, there were produced three new planes parallel to the first ones. . . . . Meanwhile, the surface remained smooth, and it was impossible to recognize the lines of division (*les fentes*) of the three planes, either because the upper stratum of ice was amorphous, or because the phenomenon of regelation had taken place.

If the process of freezing is continued, and the ice forms on the walls of the cask, so as to enclose a mass of liquid in its interior, the ice then raises itself at the upper surface in the form of a crater, and exposes (*donner accès à l'eau intérieure*) the water within, which in its turn becomes solidified. This phenomenon seems due to a double cause. At first, the ice, in descending below zero, contracts and presses on the liquid nucleus within; in the second place, the ordinary water containing foreign substances in solution becomes more and more concentrated as the ice forms; its point of congelation is therefore retarded in proportion as the formation of ice is advanced, and it continues to dilate in freezing. This reaction of the inner nucleus would become especially considerable when the gas held in solution by the water tended to be disengaged by the saturation of the liquid. This is what takes place in the following experiments:—

Bottles filled up to certain heights with water were exposed outside a window to the action of continued cold. The next day an olive-shaped liquid nucleus was soon enclosed in the mass of ice, and from the nucleus passed out in all directions bubbles of gas elongated and imprisoned by the ice. The normal disposition of the surface of the nucleus proved that the gas had acted in obedience to a hydrostatic pressure directed from within outwards. Liquid containing carbonic acid, and treated in the same manner, gave a liquid nucleus covered with long fine needles of carbonic acid, easily distinguished from those of air, which are always clariform.

When the bottle is full of water, it always bursts at a point next the window near which it is placed. Now, in this case, the congelation pushes the olive-shaped nucleus towards this point of the bottle, so that the rupture *seems to me to be due not only to the expansive force of the ice, but also, and especially, to the pressure of the liquid nucleus and the expansive force of the gas which it disengages. I have proved, in fact, that bottles of distilled water placed under the same circumstances do not burst at all.*

I have also exposed during the night to a temperature of 7 deg. or 8 deg. below zero, a flask containing carbonic acid dissolved in water up to the point of saturation. The next day the mass was completely congealed, with long vertical and attenuated filaments of gas. Moreover, the upper part presented horizontal layers superimposed to a thickness of eight centimètres, with a crateriform elevation in the centre.* These layers had been formed by the reaction of the inner water, which burst the already formed surface in order to undergo congelation, so that the upper stratum was really the most recent one.— *Comptes Rendus; Scientific Opinion.*

ICE-MAKING MACHINE.

A PNEUMATIC ice-making machine, the invention of M. E. Carré, of Moislans, has been exhibited by Dr. Roscoe at the Royal College of Physicians last week, and found to fulfil all the invention promised. It is a simple air-pump, under the receiver of which the carafe, decanter, or vessel of water to be frozen is placed. The action of the pump necessarily causes the rapid evaporation of the water, and consequent reduction of temperature. So far, all is very simple; but now comes the most ingenious part of M. Carré's invention. To remove the atmosphere of aqueous vapour resulting from the exhaustion of the air, it is made to pass over the surface of some strong sulphuric acid, kept in agitation by a peculiar contrivance, so that a large surface is presented to the vapour. By this means, rapid absorption of the vapour takes place, and, consequently, rapid evaporation and reduction of temperature; so that, in about three minutes, a decanter of water is reduced to the freezing point. An apparatus for congealing one decanter at a time costs only £4. There is but one drawback to the very extensive use of this ingenious apparatus, and that a little instruction would quickly remove. The sulphuric acid soon becomes too dilute to absorb the vapour of water. A chemist knows how to deal with this dilute acid, which only requires to be evaporated to be ready for use again. But with ordinary people, it will always be a difficulty and a loss.—*Mechanics' Magazine.*

PRESERVATION OF ICE.

A CORRESPONDENT of the *Times* has communicated to that journal the following interesting facts and experiences as to the Preservation of Ice in Hot Weather.

" As one of the originators of the scheme of importing Wenham Lake ice from America, and one formerly largely interested (although now no longer so) in that immensely successful enterprise; and as the original instructor of the English public in the use as well as in the preservation of 'lake ice,' as distinguished from 'rough ice,' I may pretend to some knowledge of the subject; and I beg to say that, although the cost of a refrigerator or ice-box may be reduced to a *minimum* by coarseness of manufacture and absence of finish, economy may be carried too far, and no box will prevent or even retard the melting of ice which does not combine the following conditions:—1. It must have double sides, bottom, and lid, with the space between the two casings filled by some non-conducting substance capable of being closely packed, in order to prevent the action of the external temperature. 2. The inner lid or cover must be practically, if not hermetically, air-tight, in furtherance of the same result. If external air enters, it will bring its own temperature with it.

3. There must be a drainage pipe at bottom to carry off instantaneously every drop of water formed by the melting of the ice, and this pipe must either be fitted with a 'trap' or curved in such a way as to prevent air from coming in where the water goes out.

"The necessity for excluding air is already explained; but it is very hard to make people believe that it is even more indispensable to carry off every drop of water. Ice has no such enemy as water. Expose a piece of ice weighing, say, 25 lb. to the air at a temperature of 75 deg., but so placed that it is perfectly drained, and at the end of twenty-four hours it will scarcely have disappeared. Wrap the same piece of ice in three or four thicknesses of blanket or flannel, and place it in a small tub exposed to the same temperature as before, and as the water filters through the blanket the ice will 'stand in its own water,' and be all dissolved in five or six hours. Wrap the same piece of ice carefully in a blanket, and place it on a grating, or on four crossed sticks, so that no water can accumulate underneath, and at the end of three or even four days it will not have entirely melted.

"Ice has two 'natural enemies,' warm air and water, but the latter is by far the more deadly. Water at 40 deg. will melt ice with ten times the rapidity of air at 80 deg.

"Dry sawdust is not a bad protector, and for ice in large quantities is the best; but for blocks not exceeding 50 lb. 'there is nothing like flannel.'

"From what I have already said, it is evident that your correspondent who recommends putting ice in a tin saucepan before burying it in sawdust is ignorant of the effect that will be produced by the water which it will at once begin to form. If he must use a saucepan, let him first punch a hole in its bottom.

"Pure ice is now in such universal use, both as a medicine and as a necessary rather than a luxury, that it will hardly be believed that the presence of cholera in 1849 almost stopped its use; and people will smile when I say that Mr. Staples, of the Albion, in 1846, told me that if he were to put lumps of ice in dishes on the table at any of his public dinners, with a view to the guests putting it in their wine, it would destroy his reputation."

[The first account of the Wenham Lake Ice Trade appeared in the *Illustrated London News*, 1844, with several engravings: it was re-printed, by permission, by Mr. Macculloch, in his *Dictionary of Trade and Commerce*, and has been thenceforth copied into various dictionaries and cyclopædias.]

SOLAR PHENOMENON.

PROFESSOR ROSCOE has notified to the Literary and Philosophical Society, the important discovery, made independently by M. Janssen, at Guntour, in Indian, and by Mr. Norman Lockyer, in London, of the visibility of the spectral lines of the red solar prominences under ordinary circumstances. Hitherto, these protuberances or red flames have only been seen during total eclipses of the sun; but by the application of the spectroscope, in conjunction with the telescope, the peculiar bright lines which these prominences exhibit, indicative of the presence of glowing gas, can now be observed whenever the sun is visible. Although the priority of this interesting discovery is due to M. Janssen, who first observed the protuberances the day after the eclipse, the method having occurred to him whilst observing during the eclipse, yet Mr. Lockyer had suggested this particular method of examination no less than two years ago, and had succeeded in his endeavour before he became aware of M. Janssen's prior observations. From the accounts, as far as they are yet published, we learn that the bright lines appear to be identical with those of hydrogen.

Mr. Baxendell stated that this discovery would give a great impetus to the progress of our knowledge of solar phenomena, and that the importance of observations on this plan could not be over-estimated.

THE SUN'S ONWARD MOTION.

FROM the consideration of the imperfect information afforded by the stars' apparent proper motion, astronomers have been able to deduce one of the most interesting astronomical discoveries yet effected. They have learned that the sun, with its attendant system, is speeding onward through space, in a certain direction which they have been able to assign, and at a rate of no less than 150 millions of miles per annum. A law also affecting the general system of stellar motions has been guessed at, and has been considered by many eminent astronomers to be supported by sufficiently satisfactory evidence. It has been supposed that the proper motions of the stars indicate a vast series of orbital motions round a point in space which does not lie very far from the star Alcyone, the principal star of the Pleiades.—*Fraser's Magazine.*

THE ATMOSPHERE OF THE PLANETS.

THE labours of Professor Tyndall, and his compeers, have shown that it is quite impossible to judge what a planet's climate may be from the mere consideration of the planet's distance from the sun. The extent and quality of the atmospheric envelope around a planet exercise fully as important an influence on the planet's climate. The sun's heat may either be retained or radiated away as fast as it is received. If a planet has an atmosphere which is always loaded with aqueous vapour, the heat poured on the planet passes freely through this vapour to the planet's surface, but it does not pass freely away again; it is retained and stored up precisely as in a glass-house. But dry air has not this power; the reflected heat passes as freely through it as the heat directly received from the sun. There are vapours and gases which have yet more power than aqueous vapour in preventing the escape of heat. Amongst these are the gases emitted from flowers; and Tyndall estimates that "a layer of air 2 in. in thickness, and saturated with the vapour of sulphuric ether, would offer very little resistance in the passage of the solar rays, but would cut off" more than one-third of the rays which would otherwise pass away as soon as received. "It would require no inordinate thickening of the layer of vapour," he adds, "to double this absorption; and it is perfectly evident that, with a protecting envelope of this kind, permitting the heat to enter, but preventing its escape, a comfortable temperature might be obtained on the surface of our most distant planet." When we remember, on the other hand, that during the full heat of the tropical summer, the lofty slopes of the Himalayas and the Andes remain covered with snow, we see how largely a diminution in the extent of a planet's atmosphere may diminish the effect of the sun's heat. And precisely as our countrymen in India find in the Himalayas the climatic relations of the temperate zones, so the inhabitants of Venus and Mercury may enjoy a climate as genial as that of our own earth.—*Saint Pauls Magazine.*

THE MOON.

Mr. W. R. Birt has read to the British Association a paper "On the Extent of Evidence of Change on the Moon's Surface." The author of this paper remarked that the two opposite questions of fixity of, or change on, the moon's surface must be decided by observation, and not assertion. With regard to evidence on the question of fixity, such evidence—resulting from observation, and not including theoretical considerations—must be exceedingly scanty; indeed, it is difficult to conceive how the unalterable state of the surface of our satellite can be determined by observation, for if, as has been asserted, "all changes on the moon's surface have ceased myriads of ages ago," we are certainly destitute of the records of the observation of the real state of that surface at so remote a period; and, even if "fixity" of the more minute details be really established at any one point by a long series of observations, it would be no argument for its universal prevalence, since a state of quiescence might be attained at very different epochs in different regions. The author next proceeded to examine the question of change, and glanced at the attempts to perpetuate a knowledge of the moon's surface by means of maps, drawings, and topographical descriptions, remarking that it is by the study of details that a definite answer must be given. These details are numerous, embracing mountains, valleys, plains, craters, rings nearly filled with bright spots, as mountain-tops, and others less bright, but presenting phenomena difficult of explanation, dark spots with bright rims, or bounded by distinct lines, separating them from the surrounding surface. All such objects must be carefully studied before a conclusion can be drawn as to their unalterable stability or their mutation. The means of obtaining evidence on those points consist in the examination of dolineations and topographical notices on the one hand, and comparing them with the moon, by personal observation of the objects, on the other. Mr. Dirt referred to a diagram giving two aspects of the same spot, one as given—lighter than some surrounding objects—by three authorities, Lohrmann, Beer and Mädler, and Schmidt; the other as observed by himself at a recent date, in which the spot is darker than all surrounding objects. In connection with these differences of colour, he put the question, Can we decide for change? In reply, he pointed out one great disadvantage, namely, the uncertainty of the number of observations on which the earlier records rest, and showed the great importance not only of increasing our own observations, but also of soliciting the aid of others, that there may be no want of confirmatory evidence to establish the certainty of what is recorded. In the absence of confirmatory observations, Mr. Dirt considered that the evidence capable of being brought to bear on questions of change is very limited, especially as former records are more or less open to be regarded as inexact drawings or inaccurate statements when they happen to differ from present observed appearances.

THE ANEROID BAROMETER.

Dr. Balfour Stewart has read to the British Association a paper on the behaviour of the Aneroid Barometer at different pressures. Experiments had lately been made, with the view of ascertaining to what extent an aneroid may be considered a reliable instrument when exposed to considerable changes of pressure such as occur in mountain districts. By means of an air-pump the aneroids, when placed in the receiver, might be subjected to any pressure. A method of tapping the aneroids had also been devised, and by this means the experiments as to the deviation of the results given by these instruments were conducted with comparative ease, and with the greatest accuracy. The experiments were still going on.

Sir William Thomson said the aneroid had become so popular an instrument that many had satisfaction in learning that it was capable of giving results with scientific precision. Dr. Stewart had shown that in taking a barometer up a mountain of 12,000 ft. the error would only be about 300 ft., and had also shown how to correct this error. By carefully using these instruments, therefore, they had a probability of determining, with much less

probability of error, the height of a mountain of 12,000 ft.
Among the very important matters which occupied the attention
of the British Association, one which might, with very great
advantage, be followed up, would be the carrying out of experiments on the elasticity of metals, and all solids capable of being
experimented upon. He remarked on the elasticity of metals,
and even of rocks, and referred to the time taken by the earth in
consolidating—that this had taken place less than a thousand
millions of years ago. The earth was not, he considered, one-
tenth as old as the popular geologists would make it.

WEATHER INTELLIGENCE.

As regards the communication of Weather Intelligence, General
Sabine informs us that the system has been further developed, so
that the drum-signal is now hoisted at ninety-seven British stations. The news thus conveyed, it should be understood, is not
a forecast or prophecy, but a fact. Since February last similar
news has been flashed to Hamburg, and the harbour authorities
there have resolved to hoist the drum, and at Cuxhaven, whenever
intelligence implying probable danger shall be received from
London. In France also, under the direction of the Ministry of
Marine, the practice of telegraphing facts has been adopted. Besides all this, the London office makes known to Liverpool and to
Holland the existence of a certain amount of barometric pressure
between two stations within a defined area. The influence which
the distribution of atmospheric pressure exerts on the motion of
the air has been much dwelt on by Dr. Buijs Ballot, of Utrecht,
and a rule has been propounded by him for inferring the coming
direction of the wind, from simultaneous readings of the barometer at different places. For more than a year past the London
office has sought to test this rule by systematic discussion of
daily meteorological charts of the British Islands, and the nearer
coasts of the Continent. The results of this investigation are,
on the whole, encouraging.—*Athenæum*.

TIDAL OBSERVATIONS.

PROFESSOR RANKINE, in introducing to the British Association
the "Report of the Committee on Tidal Observations," said that
the system on which these investigations were carried on might
be termed harmonic analysis. The extension and improvement
of tidal observation and search of data were the objects of this
Committee. The idea they attach to harmonic is a solution in
accordance with the periodic functions of the tides. The Committee has been appointed for several years; and last year's work
was a special series of tidal observations, on a system proposed by
Sir W. Thomson. The chief result of these labours was, that
the height of the water might be expressed as the sum of a single

number of harmonic functions of the time, which are certain expressions of the solar and lunar motion. The chief tidal constituents are those whose periods are twelve mean lunar and solar hours. The revolution of the earth and moon, the earth's rotation, and the progress of the lunar perigee, are the data on which these calculations are based. Sensible tides are also found dependent on the fourth motion of the moon's parallax. The Committee's calculations were drawn from curves taken by a self-registering tide-gauge. There existed a periodic lunar variation every nineteen years, which might furnish data for the estimation of the moon's mass. Ease and accuracy in estimating the height of high tide were attained by these harmonic analyses. We might also further ascertain how much the earth's angular velocity was increased or diminished by these tides from century to century. The tendency of friction is to increase the rapidity of high and low water. Lunar fortnightly tides also formed part of the same inquiry. Mr. Roberts, of the Nautical Office, had taken great pains in these inquiries, and it was hoped that his whole time might in future be given to the work.

Mr. Webster said it was impossible to overrate the importance of the subject and the labour of it. It involved questions which amount nearly to a new theory of the universe. This report is a result of the union of the Mathematical and Physical sections. There are observations also made by a Joint Committee of the Estuaries of the Wear, Humber, and Ouse, which are the bases of these calculations. Observations now taken at different places should be taken on a uniform basis, in order that a definite theory may be laid down. The paper relates also to ocean tides; in estuaries private companies have taken local tide estimates. There are many places around the coast where local tide phenomena exist; as, for example, the four daily tides round the Isle of Wight. He would suggest the appointment of a Committee of five. The President considered this a matter for the Committee. —Mr. Adams said the system was not only applicable, but more appropriate for local than for ocean tides. Admiral Ommanney desired to testify to the ability of the Report.

TERRESTRIAL MAGNETISM.

THE completion of the reduction of the Magnetic Survey of the South Polar Regions is thus referred to by General Sabine, in his annual address to the Royal Society; the papers and maps therein mentioned, being "his own achievement." "The reduction, he says, "of the great scientific work, the Magnetic Survey of the South Polar Regions, commenced in 1839, under the auspices and at the expense of Her Majesty's Government, has been completed in the present year by the presentation to the Royal Society, and the publication in the *Philosophical Transactions* of Maps of the three Magnetic Elements in Southern Parallels,

G

commencing in 30° south, and extending far beyond the limits of ordinary navigation. These maps are accompanied by tables containing the numerical co-efficients to be employed in a revision of 'Gauss' General Theory' at the intersection of every fifth degree of latitude and every tenth degree of longitude, between 30° south latitude and the south terrestrial pole. The magnetical determinations of the Survey correspond to the epoch 1842-45. Similar maps for the corresponding latitudes of the northern hemisphere, from 30° north latitude to the north terrestrial pole, are in preparation, founded on a co-ordination of results obtained by magneticians of all countries in the fifteen years preceding and the fifteen years following the same mean epoch of 1842-45, and reduced to it. It is hoped that these maps, with an accompanying memoir, will be presented to the Royal Society before the close of the present session. There will then remain for subsequent completion, the filling up (still for the same epoch) of the space between the parallels of 30° north and 30° south latitude, for which much preparation has been made in the assemblage of materials requiring only for their co-ordination the allotment of the time needed for the due examination and treatment of so large a body of materials. Should I be so happy as to be able to complete this task also (my work on Terrestrial Magnetism has now extended, more or less, over half a century), I venture to express a hope that the great work of which the foundation will thus have been laid, viz. the Revision of the Gaussian Theory, corresponding to a definite epoch in the great cycle of terrestrial magnetism, may, when a suitable time shall appear to have arrived, be taken up and completed under the auspices of the Royal Society."

We subjoin the *Athenæum* comment :—" Who is there will not join in the wish that the Nestor of the Terrestrial Magnetism may go on to finish his admirable work ?"

REGISTERING EARTH CURRENTS BY PHOTOGRAPHY.

IN a darkened room in the Meteorological Department of the Royal Observatory at Greenich, the Astronomer Royal fixed, some two or three years ago, a sensitive little galvanometer, with reflecting mirror, to register earth currents by photography. One end of the wire of this galvanometer is connected with the earth, and the other end with a telegraph wire, which, after running several miles along a neighbouring railway, is again connected with the earth. Now, currents of electricity are continually running to and fro in the earth, and sometimes these currents enter the telegraph wires; also, when they are strong enough, they overpower the ordinary working batteries, and send unreadable messages on their own account. Such disturbances sometimes stop for a time all messages and news, and on one exceptional occasion caused a panic on the Stock Exchange by delaying news of importance. Being a source of great occasional loss to the telegraph companies, the endeavour of electricians has always been to neutralize and get rid of the currents as soon as possible. The object of the apparatus in Mr. Glaisher's department at Greenwich, is, on the contrary, to watch and examine all the movements of these earth-currents. From the description already given, it will readily be seen that as currents from the earth flow through the wire erected from Greenwich Observatory, the needle of the galvanometer shows the direction, and approximately the strength, of the current. On the little magnetic needle a mirror is mounted, and a ray of light from a steady flame, after falling upon the mirror, is reflected as a spot of light upon a sheet of photographic paper. Hence, as the needle moves, the ray of light moves to and fro upon the paper. This sheet of sensitized paper is fixed round a cylinder of ebonite, which, by clockwork, makes one revolution every twenty-four hours.

At the end of the twenty-four hours the paper is taken off the cylinder, and a fresh sheet substituted. The record upon the first sheet is then developed in the usual way, and a zig-zag dark line passing across it from end to end shows the movements of the galvanometer needle and spot of light during the past twenty-four hours. The darkened room, in which this self-registering apparatus is at work, is built without iron nails, and this metal is used nowhere in the room, so that the magnetic instruments shall not be disturbed by artificial causes. In a communication made by Professor G. B. Airy to the Royal Society, he has made known the fact that the declination magnet and the earth-current galvanometer are affected at quite, or very nearly, the same time, showing some relationship between the two phenomena, the causes of both of which are at present a perfect mystery to men of science.

NEW GALVANOMETER.

Mr. F. H. VARLEY has read to the British Association a paper "On the Construction of a Galvanometer for the Detection of Weak Electric Currents." Sir W. Thomson has shown the importance and value of using small cores upon which the electro-magnetic helices are wound, and the advantage of employing small magnets for indicating and measuring the amount of force flowing through the galvanometer coil or helix. The small magnet of Sir W. Thomson has a mirror attached to it, to reflect a beam of light, so that a small motion of the magnet gives movement to a line of light thrown upon a darkened screen. It has frequently occurred to the author that smaller and lighter magnets could be employed by calling in the aid of microscopic power. Two instruments were constructed with this view. The first consists in suspending by a single filament of silk in the hollow core of the galvanometer coil a magnet of an inverted

spur-form, made of the finest steel wire that can be obtained,
and rendering its motion apparent by viewing it through a rectangular prism, by means of a microscope, in the eye-piece of
which is placed a small scale, photographed on glass: the magnet appears as a black bar bisecting the field of view; and, as
the finest wire obtainable for this purpose appears, when sufficiently magnified, as thick as a scaffold-pole, it is obvious that
the slightest motion of the magnet is rendered conspicuous by
the image moving to or from over the graduated scale in the
eye-piece. The second form is more sensitive than the first; a
small magnet, made of flat steel polished on one face, is suspended in the usual way by a single filament of silk; a small
micro-photograph of a graduated scale is placed at such a distance from the reflecting surface of the magnet-mirror that each
division equals two minutes of arc, as nearly as possible; the
image of the scale thus reflected is sent in a line with the optic
axis of the microscope; any deflection given to the magnet
causes the image of the photographed scale to move across the
field of view. The reflecting surface moving doubles the apparent motion, giving the amount due to the angle of incidence
plus that of reflection. The movement of one graduated division
being produced by a deflection equal to one minute of arc, if
magnified sixty times by the microscope, will render a motion
equal to one second of arc apparent and measurable. When
desirable, a small scale placed in the eye-piece can be made to
give a vernier reading upon the magnified scale. The magnifying
power can be increased where desired, and most minute amounts
of motion rendered measurable. The great difficulty of using
instruments of such extreme sensibility is due to the interference
of extraneous vibration communicated to the small magnets.
This, to a great extent, can be overcome by insulating the
various parts from vibration by means of antagonizing springs,
and preventing the finer vibration from being communicated
through the wire itself by covering the wire with silk or cotton,
to act as a damper to the more minute vibrations. This instrument being more compact, and not requiring a darkened room
set apart for its special use, its application is much more general,
whilst it at the same time gives much more minute and sensitive
measurements. The instrument can be used in broad day, either
in or out of doors, and is applicable to all kinds of observation.

THE PHYSICAL COMMOTIONS THROUGHOUT THE GLOBE.

WE find the following very thoughtful paper in *The Builder :*
—" These commotions unfortunately are not yet at an end. The
rendings of the earth's crust south of the equator have only
partially relieved the internal pressure, and now they have extended into the northern hemisphere, and California has been
seriously shaken, so that San Francisco has had many buildings

thrown down, and several villages have been reduced to heaps of ruins. The island of Hawaii, in the Pacific Ocean, opposite these coasts, is slowly sinking into the ocean, on the western and southern shores, which are now several feet lower than they were in April last, when the terrible volcanic eruption took place there. Since the first attack, a second earthquake, but much less violent, has occurred at San Francisco. The disturbance of the Pacific Ocean has been of an extraordinary nature in connection with these commotions—at least, with those of August; and, considering that the bed of the Pacific is like a sunken continent whose mountain-tops alone appear in the form of its innumerable islands, there seems to be little doubt that the crust of the earth is thinner there than on the continents above water. This fact seems to corroborate the idea, that the cause of all these commotions is really cosmical, or *underlies the whole* crust of the sphere, and hence operates chiefly where the crust is thinnest; and that it is such as Hopkins, of Cambridge, represented the cause of old crust-fractures to be—an expansive force, or force operating outwards from within the earth's crust. Moreover, if it be such a force, it will tend to relieve itself chiefly by *rendings north and south*, if it be connected, as we have suggested, with the earth's rotary and centrifugal force, and especially in regions more or less extending from the equatorial. Thus, too, the idea prevalent amongst geologists that the earth is still essentially a fluid or molten though encrusted sphere,—and, indeed, *à fortiori*, it must be admitted that a spherical form indicates a fluid substance,—is one which is much more capable of explaining the geological phenomena than that of a solid sphere, particularly if, as we have suggested, the admitted or recognised tendency to expansion is derived from the rotation of the sphere.

"The problem of the influence of a varying rotation on a molten and encrusted sphere (though there is no actual evidence of the earth's rotation being at present either on the increase or the decrease) is one of peculiar interest and importance, and is capable of affording curious explanations of the present states of other planets as well as our own. For example, the greater the rapidity of rotation, the lighter in specific gravity *ought* the fluid and encrusted sphere to be, as well as the more expanded in dimensions; and the less the rapidity of the rotation, the denser and the less expanded ought they to be. Now, it is evident that on a general view, this is the fact,—namely, that it is precisely those planets which rotate with the greatest rapidity that are the lightest in specific gravity, as well as the most enormous in circumference; and it is not difficult to explain what may not at first sight seem so evident. Jupiter rotates once every nine hours; and hence it is, probably, that his great molten mass is so levitated, expanded, and centrifugalized by this tremendous rapidity of rotation, that the specific gravity of his vast sphere is reduced to something like that of water, although, for all that, he may thus be an encrusted sphere. Saturn, with his centrifugal rings, his magnificent dimensions, and his specific gravity like that of cork, also rotates with immense velocity. On the contrary, Mars, Venus, the Earth, and Mercury,—all small in dimensions,—are comparatively dense in substance: hence, as the earth at least does, they probably all rotate comparatively slowly. There is no great planet of anything like the density of the smaller ones."

THE GREAT MELBOURNE TELESCOPE.

THIS gigantic Telescope (*see vignette*) which has been constructed by Mr. Grubb, is of the form known as the Cassegrainian Reflector, and is mounted upon what Mr. Grubb calls "the German system improved." We select and abridge the following leading details from the *Quarterly Journal of Science*, October, 1868:—

The mirrors, two being supplied in case of accident, are 4 ft. in clear aperture, 4½ in. thick, 30 ft. 6 in. in focus, and rest in their box on Mr. Grubb's system of hoops; the whole system of suspension and levers weighs altogether nearly 2 tons.

Of the tube, 7 ft. is made of boiler-plate iron, ¼ in. thick, to which is attached by flanges and bolts a skeleton tube, 21 ft. long, of steel bars, 3 in. wide at bottom, 1½ at top, and 1-6th of an inch thick, wound spirally round rings of carefully turned angle iron, and riveted at the joints, forming a spiral lattice of amazing strength, stiffness, and freedom from tremor. The 7 ft. of boiler-plate tube weighs 1,300 lbs., and the 21 ft. of the ventilated tube, only 1,370 lbs.

At the upper end of the tube, about 25 ft. 6 in. from the large mirror, is bolted a very stiff hollow arm of steel-plate, on the extremity of which is a V-shaped gun-metal casting, in which slides an arm carrying the small mirror of 8 in. diameter. This arm is acted upon from behind by a screw, from a pulley on the shaft of which wire cords are carried over iron guide-wheels down the side of the tube, where they are wound round a wheel, to which motion can be given by the observer, for the purpose of focusing.

The polar axis is made up of four distinct parts, viz., a cube 3 ft. square, to which is bolted on one side a cone 6 ft. long, which terminates with a bearing 12 in. diameter, resting in a peculiar "plumber-block" on the polar pier; on the opposite side a short toe-piece, which carries on parts prepared for them the two hour-circles, sector, and clamp, and terminates in a bearing 6 in. diameter resting in a Y-block in the equatorial pier; and on a third side a bell-shaped casting about 2 ft. long, which terminates in a slide carrying one bearing of the declination axis, the other being in the side of the cube opposite the bell.

The declination axis is 24 in. diameter at the bearing next the telescope, and 12 in. at the other, the bearings being 6 ft. asunder, and the axis itself about 9 ft. long. The declination axis carries at one end the telescope strapped into its cradle, and at the other counterpoise weights amounting to over two tons.

Up to the present time this has been the great objection to this form of equatorial; for the axis had necessarily to be put into the entire collared bearings without any possible relief of friction, the force necessary to move the telescope, if of any considerable size, became so great as to oblige the constructor to make the bearings of the declination axis very small, and consequently rendered the support of the telescope weak and unsteady.

Only in this telescope, and in one other (also constructed by Mr. Grubb) which is now mounted at Dunsink Observatory, near Dublin, (the object-glass being presented by the late Sir James South to the University of Dublin), has this difficulty, one of the very few objections to this form of equatorial, been overcome.

The support of the specula is one of the most important points in the mounting of the large reflector, and one which, if certain difficulties were not disposed of, would be a complete bar in the way of mounting them equatorially. Two things are essentially necessary:—

1st. A system of back-supports on which the speculum may lie in a state free from strain of any kind—as if in fact it were floating in mercury.

2nd. A system of lateral support which will preserve the mirror free from strain when turned off the zenith, and will not constrain any slight movement of the speculum.

Now, as regards the first condition, nothing could fulfil it better than the system introduced long ago by Mr. Grubb, which was made use of on such a grand scale by the late Lord Rosse, as well as by others, and this system, therefore, with some modification, was adopted.

The ultimate points of the tertiary system are gun-metal cups, which hold truly-ground cast-iron balls with a little play; and when the speculum is laid on there, it can be moved about by a person's finger with such ease as to seem to be floating in some liquid.

The system of lateral supports was devised by Mr. Grubb specially for this instrument. One objection raised to the mounting of those large telescopes (reflectors) equatorially, was that such heavy mirrors could only be supported laterally in a flexible hoop, to preserve them from strain; and, consequently, that it was impossible to mount them equatorially, when they would sometimes be turned upside down, and would roll out of their hoop. Mr. Lassel got over this difficulty by making his telescope revolve in its cradle, so that he could always keep one diameter vertical; but this plan, apart from its inconvenience and clumsiness, destroys all power of adjustment.

The driving clock is, as may be readily supposed, of rather ponderous dimensions, but is still, considering the size of the

telescope it has to move, as compact as possible, fitting in a
niche 2 ft. square and 3 ft. high. Its regulating power is that
of a governor, the balls of which, when the desired speed is
attained, fly out and bring two leather-pointed screws into contact with a circular disc, which serves by the friction thus produced to prevent any acceleration of rate. No clock is able to
produce any correction in speed until after an error has been
committed. Consequently the question becomes, What clock is
it that corrects these errors most efficiently and most instantaneously? That this clock corrects in speed efficiently is proved
by the fact that doubling the driving-wheel (viz., that weight
which will just keep the friction-screw in contact) produces an
acceleration in speed of only 1-360th part, whilst the great
vis inertia of the balls themselves (weighing about 20 lb. each)
prevents any sudden oscillation, so to speak, in rate; and that
this regulating action takes place more instantaneously than in
other clocks, may be readily understood by inspection of the
construction, when it will be seen that the actual movement required in any of the members of the governor to correct an excessive difference of power like the above, amounts to something
probably under 1-1000th of an inch, instead of having to move
by some considerable quantity a complicated system of link-work
connected with fans, breaks, water-regulators, air-regulators,
&c.

The general effect of these complicated systems is that, when
any error is made, the ensuing correction is too great, which
necessitates a counter correction in an opposite direction, and so
an oscillating effect is produced, which gradually subsides until
matters again attain their equilibrium. An experiment has been
tried with this clock which will form a record of its work, and a
comparison for other clocks.

The clock was used to give motion to a long ribbon of paper
on which lines were traced by a pencil worked by a free pendulum in a direction at right angles to the motion of the paper
itself. The result was, of course, a waved line, the length of each
wave being a measure of the second as given by the clock. During
the process at certain points the driving-weight was increased
from 1½ cwt. to 2, 2½, and 3 cwt., but the difference of rate is
insensible on the diagram, although capable of being measured
to the 1-100th of a second. If this experiment were tried with
other clocks, it would form a valuable record of their work, independent of the testimony of their constructor. The shaft
which is carried from the clock to drive the telescope is severed
in one place, and a system of six differential wheels introduced
by which the rate can be rapidly altered 1-20th part, this being
the mean of the extreme differences between sidereal and lunar
rate; whilst a small lever and graduated arc, in front of the
clock, acts on the governor-spindle, moving it up or down, and by
this means altering the working angle of the balls, to make the

final correction per lunar rate pro tem. Both of these adjustments can be made while the clock is in action.

The preparation of the specula was carried on in buildings specially constructed, wherein the mounting was both manufactured and erected. The alloy used was rather higher than what has hitherto been used for such large mirrors. The proportion was, copper 32, tin 14.77. The mirrors were cast on Lord Rosse's "bed of hoops," but with several important modifications of his process. The annealing was conducted in a circular oven, and took 23 days. Immediately on being removed from the oven, the specula were placed on the machine, and rough-ground on front, back, and edge, during which process they rested on several thicknesses of cloth; from this they were removed to their own boxes, where they rested on the system of levers before described. The finer grinding and polishing was then proceeded with, and the specula were never after raised from their supports.

For the purpose of trial, as it would have been very troublesome to be obliged to remove the speculum to the telescope on every occasion, the machine is so constructed that by turning a handle, the whole frame carrying the speculum is tilted into a vertical position by the endless screw and worm, and a dial-plate and artificial star being placed at 800 ft. distance, the mirror, used as a front-view telescope, is ready for trial in five minutes, which saves some two days' work on every occasion, besides avoiding risk. So efficiently and with such certainty did this machine do its work, that on no occasion, having once ascertained what was required, did it fail in providing the requisite alteration of figure, while as compared with other machines the ease and steadiness of its working were remarkable.

The photographic apparatus embraces every improvement proposed by Mr. W. De la Rue, and more especially those suggested by some experiments made with a rough trial apparatus fitted to the great telescope, with which photographs of the moon have been taken, exceeding in beauty and sharpness any hitherto accomplished in this or any other country.

The spectroscope has been made with a totally different style of mounting to those in general use, with a view of making it a more practically useful tool; whilst the prisms which are adapted to it are made of one very dense flint of about 90 or 100 deg., with prisms of light crown cemented on each side of angles of about 20 to 25 deg. This gives a very large amount of dispersion with very little deviation, whilst the crown prisms protect the surfaces of the flint from tarnish, which has proved such a drawback to the use of heavy flint, for although it can be removed by friction and nitric acid, yet every time this operation is resorted to it leads to a depreciation of surface.

The contract signed in the beginning of the year 1866, included all the above-mentioned parts, with the exception of the photographic apparatus and the spectroscope, and amounted to

£4,660; the instrument to be completed in twenty months, and although that time was slightly exceeded, yet, as stated by the committee, the delay arose solely from the state of the weather precluding all observation for trial of the specula.

The many large telescopes before constructed by Messrs. Grubb are well known to the scientific world; but this exceeds all in the magnitude of its proportions and the perfect mastering of all the engineering difficulties. The following are the weights of some portions of the telescope :—Speculum in box, 3,500 lb.; tube (7 boiler-plate, 1,000 lb.; lattice, 1,379 lb.), 2,670 lb.; polar axis, 3,200 lb.; declination axis, 1,500 lb.; cradle, 1,100 lb.; counterpoise, 4,700 lb.; small portions, 1,500 lb.; making a total weight of 18,170 lb. These are all parts of the moving mass. There is, besides, a weight of about 3,000 lb. of portions not movable.

Lastly, in their very interesting Report on this magnificent instrument, presented to the Royal Society in April last, the committee pay a well-earned compliment to Mr. Grubb. After minutely examining and testing it, they express their admiration of the perfection of the telescope, and of the mechanical contrivances for its adjustment; and they add that, considering the contract price, and that special works had to be erected for the purpose of constructing the instrument, they are convinced that Mr. Grubb has been influenced much more by the desire of producing a scientific masterpiece than by any prospect of pecuniary advantage.

DISTANCE-MEASURING TELESCOPE.

A WELL-arranged distance-measuring telescope has been constructed by Messrs. Elliot Brothers, the opticians, of the Strand. The measurement of distances with this telescope is accomplished by means of two wires placed horizontally in the eye-piece, which, by a beautiful mechanical arrangement, have a simultaneous movement from the centre or axis of the telescope. The object to be observed is thus always in the centre of the field of view, enabling the observer to see it with the greatest clearness and precision, and under the most favourable conditions for estimating distances. It has been tested most accurately by the Ordnance Select Committee, and pronounced to be the best telescope for that purpose. In using this instrument, the two wires are placed in contact by turning a graduated ring on the eye-piece, from left to right; the edge of the fixed index is then at 0. The telescope is then focused to the object under observation until it is seen distinctly, then the divided ring is turned until the object is seen exactly between the two wires. The required distance is then shown upon the divided ring, the face showing, in yards, the distance of infantry, or any other object of a similar height; the top part of the ring showing the distance, in yards, of cavalry, or any other object of a similar height, the average height of infantry being taken at 5 ft. 11 in. ; the average height of cavalry at 8 ft. 10 in. The utility, power, and portability of these telescopes will secure for them a large demand, especially when they are stamped with the approval of the Ordnance Select Committee.—*Mechanics' Magazine.*

PHLOGISTON.

THE *Philosophical Magazine* contains an elaborate paper by Mr. Rodwell on the "Theory of Phlogiston," in which he shows that the idea of an agent, finally called Phlogiston, did not originate in any intellectual exploit, but arose by a slow process of evolution from the ancient hypothesis of a *subtilis ignis* taught by Zoroaster and others many ages before the Christian era, and which still survives in the sacred fire of the Parsees. Fire among the Magi was supposed to represent the Deity, shown in one direction by the destructive power of fire, and in the other by the beneficent action of heat and light; and the Greek philosophers, and subsequently the alchemists, adopted many of these ancient ideas. The physical philosophy of Descartes somewhat resembles that of Epicurus and Aristotle. The physical system of Descartes is a dynamical one; and there is a partial reproduction of it in the quasi Cartesian hypothesis of molecular vortices by Rankine, and in the dynamic treatment of electrical phenomena proposed by Clerk Maxwell. The first person who burst the bonds of the ancient schoolmen was Robert Hooke, who, in his theory of combustion, published when Stahl was only five years old, gave the explanation of that process now generally accepted, and which was only revived by Lavoisier. This theory, however, was lost sight of for more than a century, having been displaced by the theory of phlogiston, which was supposed to be very subtle matter, that neither burns nor glows, nor is visible, but is capable of being agitated by an igneous motion, and of communicating that motion to material particles, thereby constituting fire. The existence of phlogiston was inferred merely because some such supposition was believed to be necessary to account for various chemical phenomena. But, finally, the phenomena were found to be explicable without any such supposition, and Lavoisier showed by exhaustive experiment that the theory itself was untenable. In all bodies, however, with which we are acquainted, there is a certain quantity of heat, the point of absolute zero being, from the best experiments, inferred to be nearly 500 deg. below the freezing-point.—*Illustrated London News.*

THE SPRENGEL AIR-PUMP.

THE Sprengel Air-pump is a very simple and effective instrument, which has latterly come into extensive use. It con-

sists substantially of a glass tube, with a funnel at the top containing quicksilver. If the quicksilver be permitted to flow down the
tube in drops, each drop will act as a piston and carry some of
the contained air before it; and a vacuum will thus be produced in
the tube or in any vessel with which it may be connected side-
wise by a pipe. The mercurial column will, in fact, produce a
vacuum like that called a Torrecelian vacuum, and which is obtained by filling a tube over 39 in. long and closed at one end,
with mercury, in the manner of a barometer; and then, by
inverting the tube, the mercury will in part run out, and a vacuum
will be formed above the mercury. A column of mercury about
30 in. high. produces a pressure which balances the pressure of
the atmosphere.

REAL IMAGE STEREOSCOPE.

MR. S. C. MAXWELL has read to the British Association a
paper on a real image Stereoscope, with illustrations on solid geometry. In ordinary stereoscopes the observer places his two eyes
opposite two lenses, and sees the virtual images of two pictures
apparently at the same time. In the real image stereoscope the
observer stands about 2 ft. from the instrument, and looks at a
frame containing a single large lens. He then sees, just in front
of the lens, a real and inverted image of each of the two pictures,
the union of which forms the appearance of a solid figure in the
air between himself and the apparatus.

ON THE COLOURS OF SOAP-BUBBLES. BY SIR DAVID BREWSTER.

IN repeating the beautiful experiments of Professor Plateau
"On the Equilibrium of a Liquid Mass without Gravity," the
colours of the soap-bubble were presented to him upon soap
film plane, convex and concave; but the changes of form which
they underwent, and their motions upon the film itself, were so
incompatible with the common theory of their formation that
he was led by a few experiments to discover their origin and
mode of production. The paper proceeded to give an account
of experiments which, Sir David remarked, were sufficient to
establish the almost incredible truth, that the colours of the
soap-bubbles are not produced by different thicknesses of the
film itself, but by the secretion from it of a new substance flowing over the film, expanding under the influence of gravity and
molecular forces into coloured groups of various shapes, and
returning spontaneously, when not returned forcibly, into the
parent films.

Sir David Brewster stated that he was sorry that he had not
brought his diagrams; but he would be glad to answer any questions regarding the subject of the paper. Several inquiries were
made as to the nature of the soap used, and whether glycerine

might not be added with advantage. Sir David Brewster replied to these questions, stating that the experiments could be made by any person in the course of a few minutes, and that all the phenomena described were emitted with ordinary soap-bubbles. A mixture of glycerine made the films last much longer.

Sir Wm. Thomson pointed out that the mechanical questions involved in the seemingly simple operation of blowing soap-bubbles were the greatest enigmas to scientific men. The extraordinary expansion and adhesion combined in the vapour spheres were well worthy of the fullest investigation.—*Proceedings, British Association.*

NEW PHENOMENA OF LIGHT.

" ON a New Series of Chemical Reactions produced by Light," is the title of a remarkable paper presented to the Royal Society by Dr. Tyndall. The *Athenæum* Report speaks very highly of the facts it discloses. " Vapours of volatile liquids are introduced into an exhausted glass tube, and subjected to the action of concentrated sunlight, or to the concentrated beam of the electric light. This is the method; the effects produced varying with the vapour employed, may, without strain of speech, be described as wonderful. The method has another advantage, for, as Dr. Tyndall remarks, the power of the electric beam to reveal the existence of anything within the experimental tube, or the impurities of the tube itself, is extraordinary. When the experiment is made in a darkened room, a tube which in ordinary daylight appears absolutely clean is often shown, by the present mode of examination, to be exceedingly filthy. With vapour of nitrite of amyl, a shower of liquid spherules was precipitated on the beam, thus generating a cloud within the tube. With a modification of the beam, the precipitation was so rapid and intense, that the cone formed by the beam, before invisible, flashed suddenly forth like a solid and luminous spear. By proper management of the light, the vapour within the tube may be made to appear of a rich pure blue colour, equal to that in the skies of the Alps. With iodide of allyl, the vapour column revolved round the axis of the decomposing beam, drawn in at certain places like an hour-glass, while delicate cloud-filaments twisted themselves in spirals round the bells of the apparent hour-glass. With iodide of isopropyl another change took place : the vapour formed globes and cylinders, which were animated by a common motion of rotation, disturbed at times by a paroxysm, in which beautiful and grotesque cloud-forms were developed, some representing a serpent's head, others buds which seemed to grow into flowers, and all of a gorgeous mauve colour. With hydrobromic acid the cloud resolves itself into a series of discs and funnels, then parasols and rings of a very pale blue

colour, and all rotating as in the former instance. With hydrochloric acid the cloud requires twenty minutes for its full development, but then it appears in sections, each possessing an exceedingly complex and ornate structure, exhibiting ribs, spears,
funnels, leaves, involved scrolls, and irridescent fleurs-de-lis.
With hydriodic acid another modification is seen, having a family
likeness to the two immediately preceding, but with marked
differences of development, for the green and crimson produced
were the most vivid that Dr. Tyndall has yet observed. The
development of the cloud, as he describes, was like that of an
organism, from a more or less formless mass at the commencement, to a structure of marvellous complexity, at which the
Professor "looked in wonder for nearly two hours."

ABERRATION OF LIGHT.

PROFESSOR KLINKERFUES has published a paper, in which he
says that the earth's motion creates a double influence upon the
rays of light. The first is known as physiological aberration,
and causes stars to appear deviated 20½ seconds, when the trans-
latory movement on the globe is perpendicular to the direction
of the star. A similar illusion affects our judgment of the
direction of raindrops when we travel by rail during a shower.
The second influence is a physical aberration, which the motion
of the earth imparts to the rays coming from celestial bodies.

PROPAGATION OF LIGHT IN WATER.

A VERY able paper "On the Alteration Produced by Heat in
the Velocity of Propagation of Light in Water" has been contributed to Poggendorff's *Annalen* by M. Ruhlmann, whose conclusions correspond, on the whole, with those previously obtained
by Baden Powell, Gladstone, and Dale, in their investigations on
this subject. It was shown by these physicists that all substances by increase of temperature exhibit a diminution in the
refractive index, and that the magnitude of the variation differs
in different substances—being least in the case of water, and
greatest in that of bisulphide of carbon.

LUMINOSITY OF PHOSPHORUS.

DR. MOFFAT has read to the British Association an account of
several interesting experiments which he had made on the Luminosity of Phosphorus. From these experiments it was shown
that phosphorus in a luminous state produced phosphorus and
phosphoric acids, and ozone also; that it was non-luminous in a
degree of temperature below 39 deg. (F.), and that it was luminous above 45 deg. (F.), but the temperature of luminosity and
non-luminosity varied with the pressure of the atmosphere, and

also with the direction of the wind. A series of experiments, extending over four years, had been made on the luminosity of phosphorus in connection with atmospheric conditions, and from the results it would appear that the equatorial or sea wind is that of phosphorescence and ozone, and that the polar or land wind is that of non-luminosity and no ozone. As the ocean is the reservoir of ozone, Dr. Moffat asks if it is not probable that its phosphorescence is the chief source of its development—a probability strengthened by the fact that the polar and land winds, in the shape of the N.E. and S.E. nodes, seem to modify its development as the land-current does. From observations made at sea, between lat. 53 N. and 39 S., Dr. Moffat was led to believe that, were it not for the modifying influence of the polar or trade winds, ozone at sea would be a constant quantity.

Dr. Balfour Stewart said there was one point on which he did not quite follow Dr. Moffat, namely, with regard to the action on metallic bodies. Whenever there was a change of temperature, currents of air were very apt to take place, and they twisted or altered the position of a delicately suspended needle. Steel, if delicately suspended, even although not magnetic to begin with, would become magnetic from the action of the earth.

Sir William Thomson remarked that the phenomenon of luminosity stored up in the ice and produced after the melting of the ice was certainly one of the most startling ever met with in physical science—the luminosity having been induced by the previous presence of non-luminous phosphorus at a low temperature. It was a most beautiful and wonderful result. He was sure the publication of Dr. Moffat's paper would induce persons in all parts of the world to repeat the experiments.

SOURCES OF LIGHT IN LUMINOUS FLAMES.

PROFESSOR FRANKLAND has discoursed to the Royal Institution on this subject. After referring to and reading chemical text-books, which assert that all our artificial lights depend upon the ignition of solid matter (carbon particles) in the intense heat developed by the chemical changes dependent on combustion, he proceeded to exhibit several examples of luminous and non-luminous flames, the difference between them appearing to be that luminous flames contain those solid, suspended particles—this view being further confirmed by the facts that ignited solids give much more light than gases heated by the same temperature, and that a lightless flame becomes luminous when solid particles are placed in it, as in the case of the lime-light. Professor Frankland then showed that doubt was thrown upon this view by certain undoubted exceptions (the combustion of arsenious acid in oxygen, and of oxygen in bi-sulphide of carbon, &c.), and also by the production of light from the non-luminous hydrogen under certain conditions of combustion. After stating that at

the moment of combustion in free air oxygen and hydrogen ex-
pand to about ten times their original volume, he performed the
experiment, but little light being evolved; but when this expansion was prevented by the gases being inclosed in the strong
glass vessels of the Cavendish apparatus, the light was very great,
proving that when the density of the gases increased during combustion the light also was augmented. The Professor then illustrated the second condition of luminosity, the temperature, illustrated by numerous experiments, after which he referred to
several facts indicating the weakness of the argument of the
necessity of solid particles in flame—viz., that the soot from a
gas flame is not pure carbon, but always contains hydrogen; and
that the luminous portion of the flame is perfectly transparent.
It is, he said, to the behaviour of the hydro-carbons under the
influence of heat that we must look for the source of luminosity.
These gradually lose hydrogen, whilst their carbon atoms shrink
together and form compounds of greater complexity, and consequently of greater vapour density, the light increasing in proportion with the density. He then referred to some of the dense
vapours existing in a gas flame, especially to that of pitch, which
is volatile when it distils over from gas retorts. The proved influence of atmospheric pressure on gas and candle flames is, he
said, strongly confirmatory of the theory that their luminosity is
due to incandescent dense gases. Among the numerous experiments, the most striking was the combustion of hydrogen and
oxygen in very strong vessels, the luminosity of the flame being
increased or diminished as the pressure upon the gases was increased or diminished.

ATMOSPHERIC OZONE.

At a meeting of the Literary and Philosophical Society were
read the following remarks on Mr. Baxendell's Laws of Atmospheric Ozone, by Professor W. Stanley Jevons, M.A.:—

" In reading the remarks of Mr. Baxendell on atmospheric ozone,
it occurs to me that a very simple explanation can be given of the
connexion he detects between the height of the clouds and the
amount of ozone at the surface—two facts which seem at first
sight entirely unrelated. The quantity of ozone which reaches
the surface will depend on three circumstances:—

" 1. The thickness of the current of air touching the surface.

" 2. The proportion of ozone existing therein.

" 3. The degree in which this current is rendered uniform by
constant mixture.

" The balloon observations of Mr. Glaisher proved what was previously inferred by meteorologists, that the atmosphere usually
consists of several strata of air which are separated by distinct
boundaries, and do not freely mix. Hence it is only the ozone
in the lowest stratum which is usually available at the surface, and

its quantity will be proportioned, *ceteris paribus*, to the thickness of that stratum. It is the height of the first layer of clouds which usually defines the upper limit of this stratum. For during my own observations, both in Australia and England, I have often noticed that smoke from a great town, or from extensive bush fires, rises only to a definite height, and seems to form the basis, as it were, of the cumulous clouds, which are the upward terminations of ascending currents. Now as, in May, the height of this stratum is, according to Mr. Crosthwaite's observations, greater than in any other month, there will be a larger mass of air which can successively come in contact with the surface and furnish ozone. But we shall only have the full benefit of this ozone when an active process of mixture is going on. At night the air in contact with the earth is cooler than that above, and therefore tends to lie in a stagnant layer, which can often be detected by the mist or smoke which it contains. This layer will be rapidly exhausted of ozone, and will be filled with the organic exhalations from the earth. Hence arises the comparatively unhealthy character, and in some climates the poisonous nature, of night air; and it may constantly be observed that a moderate wind may be blowing above our heads, as shown by the motions of the clouds or the wind felt on a mountain top, without breaking up the stagnant layer on the surface. This is one reason why the air is calmer at night than in the day. But during high winds the gusts penetrate to the surface and prevent any stagnation, so that, as I apprehend, during stormy weather, the deficiency of ozone in the evening would not be observed.

In the day-time, on the contrary, the sun's heat occasions a perpetual circulation or convection in the lowest mass of air up to the level of the cumuli. Every portion of the air is thus successively brought to the surface, and organic substances are carried off and oxidated. During my own observations on ozone, I felt strongly the imperfections of the method of measurement alluded to by Mr. Bazendell, and I thoroughly agree with him that the mysterious variations of ozone will not be understood until not only the quantity of air brought into contact with the paper be measured or regulated, but the varying source and magnitude of supply be considered.

Electrical Science.

SIR CHARLES WHEATSTONE.

THE following is a copy of the letter addressed to Sir Charles Wheatstone by the President of the Italian Scientific Society of "Forty," announcing that Sir Charles had been appointed a member of that Society in the place of the late Professor Faraday, and that a Gold Medal had been awarded to him. The President, in his letter, says :—" I will not here pass in review the various memoirs in physics which you have published in the *Philosophical Transactions*, since all carry the impression of the inventive genius which ever distinguishes all that you have done. I cannot, however, refrain from calling to mind that to you we owe the discovery of the method, as ingenious as it is original, for measuring the velocity of electric currents and the duration of the spark. The applications of the principle of the rotating mirror are so important and so various that this discovery must be considered as one of those which have most contributed in these latter times to the progress of experimental physics. Not less ingenious was the invention of the stereoscope and of the modes by which binocular vision is effected, which enables us to obtain the perception of the relief from the simultaneous observations of two plane images. Also the memoir on the measure of electric currents, and all questions which relate thereto, and to the laws of ohm, has powerfully contributed to spread among physicists the knowledge of those facts, and the mode of measuring them with an accuracy and simplicity which before we did not possess. All physicists know how many researches have since been undertaken with your 'rheostat,' and with the so-called ' Wheatstone's bridge,' and how usefully these instruments have been applied to the measurement of electric currents, of the resistance of circuits, and of electro-motive forces. And here it would be impossible to leave out of view that to you we principally owe the practical invention and the true realisation of the electric telegraph. Finally, I would call to mind your recent researches on the augmentation of the force of a magnet by the reaction which its own induced currents exert upon it. All these great acquisitions, procured by you to physical science, render you well worthy of this distinction from the Italian Society of Sciences. Preserve yourself in health and activity, and your country and all your admirers and friends are certain to find in the discoveries still to be added while you continue to work some compensation for that immense and irreparable loss which natural philosophy has received by the death of Faraday."

NEW SECONDARY BATTERY.

M. G. Plante's new "Secondary Battery" consists of a novel and peculiar arrangement for multiplying the power obtaining from a weak source. It is not unlike in form an ordinary "condenser," and the results produced from it are somewhat identical. His secondary "quantity" battery consists of a rectangular gutta-percha vessel, provided with lateral grooves, and containing a series of lead plates immersed in dilute sulphuric acid. As in a large condenser, the odd plates are joined in one series, and the even plates in another series; to either end is attached one of the poles of a weak source. In the secondary "quantity" battery, tried by M. Plante, he used six plates only with a source of two small nitric couples, and on breaking the circuit, by means of a commutator, the current obtained from the "secondary" battery is then of sufficient strength as to create temporary incandescence in a platinum wire 1 millemetre thick, by 8 centimetres long. By increasing the size and number of the plates, more powerful calorific effects, such as incandescence of iron and steel rods, may be obtained by charging the battery with two or three Bunsen's elements. An arrangement, termed by him his "secondary tension battery," produces also most powerful results. The arrangement consisted of forty secondary couples, each couple of lead plates being, in a narrow gutta-percha vessel, immersed in dilute sulphuric acid, the pole of each vessel being connected to a peculiar commutator, so that the plates could be joined up as an arrangement of tension or as one of surface. This battery was charged with three couples of Bunsen's medium size battery. On applying the current from the secondary circuit, powerful calorific effects were obtained. A platinum wire, 2 metres long, and one-quarter millimetre thick, was rendered incandescent for a few moments, and the voltaic arc was also obtained. M. Plante remarks, concerning this, that there is not, as in the case of induction, the direct production of one physical effect by another physical action, but the final result is none the less an accumulation or a modification of electrical force, which can be utilized in certain circumstances.

LIQUIDS FOR BATTERIES.

M. DELAURIER has communicated to the Academy of Science a note upon a new exciting liquid for galvanic batteries. He states that "in order to obtain very powerful batteries disengaging no deleterious gas, and of very cheap maintenance, I have proved the problem of transforming azotic acid into sulphate of ammonia, under the influence of sulphuric acid and hydrogen." This is done by the use of protosulphate of iron. The proportions in these liquids are twenty parts in weight of protosulphate of iron dissolved, sheltered from contact with the air, in thirty-

six parts of water, to which is added, with stirring, seven parts
of diluted (equal parts) sulphuric acid, then in the same manner
one part of diluted (equal parts) azotic acid.   M. Delaurier
states that the liquid produced is the most energetic and most
economical that he knows for an exciting liquid for iron, zinc,
and other metals, without any disengagement of hydrogen or
binoxide of azote. In the use of this liquid with nitric acid in
Bunsen's pile, M. Delaurier states that the action goes on without any exterior emanation of nitrous gas, without the emission
of hydrogen in the interior, and consequently that the platinum
does not polarize.

<hr>

NEW VOLTAIC BATTERY.

A NEW voltaic combination of great power is the invention of
Messrs. De La Rue and Hugo Müller, and has been designed
for Mr. Gassiott.   The elements consist of small cylinders of
pure zinc and chloride of silver.   The latter is cast upon a thin
silver wire which forms the conductor.   The exciting liquid is
merely a dilute solution of common salt.   In the battery shown
the cylinders were only 3in. long, and about the size of goose-
quills, arranged in two-ounce phials cut down to two-thirds of
their length, but a series of ten such couples decomposed water
with great rapidity.   By the chemical action taking place in
the cell the chloride of silver is reduced and chloride of zinc
formed.   The action proceeds so long as any chloride remains,
for the reduced silver adheres to the wire as a spongy mass,
which allows the liquid to permeate to any unreduced chloride.
The first cost of such a battery will be considerable, but as the
only loss will be a little zinc, it will be very economical in working.   Mr. Gassiott, it was said, is having a battery of 1,000
pairs constructed, of which, no doubt, the scientific world will in
good time hear and see much.—*Mechanics' Magazine.*

<hr>

ELECTRICITY ON MOUNTAINS.

M. HENRI DE SAUSSURE, a descendant of the great natural philosopher, has published an interesting paper in the *Bibliothèque
Universelle* on a phenomenon which has but recently attracted
attention.   Having reached the summit of the Piz Surley, a
mountain composed of crystalline rocks in the Grisons, and 3,200
metres in elevation, M. De Saussure and his party laid their
alpenstocks against a little cairn of dry stones which crowns the
summit, and prepared to take their repast.   Almost at the same
instant the narrator felt at his back, in the left shoulder, a very
acute pain, like that produced by a pin slightly pressed into the
flesh, and when he put his hand to the spot without finding any.

thing, a similar pain was felt in the right shoulder. Supposing
his overcoat to contain pins, he took it off, but the pains increased,
extending from one shoulder to the other across the whole back.
They were accompanied by pricking sensations and sharp shooting pains, such as a wasp crawling over the skin and stinging all
the time might produce. The pain next assumed the character
of a burn, and M. De Saussure actually fancied that his flannel
waistcoat had caught fire, and was about to throw off the rest of
his clothes, when his attention was arrested by a sound remind-
ing him of the reverberations of a tuning-fork. These sounds
came from the sticks which, resting against the cairn, sang loudly,
emitting a sound like that of a kettle the water in which is about
to boil. All this lasted about four or five minutes. M. De
Saussure at once guessed that his sensations proceeded from a
flow of electricity taking place from the summit of the mountain.
No spark, however, was obtained from the sticks; they vibrated
strongly in the hand, and sounded very loud. Some minutes
afterwards he felt his hair and beard stand out, causing him to
feel the sensation resulting from a razor passed dry over the
bristles. A young Frenchman who was of the party, cried out
that he felt the hair of his moustache growing, and that strong
currents were flowing from the tips of his ears; and they soon
flowed from all the parts of the bodies of those present. As
they descended the mountain the humming of the sticks and the
other phenomena diminished and eventually ceased. The sky
was cloudy, and the travellers had been overtaken at the time by
a shower of thin hail and sleet. On the same day a violent
storm broke out in the Bernese Alps, where an Englishwoman
was killed. Sleet, frost, and an overcast sky appear to be the
conditions necessary for the production of the phenomena above
described. Many of the guides have never observed them, and
others recollect having remarked them only once or twice.—
Galignani.

ELECTRO-MAGNETISM.

THERE are many who look forward to a time when Electro-
magnetism will supersede steam as a motive power. It is quite possible that electricity may be employed as a motive force, but the
amount of mechanical work which it is capable of doing is comparatively small, and it must be enormously expensive. There
are circumstances, however, in which cost is a matter of small
consideration, and in these electricity may sometimes be turned
to useful account. A lady may wish for a sewing machine, but
the labour of driving it is more than she can exert. For such,
M. Cazal has invented an electro-magnetic machine, which we
have no doubt will be competent to do the work with light
fabrics. The details of this machine we must defer until a more
complete account than we have yet met with comes to hand. For

ornamentation, too, electricity is coming into use. You may see at a fashionable ball at Paris, a lady, on the top of whose head sits a butterfly or a humming-bird. The fly and the bird flutter their wings in the most natural way possible. How is it managed? Why, within the chignon are concealed a small battery and a minute Rhumkorff coil. On the bosom of another may be a brooch, with a head upon it, the eyes of which turn in all directions. This, too, is accomplished by the use of a battery and coil so minute as to be concealed within the brooch itself. These, and more useful things, small batteries for curative purposes, easily carried about the person, are the inventions of M. Tronvé. The batteries of zinc, excited by solution of sulphate of mercury, are enclosed in vulcanite cells, so that the exciting solution cannot escape to the damage of the wearer.—*Mechanics' Magazine.*

Experiments have been tried in Sheffield, on the action of Electro-magnets upon iron in a state of fusion. In a very large blast furnace an electro-magnet was fixed in front of an opening made on one side of the furnace; the magnet was excited by a Smee battery. On exciting the magnet, we are told that the metal immediately bubbled up. According to statements furnished, we are told that the operation was accelerated with an economy of combustibles and an improvement in the quality of the metal. We should be glad to have farther details of these experiments, so as to give some further idea of the interesting *rôle* played by electricity and magnetism in a new subject.

ELECTRICITY THROUGH A VACUUM.

AN account is given in the *Comptes Rendus* of an apparatus, devised by MM. Alvergniat, for showing that electricity is unable to pass through a vacuum. A vacuum is produced in a glass tube by means of the mercurial air-pump we have already described, and the tube is made hot by a charcoal fire, while the suction of the pump is still continued. Two platinum wires have previously been cemented into the tube about two millimetres apart, and after the tube has been kept hot for some time, the opening communicating with the pump is sealed hermetically. An electric spark cannot be passed from one wire to the other, the vacuum constituting an insuperable barrier.

MAGNETISM AND HEAT.

AN experiment illustrative of the power of magnetism to generate heat has been shown by M. Louis D'Henry. If a magnet with poles pointing upwards be rotated rapidly on a vertical axis below a small copper plate, on which a glass flask is placed, the air contained in the flask will be heated, and its expansion may be made visible by any suitable arrangement, or a copper vessel of water might be substituted for the flask and plate, and by a sufficiently rapid rotation the water might, no doubt, be made to boil. The mutual convertibility of heat, electricity, and magnetism has now been conclusively demonstrated, and they are all recognized as merely certain forms of force, though what that force is we cannot tell.—*Illustrated London News.*

ELECTRIC CLOCK.

An electrical clock in the rotunda of the Philadelphia Merchants' Exchange has a running gear of the simplest description, consisting merely of two cog-wheels and a ratchet-wheel. The driving power is supplied by a weak galvanic battery, the currents from which, transmitted through two galvanometer coils placed one on each side of the clock-case, act upon steel bar magnets set within the pendulum-ball. The latter swings between the two coils, so that when one of them is "positively charged" the ball is attracted until by contact it becomes similarly electrified, and consequently repelled, then swinging over to the "negative" coil, it becomes negatively charged, again repelled, and thus the vibrations are kept up indefinitely, or as long as the battery continues working. The alternate positive and negative charges are made and broken by a simple slide bar moved by a wire pin on the pendulum rod.

ELECTRIC LIGHT.

Professor Edlund has communicated to the Royal Academy of Sciences at Stockholm, a paper containing an investigation of the laws which govern the production of the Electric Light. The voltaic current, it is well-known, produces heat and light, decomposes compounds which are conductors of electricity, excites magnetism, and induces currents in adjacent conductors. The heat produced by such a current is proportional to the square of the intensity of the current multiplied by the resistance. The entire quantity of heat generated is proportional consequently to the electro-motive force divided by the resistance of the current, and this sequence holds so long as the current performs no other work except the generation of heat alone. In the luminous arc which constitutes the electric light material particles are detached by the current from one pole and transferred to the other, and by the mechanical disintegration of the poles an electro-motive force is produced which sends a current in an opposite direction to that of the principal current. The electro-motive force in the luminous arc is independent of the intensity of the current, and the resistance of the arc is proportional to its length and increases as the intensity diminishes. The work performed by the current in the luminous arc is proportional to the intensity so long as the electro-motive force of the battery remains constant.—*From the Illustrated London News.*

SOME recent correspondence between the Trinity-house and the Board of Trade shows that the electric light at Dungeness can now be worked by either of the two engines, so that no disturbance occurs when one requires repair. The services of the high-class engineers and firemen have been dispensed with, and the Elder Brethren have since been enabled to do that which the connexion of the men with the trades unions prevented,—viz., to have their own ordinary keepers trained to drive the engines, as well as to attend to the lamps, a steady old experienced keeper being placed at the head of the establishment. The magneto-electric apparatus shown at the Paris Exhibition presented several improvements. The working by either of two machines showed that the power of the light can be duplicated in thick weather; and the engines were utilized for working the pumps of an air fog-trumpet. The electric light was compared with the flash of a first-order revolving oil apparatus belonging to the French authorities; and at 15 miles' distance the Trinity-house engineer, Mr. Douglass, estimated the power of the fixed electric light at twice that of the flash of the oil light. The superiority of penetrating power of the electric light in fog was shaken by some experiments made by the Royal Engineers, but it turned out that this result, so different from all other experience, arose from a settlement in the wood-work supporting the electric lens, causing the lens to be out of its proper position.

THE ELECTRIC WAND.

UNDER this title the London Stereoscopic Company have brought out a most amusing novelty. It consists simply of a glass cylinder, the handle of which is insulated, and when dry and warmed, has to be rubbed with a small pocket Leyden jar, so to speak, which is also carefully insulated from the hand of the operator. The electricity thus collected is discharged in an infinite number of amusing ways along glass-plates covered with tin-foil, which it lights up with the brightness of lightning. It is made to perforate cards, to light gas, to ring the little electric bells which are sent with the apparatus, to make pitch balls dance about, and all the hairs on a little image's head to stand on end. Indeed, the simplicity of the apparatus, which is designed by Professor Varley, can hardly be too much praised, and its great merit consists in the fact that while a child can use it, even the youngest child runs no manner of risk. The very small electrical generator is made so weak that any person can take its little shocks. Indeed, these small shocks form part of the amusement of the wand. With the wand are what are called metallic fireworks for parlour use. They are not fireworks in the ordinary acceptation of the term, for they are not explosive, and can only be slowly lit. They are composed of the constituents of all kinds of metals, which, when slowly ignited, burn with every kind of hue that the prism is capable of producing. All are harmless, and some of their colours are of the most wonderful brilliancy,—blue, purple, orange, green, red, white, violet, &c.—*Times.*

FIRE DAMP IN MINES.

A new invention—by M. Delaunier, of Paris,—for destroying Fire-damp in Mines, has been laid before the Academy of Sciences. It consists of a copper conductor, broken at intervals, but joined by very fine gold wire soldered to the copper ; the gold wire being surrounded by flour of sulphur, which ignite easily. By passing strong currents of electricity through the copper wire, the gold wire becomes red hot, and thus ignites the sulphur, which burns any noxious gases which may be present. It will, of course, be understood that the electric current is made to pass through the apparatus before the descent of the miners into the mine. The Academy of Sciences have, it is stated, reported very favourably on M. Delaunier's invention.

SUBMARINE CABLES.

THE India-rubber, Gutta-percha, and Telegraph Works of the Silvertown Company, opposite Woolwich, present the several works in relation to their capacity for Submarine Cable Manufacture. Here one can trace the whole processes which an ocean-cable undergoes from the time the first rough clumps of coarse and dirty gutta-percha arrive at one end of the factory, till it is sent out at the other in the form of a finished cable fit to be laid anywhere, and transmit anything. The process of preparing the gutta-percha is peculiar, but, on the whole, only amounts to this, that there are some machines which tear the blocks of gum to fine shreds, others which boil it, others which literally masticate it when dry, and others which masticate it under boiling water. It is boiled, melted, kneaded, pressed, and then masticated again, and after all these preparations comes out at last a pure gum, without a speck of dirt or foreign substance in it. It takes a long time to effect all these various manipulations, but as they go on continuously from machine to machine, it seems to lookers-on as if the gum was cleaned as fast as it is delivered, though in some parts of what may be called the manufacture the gutta percha remains for hours, and even days, undergoing one particular stage of cleansing. When perfectly fit for use, and neither too soft nor too hard, it is packed in cylinders, and thence compressed by screw levers through small dies. Through these dies the conducting copper wires are drawn by machinery, and as they pass the warm sticky gum adheres to them, and this gives them their first coat. After this operation, when the insulation

is perfect, the rest is mere matter of routine. Coat after coat
is added in the same fashion, according to the thickness of the
insulation wanted, or more properly speaking, accordingly as the
parties ordering it want a costly and good cable, or a cheap
and uncertain one. Between each layer of gutta-percha is laid
a coating of what is called compound—that is to say, a mixture
of gutta-percha and tar, through a tank of which the covered
wires are drawn, and thence pass and repass through long troughs
of cold water, to harden and consolidate all the coats. But
during every stage of these long and varied operations the electrical tests as to "conductivity" resistance, and insulation are
going on incessantly, so that the slightest flaw in manufacture
can be detected—flaws so slight that it would here be almost
impossible to convey an idea of their minuteness to those not
professionally acquainted with the working of submarine cables.
After the centre core with the conducting wire is thus made it
is a mere question of where it is to be laid as to how it is to be
protected by its outer covering of hemp and wire, though during
this stage the electrical supervision has to be, if possible, more
keen than ever, for a faulty, broken wire piercing the insulation
is the great dread of all electricians. These stages of manufacture were illustrated at the Silvertown works by two cables
which are being made there—one for the West Indies, between
Cuba and Florida, the other for the North Sea, between England
and Holland. The former has been designed by Sir Charles
Bright, M.P., and is a beautifully constructed specimen of good
submarine cable. It has seven conducting wires woven into one.
It is coated with three coats of gutta-percha and compound, and
afterwards bound in with 12 No. 9 wires, covered with hemp,
and afterwards closed in with two coats of Bright and Clark's
silicia compound. This cable tests higher than any other ever
manufactured, and, as compared with the first Atlantic cable, its
ratio of excellence may be considered as about six to one. A
cable already exists between Cuba and Florida, but it is not sufficient for the business it has to transact, and out of the surplus
profits which this first cable has earned the second is now being
made and paid for. This second line, however, is only one link
of a great chain which the International Ocean Telegraph Company
in America are about to establish, coupling up all South America
with Cuba and the West India Islands into one great junction
near Florida, and thence over North America and through the
Atlantic cable to England. This scheme, which is in view, is a
magnificent one, and if fully carried out will leave Australia
almost the only important country in the world with which we
are not directly connected by submarine wires.

Another cable at Silvertown is being manufactured under the
skilful care of Mr. Richard Culley, for the Electric and International Company. It is to be one of the largest in the world,
though not larger than that which it replaces between this coun-

try and Holland. It is to be laid between Dunwich, near Lowestoft, across to Zandvoort. The weight of this cable is nearly 20 tons a mile, but it contains four separately insulated conductors, each of seven wires, and all bound together with the most massive galvanized iron wire.—*Abridged from the Times.*

ELECTRIC SIGNALS.

SOME interesting experiments have taken place in France on board the armour-clad ship *Heroine*, and also on board the yacht *Prince Jerome*, upon the use of the electric light for signalling purposes. The results were considered of so satisfactory a nature that the experiments were ordered to be continued. The machine was that furnished by the Alliance Company, a machine that is now so well known. The power of the light obtained was equal to 200 Carcel burners—the Carcel being equal to eight candles; it therefore follows that the electric light possessed a brilliancy equal to 1,600 candles! In the direct line of the light it was stated as possible to read at the distance of 1,400 metres (1,531 yds.) an ordinary newspaper. It was found that signalling by means of short and long flashes was the most easily to be carried out. And by that means signalling was rendered easy. The commissioners on the subject report, "The apparatus experimented upon shows a very powerful focus of light, perfectly suited to night signalling, or for throwing a light over a coast or a ship. It can be considered as a veritable floating light, and would then be most useful on board the flagship of a commander-in-chief." Of the peculiar value of this electric light and its intense illuminating power, it is stated as a fact that the yacht *Prince Jerome*, fitted with this light, was enabled to steam by night through the intricate navigation of the Bosphorus, when the yacht belonging to the Viceroy of Egypt was obliged to wait until daylight.—*Mechanics' Magazine.*

CURIOUS ELECTRIC CABLE PHENOMENON.

FOR some weeks the trans-Mississippi telegraph cable between Memphis and Little Rock was out of order. At intervals no communication could be had for hours, and all at once the cable would revive, and the fluid passed through it as usual. At length it ceased to convey messages. The cable was one of the best ever laid in America; it was manufactured originally for the Red Sea. The cable was taken up, and it was discovered that four inches of the conducting wire had been completely burnt out; it is supposed that a shock of lightning had passed through the cable, and destroyed the four inches of wire. It is curious that after this destruction messages were occasionally conveyed by the cable, and it is conjectured that a slight con

nexion was formed between the burnt ends of the conductor by
moisture, which had penetrated the cable in sufficient quantities
to keep up the circuit, there being a battery at the Memphis end
strong enough to drive the electricity through at intervals. Cannot some mode be established whereby communications can be
passed through large bodies of water without a cable! It is certain that messages passed to and fro across the Mississippi without a metal connexion.—*New York Journal.*

TELEGRAPHIC COMMUNICATION.

SOME interesting facts in the history of telegraphic communication are given by Herr Neumann in an official Report recently
published at Vienna. It appears from this report that a Russian
telegraphic agency exists at Pekin, through which messages are
transmitted from the Atlantic to the Pacific Ocean. The line is
now being carried to Behring's Straits by Okhotsk, and if the
plan of the United States Government to establish a telegraph
through the territory lately ceded to it by Russia is carried out,
there will soon be an uninterrupted line of telegraphs round the
earth. The longest line in the world is that between San Francisco and St. John's, Newfoundland, a distance of 900 geographical miles. Herr Neumann annexes to his report some statistical
tables, showing that the total length of the European telegraphs
is 53,340 geographical miles; of those in America, 14,239; of
those in Asia, 4,736; of those in Australia, 1,842; of those in
Africa, 1,504—exclusive of submarine lines. There are in all
1,200 telegraph stations, and the number of persons employed
in them is about 38,000. The total weight of the wires is calculated at 1,300,000 cwt., and there are about 7,500,000 telegraph
posts, which it costs 200,000*l.* a year to keep up. The capital
expended in the construction of the various telegraphs is
17,500,000*l.*

COST OF TELEGRAMS.

IN France the greatest distance over which a message can be
transmitted is about 600 miles; in Prussia, about 500; in Belgium, about 160; and in Switzerland, about 200 miles. The
charge for a message of 20 words over the greatest distance
in France, is 1*s.* 8*d.*; in Prussia, 1*s.* 6*d.*; in Belgium, 5*d.*; and
in Switzerland, 5*d.* In Great Britain 2*s.* is charged for the transmission of a message over 500 or 600 miles, and 1*s.* 6*d.* for any
message sent 160 or 200 miles.

THE NEW ATLANTIC CABLE.

THE manufacture of the new Atlantic telegraph which is to
be submerged between Brest and a suitable terminus on the

shores of the State of New York, is progressing satisfactorily. The new cable is almost identical in construction with those which were completed in 1866; the only difference being that the diameter of the conducting copper core is slightly greater, and the outside wires are of homogeneous Bessemer steel, galvanized, having a breaking strain of about 1,000 lb., while the wires outside the existing Atlantic lines have a breaking strain of only about 800 lb. The new cable will be laid in two lengths—one from Brest to St. Pierre, in deep sea, of 2,325 miles, not including slack, and the other from St. Pierre to the terminus, of 722 miles in length, not including slack. The latter section will be similar to the Persian Gulf cable, as it will have to be laid in comparatively shallow water, and its exterior wires will be protected with Bright and Clark's patent siliceous compound, which consists principally of powdered flint and pitch. The construction of the shore ends will be similar to that of the existing Atlantic lines, and will gradually become thinner until they assume the deep sea dimensions. In order to avoid the dangers of injury from rocks and icebergs, the new line will be laid to the south of the present cables, below the southern edge of the Great Bank, so that it may be laid in deep water. Sir James Anderson, who will command the Great Eastern during the expedition organized for the submergence of the line, has made the following observations regarding the Newfoundland Banks: "By keeping in the 500 fathom line upon the Milne Bank, and around the southern edge of the Grand Bank, there is no possibility of ice or any other agency that can be suggested injuring the cable. The northern edge of the Grand Bank was avoided because it is uncertain at what depth the icebergs ground. They are said, upon good authority, to ground at times in 90 fathoms. It is not certain at what depth the vessels employed in the seal trade may sometimes choose to drop an anchor for the purpose of keeping in the track of ice floes. These dangers are avoided by the track chosen for the proposed cable, and the track from the southern edge of the Grand Bank to St. Pierre, and thence to the place of landing in America, is entirely free from any danger from ice, and does not cross any anchorage resorted to by the fleet of fishing vessels."

The breaking strain of the new steel cable will be 7¼ tons, and the strain required for submersion need not be more than 14 cwt. If at any time it be necessary to haul up any portion of it already laid the strain need not exceed a ton and a half in the deepest water. The weight of copper forming the conductor of the existing Atlantic cables is 300 lb. per knot; in the new cable it will be about 400 lb.

AMERICAN TELEGRAPH BUSINESS.

THE number of miles of line worked in the United States in 1848 was estimated to be about 12,000. In 1858 the number had increased to nearly 40,000, and before the close of this year there will be completed and in operation about 120,000 miles of telegraph wire. Supposing the same rate of increase to be kept up, it might be supposed that 250,000 additional miles of wire, which would be necessary, might easily be put up by 1878. But there is a serious obstacle in the way. Even now, along every avenue, turnpike, and railroad leading from New York, the way is bordered with a forest of telegraph poles, cross-arms, and wires. It has been found by experience that every addition to the number of wires upon a range of poles adds to the difficulties of working and increases the danger of interruptions. In severe storms they present so much surface to the gale that they are frequently blown down, especially when snow and ice adhere to them. For these and other reasons it is evident that the wires cannot be indefinitely multiplied in number, nor, indeed, be greatly increased, without creating serious difficulties. It is not likely that the speed with which messages can be transmitted under the present system is susceptible of any considerable improvement. A good operator can send 2,000 words per hour, and more will fall below this rate than go above it. The most practicable way of getting over the difficulty appears to be the adoption of a species of shorthand. Some experiments have been made in England with figures standing for words, and even for common phrases, and a number of short messages were sent in less than half the usual time. This system may afford great aid to the telegraph, but it is a little doubtful if it will be found applicable to the messages which have to be sent to the press.—*New York Tribune.*

THE ELECTRIC LIGHT APPLIED TO PHOTOGRAPHY. BY DAVID WINSTANLEY.

IN a recent article on this subject, it is stated, *à propos* of the electric lamp of Mr. John Drowning, that perhaps the writer's own experience of such instruments had fallen on evil days, inasmuch as he entertained the idea that huge, bulky, and expensive pieces of apparatus were necessary for the production of the electric light, and that the light when produced, though bright in itself, had its reputation sadly dimmed by its flickerings and occasional outgoings. Whether our experience happen to fall on evil days or not, it is quite impossible for any person to be thoroughly up in all the improvements that have taken place in every branch into which our art may be divided. The writer of the present article, whose life during the past three or four years has been spent almost entirely amongst artificial lights of nearly

every imaginable variety, including the electric light as produced
by voltaic and induced currents, and regulated by many varieties
of lamp, automatic and otherwise, would respectfully offer a few
remarks supplementary to the editorial article alluded to.

In the first place, a few words on electric lamps and lights
generally :—

1. The construction of the electric lamp, as the regulator of
the electric light is technically called, has nothing whatever to
do with the intensity of the light itself, which depends upon the
battery power or magnetic force employed.

2. All sources of artificial light, whether electric or otherwise,
which are intended for such optical purposes as photo-enlarging,
photo-micrographing, lantern exhibitions, &c., must necessarily be
virtually motionless. The magnesium light, it may be said, is
not motionless, but, on the contrary, is constantly moving through
the space of an inch or thereabouts. In the midst of this movement, however, which takes place with some amount of periodicity, the light does occupy always one and the same place, and,
in virtue of there being a position in which we may depend upon
finding the light from time to time, its utilization for optical purposes becomes practicable. In the case of a candle, however,
each moment of its consumption sees the flame descend to a position lower than that which it occupied a moment before; and,
therefore, were its flame sufficiently luminous or sufficiently
actinic for any of the before-mentioned optical purposes, the candle
itself would be inadmissible, unless supplied by some clockwork
arrangement by which its whole body might be raised to such a
speed as would exactly compensate for the descent of the flame.
In the case of an electric lamp, however, such a clockwork arrangement, if necessary at all, would cost as much and be quite
as complicated as a thoroughly efficient lamp itself, because of
the varying extent of the incidental consumption of the carbons
caused by inequalities in their composition, and by variations in
the electric force passing through them.

3. After allowing sufficient space to secure the carbons in their
holders, and adding to this amount twice the length of each carbon to be used, every inch of altitude possessed by the regulator above the total of these two sums is a degree of elevation
entirely unnecessary for all ordinary and many extraordinary
purposes.

4. A variety of brake employed in an electric lamp, which
brake depends for its action upon friction, if not supplied with
adjusting screws whereby the said friction can be increased or
diminished at pleasure, is constantly liable to derangement, which
will at least interfere with, and in some lamps may entirely suspend, the regulating action.

5. Where a reflector of any kind is to be used, it is desirable in
all instances that the greatest possible amount of light should be
reflected, and the least possible amount lost, so far as it can be

done without in any way interfering with the object for which the
light is intended.

6. It is desirable that the lamp should, as a rule, be able to
carry carbons of sufficient length to last a whole day.

In the second place, to return to the particular lamp described
in the article mentioned :—Like many other electric lamps which
have from time to time been proposed, Mr. Browning's is, judging from the illustration by which its description was accompanied, entirely destitute of any apparatus by which the lower
carbon can be raised, and, consequently, its spark is a constantly
descending one.

Although in the particular lamp now under consideration "the
altitude is rather under eleven inches," the elevation, in proportion to the length of carbons which may be used, is unnecessarily
great.   From the diagram of the apparatus as given, it would
seem that the longest carbons which can possibly be used in Mr.
Browning's lamp would each be but little over three times the
length of what is shown as being in use in the illustration, whilst
the elevation of the lamp itself, when occupying the least space
which it is possible for it to occupy, would be something over
twice the length of the whole available carbon ; in other words,
though the lamp stands some eleven inches high, it is only capable
of using carbons each some two and a half inches in length,
and of these five inches of carbon it would be neither safe nor
expedient to consume more than four.   Therefore, with a current
of electricity anything like powerful enough to give a really useful
light, four inches of carbon, narrow enough to retain their points
under the influence of such a current, will last only about two
hours, at the end of which time fresh carbons must be inserted.

Judging from the diagram, it would seem that the upper carbon
is arrested in its downward course through the influence of the
electro-magnet, by being what we may call "skewifted" in its
bearing : and, unless the lamp be supplied with a screw by which
the extent of this deflecting pressure can be regulated at pleasure,
which in this instance does not appear to be the case, its automatic action will always be liable to be interrupted.

The lamp, taken as a whole, is an ingenious arrangement ; and,
for the regulation of an electric light intended for general, and
not for exact, optical illumination, and requiring only to burn for
a comparatively short time, it is, I have no doubt, as efficient a
cheap lamp as is to be found in the market.   As an auxiliary to
electro-magnetic apparatus worked by steam-power for the production of a long-continued and powerful light, to be used in front
of a condensing lens, the present lamp does not appear to be
adapted ; but, probably, this is a use to which its inventor never
intended that it should be applied.   Such, however, it seems to
the writer, is the use to which lamps must be applied in order to
make them available for any commercial photographic purpose.—
British Journal of Photography ; quoted in Scientific Opinion.

Chemical Science.

HEAT WITHOUT COAL.

RECENT scientific discoveries should do much to lessen the alarm of those who fear the exhaustion of our coal-fields. Mr. H. W. Pond, of Newark, U.S., remarks, in the *Mining Journal*, that economists have speculated on the possible discovery of some method of producing heat independent of coal, and the decomposition of water has been regarded as a probable expedient. With our present knowledge and appliances it appears not to be difficult to realize this proposition, even in competition with coal—at least in a small way. The agents are obvious—wind-power, a magneto-electric machine, oxygen and hydrogen gas-holders, and the electrolysis of water. The result would be that oxygen and hydrogen would be available for the production of the heat which would be required. For use in the arts the oxy-hydrogen furnace would, of course, offer advantages far above any other known; and results could be reached impossible with the lower temperature of the coal fire, while the flame would be free from deleterious substances common to coal. Owing to its gaseous form, and the intensity of its heat, this fuel would be manageable in many ways impracticable with coal. For instance, seams could be hardsoldered with great rapidity with the jet of the compound blowpipe, and it is probable that the joints of steam-boilers could be heated for welding in a suitable oxy-hydrogen jet.—*Mechanics' Magazine.*

PURE SILVER AND ITS PROPERTIES.

METALS are rarely seen in a state of absolute purity, and as very small amounts of contaminating substances considerably modify their physical properties, and to some extent also their chemical behaviour, the characteristics of pure metals are but little known. Our knowledge of Silver has recently been considerably extended by the experiments of Professor Christomanos, of Athens, who has obtained the pure metal by distillation. Silver was well known to be volatile to a slight extent at very high temperatures, but the Professor, by the use of a sort of bullet-mould made of well-burnt lime, into which he could direct the flame of an oxy-hydrogen blowpipe, was enabled to obtain enough of the metal to experiment with. The pure metal he describes as of dazzling whiteness. Its specific gravity is 10·575, which is a trifle higher than that usually given. It is, of course, easily soluble in nitric acid, and in hot concentrated

I

sulphuric acid. In extremely thin layers it shows by transmitted light a bluish-green colour; in somewhat thicker, from a yellow to a yellowish-brown colour. In the first case, it allows the chemical rays of light to pass, as the Professor proved in an original way. Chemically, pure silver he finds is easily soluble in a hot solution of cyanide of potassium. When in such a solution, heated to 60 or 70 deg. C., a glass rod heated to a somewhat higher temperature is immersed, an uniform layer of metallic silver is deposited, which becomes thicker the longer the rod is allowed to remain in the solution. By filling a test tube with mercury heated to 110 deg. C., and immersing it for a moment or two in the solution, a dull white coating of silver was deposited on the outside, which on the inside was seen as a brilliant silver mirror. The tube was then filled with equal volumes of hydrogen and chlorine, and carried into sunlight, whereupon combination and explosion took place. In the case of a tube left in the solution for a longer time for a thicker layer of the metal, and filled with the same gases, combination only took place slowly and without explosion. It may be that the mode of silvering glass above described may be utilized for the silvering of glass globes and other ornamental objects, which are now silvered by the somewhat complicated reduction processes.—*Mechanics' Magazine.*

GILDING.

A PROCESS for Gilding, the advantages of which, we must confess, we do not exactly see, but which has been deemed worthy of a prize by the Société d'Encouragement, is published in the Bulletin of the Society, so we transcribe it for our readers. The articles are first electro-gilded in the ordinary way. An alloy of gold and quicksilver is afterwards deposited from a solution of the cyanides of the two metals. When this coating is sufficiently thick, it is done over with a paste formed of borax, with some additional boracic acid, and sal ammoniac. It is then placed in a fire until the paste is calcined, and afterwards in a bath of acidulated water. By this process a surface of gold is obtained in just the same state as by the old water-gilding process, without any risk to the health of the workmen, and it may be burnished or left dull as desired.

The gilt wire employed in making gold lace soon tarnishes, especially in a London atmosphere. The surface of gold seldom being perfectly continuous, sulphur gets at the metal beneath, whether copper or silver, and the wire quickly blackens. To avoid this result, M. Helouis proposes to insert a thin layer of platinum between the copper and the gold. To do this, he drives a bar of copper into a ring of platinum strongly heated. The contraction of the platinum on cooling fixes it firmly to the copper, and the wire may be drawn without any risk of separat-

ing the two metals. The gilding may be done as usual. The
colour obtained is as good as that of silver-gilt wire, and, of
course, lasting, for sulphur will have no effect on the platinum.

REFINING COPPER.

THE following improved process of Refining Copper has recently
been invented by Dr. Le Clerc, of 29, Boulevart St. Martin, Paris.
Black copper in a more or less impure condition, is subjected to
the action of heat in a reverberatory furnace, as is ordinary, and
immediately it has reached a sufficiently high temperature to
begin to soften, water is projected on the heated metallic mass in
the form of very fine jets or small rain until the complete fusion
of the metal is obtained, when the supply of water is stayed.
Supposing the cupreous mass subjected to the action of heat contain copper, iron, sulphur, arsenic, antimony, lead, or tin, under
the influence of the high temperature, the water on coming in
contact is decomposed and forms sulphuretted, arseniated, or an-
timoniated hydrogen, which disengages. The oxygen acts on the
copper, iron, and other fixed metals, and forms oxides, which derive silica from the sides of the furnace and pass into the form of
scoria. A small amount of sulphuric, arsenious, and antimonious
acid is produced, which disengages in the form of vapour. Immediately the copper is completely fused, Dr. Le Clerc places a
tube of refractory clay in the metallic bath, the diameter of which
tube is proportionate to the quantity of copper forming the bath.
This being done, he projects a large amount of atmospheric air
through the tube, when an extremely active reaction is produced,
and the refining is operated in a very short time, often in a few
minutes only, when the copper to be refined is not very impure.
The air may be introduced into the midst of the copper bath in
various ways, but the most practical mode is to pass it through
the tube at the centre of the roof of the furnace. This tube may
be readily raised out of the bath, when it is not required to introduce air; and it always remains at a sufficiently high temperature
to prevent its breaking, as it would do were it subject to sudden
variations of temperature. It is essential that the treatment
should extend beyond the period of oxidation, and when the metal
on examination is found to be of a deep or brick-red colour the
refining is complete, after which the metal is treated exactly as
in ordinary.

Dr. Le Clerc gives the following as the theory :—The bath of,
metallic copper, although much purer than at the commencement
of the operation, still contains sulphur, arsenic, antimony, and
iron. Under the influence of the air introduced into the bath excessive oxidation takes place, whereby the sulphurous, arsenious,
and antimonious acids are disengaged, oxides of copper and iron
being also produced in the form of scoria. Combinations of oxi-
sulphuret of copper and antimony are also obtained more or less

arseniated. An important point to be noted, which is also
well-known, is that, by extending the oxidation, which may be
readily done, a large quantity of oxide of copper may be obtained, which has the property of giving up its oxygen to foreign
matters, and in this manner conduces to the reduction of the
copper. The production of the protoxide of copper in super-
abundant quantity necessarily causes a considerable loss of copper, which combines under the form of complex silicate of iron
and oxysulphuret of antimony, more or less arsenical, formed by
the influence of the air insufflated into the metallic bath. For
this reason, when, after having introduced the air, the metal assumes a deep brick-red colour, Dr. Le Clerc takes the precaution
to add to the bath a mixture of from two to five per cent. of
charcoal and lime, and to well stir them. In this manner he obtains the complete reduction of the silicate, and all loss of copper
is thus entirely prevented. Immediately following the action of
this mixture the refining process is proceeded with in the ordinary
way.—*Mechanics' Magazine.*

NEW ALLOYS.

Two new Alloys of tin and lead are described by M. Piho.
While containing less tin than is used in common pewter, they
are said to possess most of the advantages of that useful alloy.
They are not acted upon by vinegar, sour wine, or salt water.
The first is made by melting 1 part of tin with 2·4 parts of lead.
The lead is first melted and skimmed, then the tin is added, and
the mixture is stirred continually with a wooden stick until it
begins to cool, to prevent the lead from settling to the bottom.
This mixture has the density of 9·64, and its melting point is
320 deg. Fahr. It may be rolled cold, and the plates do not
crackle when bent. It takes a very good polish, and tarnishes
but little on exposure. It will mark paper like lead, and is so
soft that it may be scratched with the nail, but it will not foul a
saw or file.

The second Alloy is made by melting together in the same way
1 part of tin with 1·25 parts of lead. This alloy is less elastic
and harder than the foregoing. It is rather brittle, less malleable
than the former, and fills up a file. Neither of these alloys was
acted on by boiling with acetic acid for half an hour, and standing in the acid for twenty-four hours longer, nor had salt water
any action upon them ; hence, they may be useful for some kinds
of utensils.

The analysis of a few Alloys in use in Paris for spoons and
forks, and said to be unusually good in appearance, has been
made by a German chemist. The first, a beautifully yellow
alloy, is really aluminium bronze, and is composed of 69·3 parts
of copper and 10·5 parts of aluminium. The second is also a
gold-resembling alloy, and has been named "oreido." This is

composed of 79·7 parts of copper, 13·05 parts of zinc, 6·09 parts of nickel, with 0·28 parts of iron, and 0·09 parts of tin. The two last-named are no doubt accidental ingredients. The third is a beautiful white metal, very hard, and taking a beautiful polish. It is composed of 69·8 parts of copper, 19·8 parts nickel, 5·5 of zinc, and 4·7 of cadmium.

CRUCIBLES.

MR. GORE communicates to the *Philosophical Magazine* an excellent way of making Charcoal Crucibles, &c. He first shapes the articles out of wood, and he finds that lignum vitæ, kingwood, ebony, and beech answer best. After the vessel has been formed, the wood is carefully dried in a warm place. The articles are then enclosed in a copper tube retort, having two exit tubes for the escape of gas. This retort is heated slowly at first, and finally for some time to bright redness, to completely carbonize the wooden vessel. It is necessary, Mr. Gore says, to turn the retort continually, and so distribute the heat, that none of the tarry matter evolved may condense upon the articles; otherwise, he tells us, their shape and dimensions may be curiously altered. The heating is to be continued until no more gas is evolved, and care must be taken not to heat too rapidly, or the article will fall to pieces. Charcoal made in this way from lignum vitæ is remarkably hard, and the texture is so close as to make it apparently quite impervious to liquids; even after immersion in the strongest hydrofluoric acid the surface had no acid taste. Rods made of this lignum vitæ charcoal conduct electricity admirably, and would probably, Mr. Gore says, answer well for pencils for the electric arc.

NEWLY OBSERVED PROPERTIES OF PARAFFIN.

PARAFFIN, generally regarded as one of the most stable and unchangeable of bodies, has been found by Bolley to be subject to alterations which are worthy of attention. In the first place, it seems to be liable to oxidation, like ordinary organic fats. For example, Bolley took a specimen of paraffin, fusing at 53 deg. Centigrade, and having the centesimal composition—carbon, 85·61; hydrogen, 14·39; and kept it exposed to air, and heated to 150 deg. Centigrade for eight days. In the course of this time the paraffin, originally white, became browner and browner, and was at last changed into a tough, doughy, blackish mass. A good deal of this mass was unaltered paraffin, soluble in absolute alcohol; but after the removal of this, a dark brown residue was left which consisted in 100 parts—of carbon, 70·01; hydrogen, 10·25; oxygen, 19·71. Other changes are produced by the continued action of heat on paraffin. Thus, in a specimen which first boiled at 300 deg. Centigrade, a strong evaporation was afterwards remarked at 150 deg. Centigrade. These facts will be of

interest to those who employ paraffin baths, and, as Bolley suggests, may have important bearings on the industry of the body.
Paraffin is most likely a mixture of various hydrocarbons with
various boiling and melting points. Those specimens which have
the highest boiling point have also the highest melting point,
and from this circumstance it is fancied that a hard paraffin can
be produced at will.

COMBUSTION UNDER PRESSURE.

PROFESSOR FRANKLAND has read to the British Association a
paper on Combustion under Pressure, illustrated by experiments.
He commenced by stating that the origin of the paper now
communicated arose from observing the way in which candles
burned at the top of Mont Blanc, and the law deduced therefrom was that the diminution of illuminating power was exactly
in proportion to the diminution of atmospheric pressure. The
Professor stated that some years ago, while he was on the summit
of Mont Blanc at night, he was struck with the want of illumination in the candles burnt in the tent in which they stopped for
the night. He had observed similar results in other elevated
regions. The diminution of the illuminating power was in all
probability due to the reduction of atmospheric pressure. If
they took an ordinary gas flame, and placed a piece of paper
with writing on it against the flame, looking steadily through it,
they would be able to read the writing as well, or nearly as well,
as if the flame were not there at all.

The commonly received opinion was that they must have incandescent solid or liquid substances in order to produce a white
light in gaseous flames. In following ont this subject, he had
been brought into contact with a number of flames which
emitted a considerable amount of light, but which did not contain any solid matter whatever. One was metallic arsenic,
burnt with oxygen gas; it emitted an intense and brilliant white
light. Bi-sulphide of carbon also emitted a very intense light—
indeed, so intense that it had been employed to take instantaneous photographs. This was produced without the possibility
of a solid or liquid matter existing in the flame while the light
was being evolved. If oxygen and hydrogen were enclosed in a
soap-bubble or other light envelope, and exploded, there was
scarcely any light produced; but if they were enclosed in a
strong vessel and exploded by means of an electric spark, at the
moment of their combustion the light would have an increased
luminosity to the extent of ten times above that in the previous
case. Ignited gas emitted light in proportion to its density.
The increase of luminosity in flames, the Professor considered to
be due to the presence of dense hydrocarbon vapours. One of
the most interesting experiments shown was that of sending an
electric spark first through air under ordinary pressure, and

then through air under doubled pressure. The result was, that the light of the spark due to the combustion of the air was very much increased. The spark was sent also through many other gaseous and vapourised substances, showing most conclusively that the greater the atomic weight of the bodies, the greater was the luminosity of their flames when submitted to combustion by the electric spark.

CHEMICAL NATURE OF CAST IRON.

THE report on the Chemical Nature of Cast Iron (read to the British Association), was important, inasmuch as it stated that although Dr. Matthiessen and Dr. Prug had made seventy experiments in the production of pure metallic iron from its various compounds, they had not succeeded in obtaining any iron perfectly free from sulphur. Dr. Matthiessen hoped, however, by continuing his researches, yet to obtain a perfectly pure sample of metallic iron.

In the course of the discussion which followed, Mr. Salton suggested that probably the presence of sulphur in iron was only another instance of the persistence of that element in the atmosphere, as shown by the experiments of Mr. W. F. Barrett, who first devised the method of detecting the presence of sulphur upon the surfaces of bodies exposed to the air by projecting upon them a flame of hydrogen, a magnificent blue flame resulting therefrom.

SOREL'S CEMENT.

THIS is formed by making oxide of zinc into a paste with a solution of chloride of zinc. This paste quickly sets into a hard mass, which may be applied for stopping teeth, and a variety of useful purposes. Dr. Tollens gives a cheaper form of the same cement, which may be used for stopping cracks in metallic apparatus, and cementing glass, crockery ware, and other materials. He mixes equal weights of commercial zinc, white and very fine sand, and makes the mixture into a paste with a solution of chloride of zinc having the density 1·26. The mixture sets rapidly, but allows plenty of time for its application. As it resists the action of most agents, it will be very useful in the chemist's laboratory.

MANUFACTURE OF SULPHUR.

A PAPER has been read to the British Association "On the Manufacture of Sulphur from Alkali Waste in Great Britain," by Dr. Ludwig Mond. The author called attention to a new industry —the recovery of sulphur from alkali waste, which had made very rapid progress during the past few years. The importance

of the subject had been very ably pointed out in 1861 by Mr.
Gossage, in a paper "On a History of Soda Manufacture." Mr.
Gossage stated that two-fifths of the total cost for raw materials
used for the production of a ton of soda-ash was incurred for
pyrites from which to procure a supply of sulphur; and it was
well-known that nine-tenths of this sulphur was retained in the
material called alkali waste, which was thrown away by the
manufacturer. A problem was thus presented for solution, which,
if effected, would cause a large reduction in the price of soda.

Dr. Mond went on to say that the problem had been brought
very near a satisfactory solution. He called attention to a process with which his name was connected. He took out a patent
in 1863 for the process, and its merits had been very fully recognised in this country. The process was carried out in the following way :—The first product of Leblanc's famous process for
the manufacture of soda, called rough soda or black ash, was
now almost universally lixiviated with water in an apparatus
which was first used for this purpose in Great Britain, and was
composed of a number of iron tanks connected in a very simple
manner by pipes and taps, &c., so as to allow the water to enter
a tank filled with black ash already nearly spent, and thence to
flow through others filled with black ash richer and richer in
alkali, until it met fresh black ash in the last tank, thus becoming an almost concentrated solution of alkali before leaving the
apparatus. The alkali waste, or insoluble residue of the black
ash remained thus in these tanks deprived of all alkali ; and as
it had been immersed in the liquor throughout the whole time of
lixiviation, it was consequently obtained in a very porous condition. The tanks were already provided with a false bottom.
The whole process of oxidation and lixiviation of the waste,
though it was repeated three times, was finished in from
sixty to seventy-two hours. When the waste left the tanks, all
the recoverable sulphur had been taken out of it, and could no
more give rise to the dreadful exhalations of sulphuretted hydrogen, or to the formation of those well-known yellow drainage
liquors which had hitherto caused the waste to be so great a nui-
sance, the one poisoning the air and the other the water in the
neighbourhood of the vast heaps of waste surrounding many
works. Almost all the sulphur left in the waste existed in the
form of sulphite and sulphate of calcium, which were both innocuous ; and together with the carbonate and hydrated oxide of
calcium, as well as with a little soda, alumina, and soluble silica,
which were all to be found in the waste, made this waste a very
valuable manure for many soils and crops. By other processes
which Dr. Mond explained, he said he obtained sulphur of a dark
colour, the waste from which was turned to advantage and made
comparatively harmless. By his processes fully one-half of the sulphur contained in the waste was recovered. The cost was small.
A plant for the recovery of 10 tons of sulphur per week would be

about £800 ; and the sulphur could be made at £1 per ton. The
recovered sulphur being very pure was not used to replace pyrites
in the manufacture of soda, but for purposes where Sicilian sulphur
or brimstone had hitherto been employed, this Sicilian sulphur
having a much higher value than the sulphur in pyrites, and
averaging upwards of £6 per ton. And so large were the quantities of brimstone used, that the British alkali trade, in spite of
its enormous extent, could only produce a small portion of the
sulphur yearly exported from Sicily, which country had hitherto
had the monopoly of the supply of this article.

NEW EXPLOSIVE COMPOUNDS.

PROFESSOR ABEL, of the Royal Arsenal, Woolwich, has recently
patented an invention which relates to an improved production of
Explosive Compounds, capable of being employed as substitutes for
gunpowder or other explosive mixtures or compounds for blasting
and for other similar purposes. It consists, first, in the employment of a mixture of gun-cotton, employed either in a filamentous
or in a pulped condition, in conjunction with a large proportion
of an oxydizing body, such as chlorate of potash, nitrate of potash,
or nitrate of soda, with a small portion of alkali, or of an alkaline
carbonate. The mixture of these substances is effected in one
way by mixing the same in a dry state, and then adding a sufficient quantity of water, in order that the whole may be converted
into a semi-fluid consistency or mass, from which, by the application of heat, the water is evaporated during continued agitation
and incorporation until the mixture assumes a pasty consistency,
and the resulting plastic mass is then granulated or moulded into
discs, sheets, or any other desired form. The product is dried,
and is then suitable for blasting or other similar purposes. Instead of mixing the substances in a dry state, the oxydizing compound and the alkali are dissolved in water, and the boiling
saturated solution is mixed with gun-cotton, either in a filamentous or in a pulped condition. This mixture is evaporated as
before to a paste or plastic mass, capable of being granulated or
moulded into any desired form. The resulting product is to be
dried, and is then suitable for employment for blasting purposes.

The proportions of the above materials which Professor Abel
has found to give good results, are from thirty to sixty parts by
weight of the oxydizing materials, with from seventy to forty parts
by weight of gun-cotton, and about one part by weight of alkali,
or alkaline carbonate; but these proportions vary according to the
precise nature of the ingredients used, and to the purposes to
which the resulting product is to be applied.

The second part of the Professor's invention consists in impreg-
nating either partially or completely with nitro-glycerine, granulated or moulded masses.

WHAT IS LAVA?

COMING no one knows whence, it might be suspected to be formed of, or at least to contain, unusual substances; but such apparently is not the case. Here is an analysis, by M. Silvestri, of Lava recently thrown out of Vesuvius :—Silica, 39; lime, 18; alumina, 14; magnesia, 3; protoxide of iron, 13; potash, 1; soda, 10; water, 2; which means that the specimen closely resembled common wine-bottle glass. In short, lava, though varying considerably in colour and solidity or friability, and occasionally containing little groups of crystalline minerals, would seem to be a sort of rough natural glass or earthenware mainly produced from sand, chalk, clay, and similar common earthy substances. —*Express*.

NITRO-GLICERINE.

A TERRIBLE explosion has taken place at Quenast, in Belgium. It appears that a waggon, accompanied by M. Grillet, of a Hamburg firm, which manufactures this dangerous article, brought 2,000 kilogrammes (4,411 lb. English) to the quarries belonging to M. Zaman, where it was to be used in blasting operations. The people employed in the works, and those in the neighbourhood, had been persuaded that this chemical product was more efficacious and less dangerous than gunpowder. That it is more efficacious we admit, but that it is less dangerous we strongly deny. The waggon arrived at the quarry, followed by M. Grillet, three soldiers, and two workmen to unload it; two carpenters were working at a little distance, and a young girl was close by. Suddenly a tremendous explosion took place. The persons just mentioned disappeared in an instant, having been blown to pieces; and the shock was felt at Loth, eight miles off. A store close by was quite destroyed, and the houses, trees, and fields, within a circle of 500 yds., were devastated. No other lives were lost, but had the explosion taken place a quarter of an hour later, when the quarrymen had assembled, the catastrophe might have been much more terrible, as 700 men are employed in the works. In the face of the terrific accidents which have taken place since the introduction of this treacherous compound, we are much surprised to learn that any can be ignorant of its dangerous nature. It is evident that although much improved since its first introduction, it yet remains a highly unsafe explosive, and one by no means to be depended upon. We thought and wrote so once—and more than once—of gun-cotton.—*Mechanics' Magazine*.

GUN-COTTON.[*]

A GREAT discovery in connection with Gun-cotton has been made by Professor Abel and Mr. Brown, of the Royal Arsenal,

[*] Schonbein, the inventor of Gun-cotton, died August 2nd last, at Sauersburg. (See OBITUARY.)

Woolwich. The last stages are the preservation from decomposition by the use of a weak alkali, and the employment of compressed charges. Used in the last-mentioned form, we have said that gun-cotton, lighted by an ordinary fuse, has about six times the destructive force of an equal weight of gunpowder. Messrs. Abel and Brown, however, after experimenting with gun-cotton saturated with nitro-glycerine, igniting it necessarily with the detonating fuse, tried the experiment of igniting gun-cotton alone in the same way. The results showed that cotton exploded in this way has, perhaps, greater destructive force than nitro-glycerine. This is a great and valuable discovery, for we have here the comparative safety of the gun-cotton with the tremendously destructive powers of nitro-glycerine. For mining and quarrying purposes it removes the only source of danger from the use of gun-cotton, for no hard tamping is necessary. It is sufficient to fill up the bore with sand.

Gun-cotton may be carried and stored with perfect safety from explosion. The detonating fuse can be inserted in the charge when required for use. As with nitro-glycerine, the explosion of one charge provokes the explosion of another. Thus, if it were required to destroy a stockade, it is only necessary to place the charges at distances round the enclosure, and explode one. The explosion of all follows with instantaneous rapidity. Moreover, as said above, when the cotton is exploded by the detonating fuse, it is not necessary to confine it. Just as with its analogue, nitro-glycerine, the explosion is so instantaneous that the shattering effects are just as great as when the charge is quite exposed. Large blocks of granite, and thick plates of iron have been shattered by exploding open charges upon them.

The Safety Gun-cotton now manufactured by Prentice and Co. is absolutely safe under ordinary conditions, merely blazing up, and not exploding, upon ignition. When, however, it is tamped well home for work, its latent energy is fully developed. On the one hand, it is safer than gunpowder, and on the other it is far more powerful. The presence of the element of safety is guaranteed by the fact that the North-Eastern Railway Company now carry the safety gun-cotton as ordinary goods in covered waggons. We believe other railways are following this sensible example, and hope it will lead to the general adoption of gun-cotton in place of gunpowder for blasting purposes. For military or sporting purposes it is not yet quite adapted, although from experiments now being carried on, we hope to see it so utilised before long. We will close with a note of warning to all nitro-glycerine users, who, with every precaution, are in a far more dangerous position than even with gunpowder. They may be exceedingly careful at first, but the very slightest deviation from the very greatest care, into which they may be betrayed by familiarity with danger, may be attended by the direst results.

WHITE GUNPOWDER.

WITHIN the last few years we have drawn attention to a great number of new explosive compounds, the majority of which have been intended to supersede Gunpowder. That they have not yet wholly succeeded in so doing is evident from the fact that for artillery and small-arms gunpowder still continues to be used. The application of some of them to blasting purposes has been more successful; for this use, their destructive energy admirably adapts them, provided their power can be controlled, as in the case of M. Nobel's dynamite, recently described in our pages. But in this and every other explosive compound, there is an element of danger ever present in the combination of the ingredients. Could we dissever this danger from the preparation as well as from the use, and produce a result equal to gunpowder at a greatly lower cost, we should certainly achieve something; but if, in addition to this, we could include advantages which gunpowder does not possess, we should unquestionably render a still greater service to the State. All these conditions we think will be found in the White Gunpowder we are about to describe, and which we have tried sufficiently for our present purpose. White gunpowder is in appearance nothing more nor less than its name implies—a perfectly white impalpable powder, resembling flour, powdered chalk, or magnesia in its appearance. With whom it originated we cannot now say; we can only go back ten years, when Mr. Henry W. Reveley, C.E., of Baker-street, Reading, found the description of the composition and its manufacture in a French newspaper. Since that time, Mr. Reveley has constantly made and used it, in preference to the ordinary gunpowder, both on account of its superior propelling power—which is at least one-third greater—and its perfect cleanliness. It produces neither smoke nor flash of flame at the muzzle on discharge, and can be used in a casemate with perfect comfort to the gunners. Mr. Reveley, to whom we are indebted for particulars of manufacture, has used it for every purpose to which ordinary gunpowder is applicable, and invariably with the most perfect success.

The manufacture of the various modern explosives—especially of the nitro-glycerine species, or the application of the gun-cotton process to vegetable granulations—is a delicate chemical operation. The results, too, are, for the most part, dangerous and uncertain, as lately proved in the lamentable accident at Portsmouth, which resulted in the death of Lieutenant Meade, R.N., and his assistant. But Mr. Reveley has made many parcels of the white gunpowder during the last ten years, and has always found them uniform both as regards strength and other properties, and he has never met with the slightest accident, although he has tested it very severely. The composition of white gunpowder is as follows :—

Chlorate of potash	.	.	.	.	. 48
Yellow prussiate ditto .	.	.	.		. 29
Finest loaf sugar	.	.	.	.	. 23

Parts by weight 100

In manufacturing this powder, the yellow prussiate must be dried in an iron ladle until it is as white as the chlorate. The ingredients are ground separately to very fine powder, and are then mixed by means of a conical sieve until they are thoroughly incorporated, but not by trituration. For small quantities, Mr. Reveley uses a common Wedgewood mortar and pestle, which must be perfectly dry and clean. The operation does not take many minutes, and with the above precautions its manufacture is absolutely free from danger. In loading it is treated in the same way as ordinary gunpowder, being pressed down by hand solid, but not hard. The charge is ignited in the usual way, either with a common cap and nipple, or in a rim or central fire cartridge. No alteration is required in the arm, but the cartridge case must be little more than half its usual length, which will give the same result as double the quantity of ordinary gunpowder, but with greater quickness, penetration, and accuracy. In actual use it does not appear to possess a bursting so much as a propulsive power, and Mr. Reveley has obtained some of the highest penetrative results in his practice with it. The economy of this powder will at once be apparent when we state that its wholesale cost is about 86s. per cwt., but as its strength is at least one-third greater than that of ordinary gunpowder, its cost may be comparatively estimated at about 60s. per cwt.

One important feature in the manufacture of white gunpowder is that it does not require to be—indeed it cannot be—granulated, which process is the great source of danger in powder-mills. The universal use of the cartridge entirely obviates any objection that may be made to white gunpowder on that score, or on the score of similarity in appearance to other substances, and, owing to its compact form, it only occupies half the usual space. We may here state that Mr. Reveley has no pecuniary interest in this invention, nor has he attempted to protect it by a patent.—*Mechanics' Magazine.*

PRACTICAL NOTES ON GREEK FIRE. BY MR. JOHN HORSLEY, F.C.S.

THIS preparation is altogether a misnomer, as the original Greek Fire was simply a fireball made of various combustible materials, which was thrown in an ignited state into an enemy's camp, whereas the so-called Greek fire is a solution of phosphorus in a very volatile spirit known as bisulphide of carbon, which when sprinkled upon substances of a loose or spongy character more or less rapidly produces spontaneous ignition, as may be seen from the following table. The constant temperature of the

particular part of the room in which my experiments were made,
for a fire was burning at the other end, was 55 deg. Fah. :—

Paper ignited in	.	.	.	.	1 min.
Wood shavings	.	.	.	.	1¼ „
Linen	.	.	.	.	1½ „
Packing or sail-cloth	.	.	.	2 „
Woollen cloth	.	.	.	.	5 „
Carpet	.	.	.	.	6 „

At higher temperatures, of course, the ignition would be much
quicker (summer weather, for instance), as phosphorus instantly
fires when a direct heat of 70 deg. is applied to it, and the
slightest friction with the finger in its dry state produces a serious
burn which heals with difficulty. I therefore particularly caution
persons against meddling with it under any circumstances.   Atmospheric changes, such as cold and damp weather, appear to
exercise a considerable influence over phosphorus in checking or
retarding its ignition, hence it is that the application of liquid
phosphorus to paper and other light substances will not then inflame them so quickly as when the weather is warm and dry, as
I have now satisfactorily ascertained; for even at a temperature
of 47 deg. some of the articles mentioned above took a considerably longer time to ignite, but the day, however, was very wet,
and there was no fire in the room during the time of conducting
the experiments, thus :—

Paper required	.	.	.	.	6 min.
Wood shaving	.	.	.	.	7 „
Packing or sail-cloth	.	.	.	11 „
Linen	.	.	.	.	20 „
Woollen cloth	.	.	.	.	34 „
Carpet	.	.	.	.	45 „

So that although no spontaneous combustion follows the use of
Greek fire, or liquid phosphorus, in the open air on a cold or damp
day, it is nevertheless quite as much, if not more dangerous than
if it had succeeded, as the slightest heat or friction, either by the
hands or feet, on a part containing phosphorus would readily inflame it like an ordinary lucifer match, and in the case of a
female, her long clothes might be speedily set on fire. In a warm
room the danger would be even greater still, therefore the utmost
caution should be used in walking about or handling any article
of furniture.   The peculiar smell of phosphorus would at once
indicate the fact of its having been used at all, and if white fumes
are being evolved, say from a carpet, instantly proceed to wet it
with water to prevent its ignition, cut out the spot, and throw it
into the yard to dry and be burnt; the same with curtains, bed
clothes, &c.; in fact, this is the only way of getting rid of any
danger. In proof of cold retarding ignition I placed in the open
air on the window-ledge of the laboratory, the temperature being
35 deg. Fah., a small saucer containing some shavings sprinkled
with liquid phosphorus.   Several hours elapsing without any

visible effect, the saucer was brought just inside the room, when the shavings instantly inflamed.

If a naked floor-board has had some liquid phosphorus spilt on it, although unlike the more spongy articles, it will not inflame to any great extent otherwise than superficially, yet it is equally liable to do so by heat or friction, to avoid which the boards must be first well wetted, and then planed as quickly as possible, throwing the shavings into the fire,—keep wetting and planing as long as white fumes appear to be evolved. A floor-board may, however, be set on fire indirectly through the medium of the carpet on which the liquid had been sprinkled spontaneously igniting. By way of showing what may be done with a naked floor-board, I sprinkled some liquid phosphorus on a small deal board ¾ in. thick, placed on a stout sheet of iron, with a thermometer at 50 deg. standing by its side, and in ten minutes the surface of the board burnt with considerable activity. Again, I took another board, sprinkled some phosphorus over it, and when dry, gently drew across it a pointed iron instrument, and the friction was quite sufficient to set it on fire, so that it cannot be denied that there is great danger attending the use of this liquid.

Some persons have recommended the use of wet sand to stay or prevent the absorption of the liquid, but that does not appear to be of much value, as by its density it still finds its way to the board, which, unless kept constantly wetted, is liable to be fired by the slightest heat or friction of any kind. If a fire has been actually produced, the best thing to extinguish it before it goes far is to have ready a solution of carbonate of ammonia (common smelling-salts), say 1 lb. of the salt to seven pints of water, adding one pint of the crude ammoniacal liquid of the gas works, and keep projecting it on to the flaming body till it is put out. This appears to be all the consolation and advice I can at present offer, but I propose continuing my experiments and investigations upon this infernal liquid, as the present excited state of society, and the country generally, demand that something should be done to counteract, if possible, the mischief being perpetrated by those misguided and cruel wretches calling themselves Fenians, who will assuredly find that this stuff is not quite so easy to manage, nor so harmless to themselves as they imagine, as the following recent incident clearly shows. A few more such occurrences will probably cause it to be laid aside altogether, and the sooner the better, for there is no question as to its danger.

"Greek fire" appears to be a dangerous weapon to carry. Two men were going along a street in Cork lately, when the explosion of a bottle in the pocket of one of them made him throw off his coat instanter. The man and his companion then made off. On a bystander lifting the coat, another bottle exploded, and before the police came up, the garment, and what seemed to be a number of documents in it, were partly destroyed.—*Ibid.*

GLASS FOR LIGHTHOUSES.

A PAPER has been read to the Institution of Civil Engineers, "On Lighthouse Apparatus and Lanterns," by Mr. David M. Henderson. It was stated that the Glass used in lighthouse apparatus was nearly all made at Saint-Gobain or Birmingham, and was of the kind known by the name of crown glass. Different mixtures had been employed for the purpose; but M. Reynaud, the director of the French lighthouse service, now gave the composition as—

Silica	72·1
Soda	12·2
Lime	15·7
Alumina and oxide of iron	traces.
	100·0

At Birmingham various mixtures had been tried, of which several examples were given, the following being about an average:—

	cwt.	qr.	lb.
French sand	5	0	0
Carbonate of soda	1	3	7
Lime	0	2	7
Nitrate of soda	0	1	0
Arsenic	0	0	3

English glass was supposed to be of the refractive index of 1·51. That produced at Saint-Gobain had formerly an index of refraction as low as 1·50, but now it was 1·54, and frequent experiments were made to ascertain that the standard was maintained.

The furnace for melting glass was generally rectangular in plan, and was constructed of the most refractory materials; and the sides were arranged so as to allow of the easy withdrawal of the pots. Six, and sometimes eight, pots were placed in the furnace, arranged in pairs, with a firegrate at each end. The flame filled the whole interior of the furnace, and, after circulating round the pots, which were covered to prevent the colour of the glass being injured by dust or impurities from the coal, found its exit by flues. Great care was necessary in the preparation of the pots, which were made of about one-half new fire-clay, and one-half old potsherds, finely ground. The length of time a pot would last depended upon (1) the quality of its manufacture; (2) its being slowly and thoroughly dried—a process occupying about six months; and (3) the care bestowed on it in the furnace, and whilst withdrawn for casting. The average number of castings from each pot was about twenty; and the time the pot was out of the furnace at each casting was about three minutes. It was mentioned that Mr. Siemen's regenerative furnaces were now in use for the manufacture of lighthouse glass with perfect success. When the metal was ready for casting, each pot was lifted from its seat, withdrawn from the furnace, and carried to the foot of a crane, the lifting-chain of which had attached to its end a clip to embrace the pot. A mouth-piece of wrought-iron was fitted to the pot before casting, to facilitate the pouring, and the workmen tipped over the pot by means of long handles.

The casting-table was circular, and was mounted on a frame, so that by means of a handle it could be turned round, and each part of its outer circumference brought consecutively under the pot of molten metal. The moulds into which the glass was to be cast were arranged round the outside of this table, and were caused to revolve slowly under the continuous stream of liquid glass flowing from the melting-pot, so that each mould was filled in succession, thereby enabling the immediate return of the empty pot to the furnace. The moulds were of cast iron, of a uniform thickness of ⅜ in., and were supported on feet cast on, the size being such as to allow ¼ in. thickness of glass all round for the grinding process. The small lens-rings and prisms were cast in one piece, but the larger ones were cast in segments. The large belts, or central lenses for fixed lights, were generally cast flat, and were afterwards bent on a saddle to the required curve in a kiln.

Sand, emery, rouge, and water were the four necessaries for glass grinding and polishing. The sand had to be applied, with abundance of water, until it lost its cutting qualities. The emery, after being ground to a fine powder, was agitated in water, and the mixture was passed through a series of vats or tubs, so that the emery was divided into as many qualities as there were tubs, the coarsest being deposited in the first tub, the finest in that farthest from the supply. The rouge, which was an oxide of iron, was prepared from the sulphate, and was separated into qualities by means of water-tubs, as in the case of the emery. The glass of optical apparatus was ground on horizontal circular tables, securely fastened to the tops of wrought-iron vertical spindles, which received motion from the main shafting in various ways. The surfaces of these tables were divided out, like the face-plate of a lathe, to receive the different sizes of "carriers," or supports of cast-iron, which were bolted to them, and were arranged to hold the lenses or prisms to be ground. Plaster of Paris was then laid on the "carriers" in bands, the bands being reduced to the exact size by turning the table round under a gauge secured to the framing of the machine. The glass was laid on these strips, and was secured in place by means of pitch, care being taken in the larger sizes, which were ground in segments, to place a thickness of pitch between each joint, so that glass did not touch glass.

MANUFACTURE OF SODA AND POTASSA.

WHEN Soda is made by what is commonly known as Leblanc's
process, coal is used to reduce the sulphate of soda. The impurities in the coal, which consist principally of pyrites, aluminous
shale, and other materials containing silica and alumina in large
proportions, act injuriously in the manufacture. Pyrites, by
communicating oxide of iron and by assisting in the formation
of sulphide of sodium, damages the soda produced, and renders
it unfit for many purposes, and aluminous shale and the other
materials containing silica and alumina cause great loss of soda
by assisting to form insoluble compounds of soda. The object
of an invention recently patented by Mr. James Hargreaves, of
Appleton-within-Widnes, is to produce soda and potassa of uniformly good and of better qualities than have been obtainable
hitherto. This he accomplishes by using coal free from the impurities above mentioned. To effect the separation of the impurities, the coal is washed as it comes from the mine (preferably
small coal) in a liquid bath, the liquid being of such specific
gravity that the coal will float whilst the impurities, being of
higher specific gravity, will sink to the bottom. The coal is
"tipped" into the liquid, is agitated by a rake, and is then
skimmed off in the pure state. The liquid bath is a solution of
sulphate of soda or of sulphide of sodium, when the pure coal is
to be used in the manufacture of soda. The pure coal is mixed
with sulphate of soda and limestone, either in a wet or dry state,
to make "black ash." When the pure coal is to be used in the
manufacture of potassa, Mr. Hargreaves uses a liquid bath of a
solution of sulphate of potassa or sulphide of potassium; in other
respects he proceeds in the same manner as in making soda.
These improvements are, of course, applicable to the coal used in
the reduction of sulphate of soda when used in the manufacture
of glass.—*Mechanics' Magazine.*

CHLORIDE OF METHYLENE.

MR. W. H. PERKIN has read to the British Association a paper
on the Chloride of Methylene, found by the action of nascent
hydrogen on chloroform.   Mr. Perkin stated there could be no
doubt that this question was one of considerable importance,
because, if isomerism exists in the monocarbon series, in all probability it would be found also to exist in the derivatives of all
other polyatomic elements.   Those considerations had induced
him to commence a fresh examination of some of the monocarbon
derivatives, hoping that an experimental comparison of their
properties might in some degree help to the solution of this problem.   He gave a description of the experiments he had made
upon the chloride of methylene, obtained from chloroform by
the action of nascent hydrogen. Chloride of methylene obtained

in that manner, possessed the same (or nearly so) boiling point as that obtained by Buthrom from the chloride prepared from the iodide of methylene from iodoform. Chloride of methylene was but little acted on by sodium.

The President of the Chemical Section remarked he was sure that every chemist would regard with satisfaction the results obtained by such a chemist as Mr. Perkin. It was quite time that their views upon the subject should be settled one way or another. It certainly differed from the ordinary brodino of methylene, and they were almost compelled to come to the conclusion that two methylenes existed, and the experiments made by Mr. Perkins went far to establish this point.

SULPHOCYANIDE OF AMMONIUM.

DR. T. L. PHIPSON, in a paper read to the British Association, alluded to the presence of this salt in large quantities in some kinds of sulphate of ammonia of commerce; and to a method of estimating the amount of this compound, as the whole of its nitrogen is not available for agricultural purposes. This method consists in determining the sulphocyanogen as an insoluble salt of copper duct at 100 deg. C. The next point considered was connected with the properties of sulphocyanide of ammonia. In dissolving in water the salt produces a greater degree of cold than any other, though in crystallizing it gives out heat, which causes the surface and interior of the liquid to take curious motions. The alcoholic solution of the salt shows the peculiar phenomena of super-saturation to a very limited degree. Finally, the author gave his analysis of supercyanogen, which showed it to be, as Vœlkel stated, C_6, H_2, N_4, S_8, O, and not that admitted by the French chemists. The salt could be purchased in London for from £10 to £13 per ton. The sulphocyanide and sulphate of ammonia were mixed in about equal proportions, and had lately entered very largely into the manufacture of artificial manures.

ALLEGED POISONOUS QUALITY OF BEEF-TEA AND EXTRACT OF MEAT.—BY BARON LIEBIG.

ALTHOUGH it is contrary to common sense to believe that the daily food of men and animals could possibly contain a substance injurious to health, it was nevertheless to be expected that the experiments made by Dr. Kemmerich on the effect of Beef-tea and its salts on animals would produce anxiety and fear in some weak minds; and indeed the article which appeared in *Once a Week*, entitled "A Word of Warning to Cooks," is a proof that such fears really existed. I believe, however, that a simple acquaintance with the experiments of Dr. Kemmerich will be sufficient to dispel them completely. The results of these experiments

ure of a very harmless character.   Dr. Kemmerich made most of
his experiments, not upon men, but upon graminivorous animals,
—viz., upon rabbits,—and only one experiment was made by him
upon a dog.   The broth was made from horseflesh, and injected
into the stomach of the animals in progressively augmented quantities, the chief results of which are as follows :—

A rabbit weighing not quite two pounds, which had received
the broth from one pound of horseflesh (equivalent to half an
ounce of extract), remained perfectly well.   It polished itself
with its paws, was very lively, and no disturbance in the state of
its health was afterwards perceptible.

A second rabbit, of two pounds weight, into the stomach of
which the extract of one pound and a quarter of horseflesh had
been introduced, deported itself in just the same manner ; its
pulse became more vigorous, its breathing slower, and it remained
lively and healthy.

When, however, the doses were increased, and the extract of
two pounds and of two pounds and a quarter of flesh were injected into the stomach of the rabbit, such quantities of concentrated animal food were evidently too much for the little
graminivorous creature, which by doses Dr. Kemmerich succeeded in killing,—a result at which nobody will be surprised.
It follows that Dr. Kemmerich could likewise have killed stronger
animals with beef-tea ; and it may be assumed that he would
have killed even a man of 140 lb. weight (seventy times heavier
than the rabbit) by a dose of beef-tea seventy times as large,—
namely, by the broth of 140 lb. of flesh, equivalent to about 4 lb.
of extract of meat.   Less than a couple of pounds of extract
would, however, scarcely have been sufficient for one of the experiments of Dr. Kemmerich on a carnivorous animal, contrasted
with the experiments on the rabbits.   He did not succeed in poisoning that animal with beef-tea.

It was a small but very strong terrier, which had taken the
broth of four pounds of flesh (equivalent to two ounces of extract), which the animal seemed to enjoy considerably. As, however, the whole quantity was too much for it, it became necessary
to inject the remainder into its stomach.   Notwithstanding the
enormous quantity of extract of meat which had been introduced
by force, the terrier remained very comfortable and lively, and no
symptom of any disturbance of its health became manifest.
Double the quantity of meat broth which killed the rabbit had
not the least injurious effect on the little dog.

These experiments and the above calculations show sufficiently
what is to be thought of the poisonous effect of beef-tea ; it belongs to the category of cases where people have eaten *pâté de
foie gras*, turtle soup, or oysters, to such excess as to cause
death ; but no sensible person will ever dream of ascribing, on
that ground, poisonous qualities to *pâté de foie gras*, turtle soup,
or oysters.

The experiments of Dr. Kemmerich are described in his *Dissertatio Inauguralis* for obtaining the degree of Doctor from the medical faculty at Bonn; and in connecting with his conclusions the meaning of the word "poison," he in fact succeeded in drawing to his work the attention of the public, which otherwise would probably have taken little notice of it.

Dr. Kemmerich ascribes the effect of beef-tea not to its aromatic and combustible ingredients, but to the potash salts which it contains, and of which it is well-known that in larger doses they exercise an injurious effect on the organism; nevertheless—and this is a matter of great importance—potash salts are an element of all articles of food; they not only form the chief ingredients of the salts of all sorts of flesh, including the flesh of fish, but likewise of all other food, and of all the food of animals. The alkaline salts of bread, vegetables, and hay consist of potash salts, and with the exception of chloride of sodium (kitchen salt), soda salts are but rarely contained therein; in fact, it may safely be asserted that without the potash salts our food would be quite unfit for nourishment.

It does not follow, therefore, that these salts, when taken in excess, like any other—even the most harmless substance—might not eventually exercise an injurious effect. It is, however, preposterous to apply the meaning which we are accustomed to attach to the word "poison" to the effects of such an excess. It is surely quite absurd to connect this meaning with substances which we daily take in our food, and which are quite indispensable to our existence.

Dr. Kemmerich himself says (p. 31),—" I do not think of the possibility that beef-tea, in the form in which it is used for household purposes, could be the cause of poisoning; it therefore does not require a medical warning to protect from poisoning with Liebig's extract of meat." He further says:—" In medical practice, wine, ether, camphor, and musk are eminent analeptica (invigorating and refreshing remedies). Compared to these giants of medicines beef-tea modestly occupies a subordinate position. If, however, it be necessary to preserve the exhausted body from protracted illness, then there is no other remedy in the whole rich store of medicine which can afford such assistance for regenerating the diseased organism as repeated doses of beef-tea."

One of the three theses defended by Dr. Kemmerich, on his promotion before the medical faculty at Bonn, is worthy of observation by the British Navy. It runs thus:—

Thesis 2. "The best remedy against scurvy is beef-tea, or Liebig's extract of meat."—*Lancet*.

WHY COFFEE IS A STIMULANT. BY M. PERSONNE.

THE changes which heat effects in the elements contained in the green Coffee-berry have been little studied; we merely know, from the researches of MM. Buitron and Fremy on the one hand, and of M. Payen on the other, that the brown bitter substance and the aromatic principle are produced by the decomposition of that part of the coffee-bean which is soluble in water, and that a large part of the caffeine disappears during the roasting. It is said that this (caffeine) is carried away with the volatile products generated in the operation.

By roasting coffee in an apparatus which allows of the recovery of all the volatile products, I have ascertained that if caffeine be carried away with the volatile products, it can only be in such small quantity as is not appreciable by weight, and cannot explain the considerable loss which takes place during roasting carefully performed. The loss is experimentally found to equal nearly one-half of the caffeine originally existing in the coffee. I have succeeded in demonstrating that the lost caffeine has been transformed into a volatile base—methylamine, or methylammonia $(C_4 H^5 N)$, which was discovered by M. Wurtz. The following are the facts which prove the change of caffeine into methylamine during coffee-roasting.

If pure caffeine be submitted to the action of heat, and the vapour be carried through a tube heated to about 300° Cent. (about the heat which is necessary for roasting), and filled with fragments of pumice-stone, which delay the passage of the vaporized matters, only a feeble decomposition occurs; the greater part remains unchanged, and the little that is decomposed gives no characteristic product except cyanogen. This experiment tends to prove that it is not the caffeine which furnishes the volatile alkaloid existing in roasted coffee. But a very different result is obtained if, instead of acting on free caffeine, we experiment on caffeine in analogous circumstances to those in which it exists in green coffee. M. Payen has, in fact, shown, that caffeine exists in that berry in the form of the tannate, i.e., a combination of caffeine with a tannin peculiar to coffee. On submitting to the action of heat the tannate of caffeine which has been prepared with tannin of gall-nuts, we obtain, as with green coffee, methylamine: this compound behaves, under the influence of a temperature of about 300° Cent. in a manner similar to the tannate of caffeine first isolated by M. Payen. The whole of the methylamine produced during the roasting of coffee is not found in the solid residue; a certain proportion escapes with the volatile matters. It is easy to extract the alkaloid from roasted coffee by distilling the extract of coffee, *made with cold water*, with a weak base, such as lime. The addition of this alkali to an infusion of coffee immediately liberates the methylamine, the special ammoniacal odour of which is readily perceptible.—*The Practitioner; Scientific Opinion.*

DIASTASIS.

If we pour warm water on barley malt reduced to powder, and macerate for a few hours in a warm place, the liquid, on being strained, will contain a white and uncrystallizable substance, called Diastasis, soluble in water and insoluble in alcohol. This curious substance has the peculiar property of converting starch first into gum, and then into sugar. In a paper addressed to the Academy of Sciences by M. Dubrunfaut, this chemist announces that he has separated the active principle of malt, which he calls maltine. It possesses the property of forming an insoluble combination with tannic acid, in which it preserves its active properties in a remarkable manner. Maltine, compared with diastasis, is infinitely more active; the malt in the breweries contains one-hundredth part of maltine, representing at least ten times as much as is necessary for the manufacture of good beer ; so that nine-tenths of the maltine are wasted. M. Dubrunfaut observes that if it were chemically extracted, it might cause an immense saving, and might be very usefully employed either in distilleries or breweries, or in glucose manufactories.

INDUSTRIAL PRODUCTION OF HYDROGEN.

MM. TESSIE DU MOTAY and MARESCHAL have patented in France a mode of preparing Hydrogen for industrial purposes. Water has been regarded by many enthusiastic people as a cheap source of fuel, and several modes of decomposing it and separating the hydrogen have been proposed. Steam has been passed over red-hot coke, by which a mixture of hydrogen, carbonic oxide, and carbonic acid is procured. The last was separated by means of lime, and the two former were then available for heating purposes. But in this case the amount of coal expended in procuring the combustible gases exceeded in heating value that of the gases produced. More recently, it was ingeniously proposed to make water indirectly decompose itself, by employing water-power to drive large electro-magnetic machines, and using the stream of electricity generated to decompose water, and thus furnish a continuous supply of gaseous fuel. In this case also the amount of fuel obtained is altogether insignificant, when compared with the force expended in procuring it. Thus the question has been left until now, and although it cannot be said that the inventors named above have solved the difficulty, they have, perhaps, as the phrase goes, made a step in the right direction. Their process is really the decomposition of the vapour of water by red-hot carbon, but they do not use steam ready generated, and thus save the expenditure of coal in one furnace. They take alkaline or earthy hydrates, mix them with coke-dust, and heat the mixture in retorts. The heat expels the water from the hydrates, and the vapour coming in contact with the carbon produces hydrogen and carbonic acid. The latter is removed by passing the mixture over an alkaline carbonate, which is thereby converted into a bi-carbonate, and the hydrogen passes on. The bicarbonate, when re-heated, gives up the extra atom of carbonic acid, and the neutral carbonate is regenerated for another operation. Another mode of procuring pure hydrogen is to drop hydrocarbon (petro-leum oil) upon lime heated to cherry redness. In this case, they say, pure hydrogen is set at liberty, and carbonate of lime is formed, but the source of the oxygen necessary for this is not obvious. There is an apparent economy in the first process of these inventors, but still the hydrogen will be a very expensive fuel, while, for lighting purposes, it would require to be "carbu-retted," which would make it dearer than coal gas. The inven-tors evidently do not realise all the obstacles in their way. Hydrogen, it is true, weight for weight, gives four times the heat of coal; but hydrogen is so extremely light, that, bulk for bulk, it is calculated that the gas will give only one five-thousandth part the heat of coal. The loss in storing and distributing, too, would be enormous. The extraordinary facility with which hydrogen passes through cast-iron, shows what would happen in gas-holders and pipes. We fear the day is very distant when hydrogen will be directly available for fuel, and can only recom-mend these processes to the attention of the Aëronautical Society. —*Mechanics' Magazine.*

BISULPHATE OF LIME.

AN invention, patented by Mr. W. T. Read, of Old Broad Street, for the better application of Bisulphate of Lime, now largely used in the brewing-trade to control fermentation and prevent acidity, &c., is attracting considerable attention. It is calculated not only to save a large amount of labour, but also to perform in a perfect manner a process that has hitherto been rather clumsily carried out. The value of our annual exportation of beer and ale is nearly two millions sterling.

METHYLIC ALDEHYDE.

DR. A. W. HOFMANN produces Methylic Aldehyde by directing a current of atmospheric air, charged with the vapour of methylic alcohol, upon an incandescent platinum spiral. Fittig and others have shown that in the hydrocarbons of the aromatic series, the higher members may be obtained by substituting radicals of the ethyle series for hydrogen. Amyltoluole is obtained by the ac-tion of sodium on a mixture of bromide of amyle and bromi-nated toluole. Otto and Morries describe the formation of a compound of mercury with naphthaline, and Lossen has examined the products of the oxidation of naphthaline. When toluene is

distilled in a current of chlorine, chlorinated toluole is formed.
When the chloride is heated with diluted nitric acid, it is formed
into hydride of benzoyle, yielding a considerable quantity of
benzoic acid. The process would enable large quantities of benzoic acid, or of oil of bitter almonds, to be produced at very small
expense.—*Illustrated London News.*

BISULPHIDE OF CARBON.

Is by far the best extractor of grease, but there are several
objections to its general use. Its vapour is very inflammable,
and is also decidedly poisonous, although there may be some
exaggeration in the stories told of its effects. One thing is quite
certain—the smell of the commercial article is extremely unplea-
sant. The bad smell, however, we learn from Millon, does not
belong to the pure sulphide, and may be easily removed. It is
only necessary to agitate the liquid well with an equal volume of
milk of lime, and then distil off at a low temperature. Litharge,
and copper, and zinc shavings, remove the compounds which give
the bad odour equally well, and may be used in place of the
lime. Simple agitation with these bodies will take away the
greater part of the smell. Purified by either of these methods,
the sulphide only possesses a faint odour somewhat resembling
that of chloroform, and it may therefore be used to remove
grease, like benzine. It must be mentioned that, left to itself,
the sulphide soon again acquires a bad odour; but this, it seems,
may be prevented by keeping a little litharge, or some copper or
zinc shavings in the bottle.

NEW ANÆSTHETIC.

ALTHOUGH laughing-gas has only quite recently and suddenly
come again before the notice of the profession, the properties of
the oxides of nitrogen have not been wholly neglected by phy-
siologists. In two papers published about four years ago, Dr.
Hermann arrived at some interesting results touching the physiological action of nitrous and nitric oxide. (Reichert, Du Bois
Reymond's *Archiv,* 1864, p. 521; 1865, p. 469.) From these
researches it would appear that while laughing-gas is very readily
absorbed by blood, it neither enters into combination with, nor
produces changes in, nor suffers changes from, the action of
blood. As our readers are aware, it is now generally believed
that the oxygen present in blood exists in a peculiar loose combination with the blood corpuscles, and is not retained by simple
physical laws of absorption. Laughing-gas, on the contrary, is
merely physically absorbed, and blood will take up rather less of
it than it will of water—that is to say, 100 volumes of blood will,
at the temperature of the body, absorb somewhat less than 60
volumes of laughing-gas. Blood saturated with laughing-gas

shows no sign of change; the spectrum appearances are the
same; the blood corpuscles are unaltered; and, according to
Hermann, the oxygen is not driven out. In the blood, and probably in the body, laughing-gas suffers itself no change. It does
not give up its oxygen for purposes of oxidation, as Sir Humphry Davy thought. It gives rise, therefore, to no free nitrogen,
but goes out of the body as it comes into the body, pure and
simple laughing-gas. Hence it is itself of no respiratory use;
and, when mixed with a quantity of oxygen sufficient for the
needs of the economy, has no more direct effect on respiration
than has nitrogen or hydrogen. From these facts, we may gather
that the mode of action of laughing-gas is that of a body having
distinct effects on certain parts of the system; and does not depend, like that of some other agents, on any direct interference
with the function of respiration. Readily absorbed by blood,
and yet with its limit of absorption soon reached; passing away
from the blood into a pure atmosphere as quickly as it passed
into the blood from the receiver in which it was previously confined; suffering no change itself, and causing no obvious gross
chemical changes in the fluids or tissues of the body, it certainly
seems peculiarly fitted as an agent for producing temporary conditions of the economy. On the muscles and hearts of frogs it
has no more effect than nitrogen or hydrogen. The American
dentist, Wells, actually applied it in practice without encouraging
success; his partner, Morton, pursuing the research, was led to
the use of sulphuric ether, and became, in fact, the great discoverer of practical anæsthesia, and the benefactor of his race.
The decision adverse to the use of protoxide of nitrogen gas
seems at least to have been hasty. Dr. Thomas W. Evans, of
Paris, has this week given a series of demonstrations of its use
at the Dental Hospital of London and at the Central London
Ophthalmic Hospital to crowded circles of dentists and surgeons,
and has produced results hitherto unknown here. Given by his
and Colton's method, the period required to produce unconsciousness has been less than 45 seconds; the operations have
been harmless; the sensations of the patient agreeable; there
has been no struggling or distress. The recovery has been almost
instantaneous, and without headache, giddiness, sickness, or
prostration, such as so frequently follow chloroform. In fact, in
many instances, three minutes after the patient has expressed a
willingness to submit to operation, he has been chatting gaily by
the chair, the tooth having meantime been painlessly extracted,
and he having passed through a period of total unconsciousness
without any disagreeable sensations.

As will be seen from the account which we give of these remarkable demonstrations, it has yet to be shown how far the unconsciousness can be protracted, as is necessary for prolonged
surgical operations. And great caution must be enjoined in using
this agent, even by Dr. Evans's method, until we have more of our

own experience to guide us. But taken with all qualifications, the results are very surprising, deeply interesting, and of great promise as supplying that important desideratum—a painless and rapid anæsthetic, suited for those who have to pass under the hands of the dentist, and for the quicker operations of surgery, and of which the effects are entirely transient.—*British Medical Journal.*

A NEW MEDICINE.

THE *Moniteur* gives an account of a tree called "Haofash," which grows on the mountains of Baria, in French Cochin China. MM. Condamine and Blanchard, two French travellers, have at length succeeded, after much fruitless research, in finding this tree. The Annamites, who gain their livelihood by selling the bark of the haofash to professional men, wait till the tree has attained its third year before stripping it of its bark—its usual height at that age being about twenty-four feet, with a circumference of a foot and a-half, or thereabouts. The operation is performed in June, when the tree has neither blossoms nor fruit; it is hewn down, and then denuded of its bark methodically, in slices about two feet long and three or four inches broad. These strips are made up into bundles weighing from thirty to forty pounds. The bark of the haofash is outwardly of an ash-grey colour, and inwardly brown. It has a strong aromatic smell, and a slightly bitter taste. When chewed, it reddens the saliva; it is a powerful styptic, and administered by the physicians of the country in cases of colic, diarrhœa, and dysentery. The dose for a decoction is generally from six to ten *grammes* in 100 *grammes* of water, boiled down to one-fifth; but sometimes they merely put a bit of the bark into hot water, occasionally rubbing the former against the rough sides of the earthern pot used for the purpose, and then make the patient drink the liquid, which is then sufficiently strong to cure a simple colic.

WATER AS A SIMPLE BODY.

MR. H. WILDE has contributed to the *Philosophical Magazine* a paper announcing the startling discovery which he believes he has made, that water, instead of being composed of oxygen and hydrogen, as had formerly been supposed, is a simple body, and is transmutable, either wholly into oxygen or wholly into hydrogen, by the action of electricity. The experiments upon which this conclusion rests are described at great length, but their results are believed by other observers to be reconcilable with the common hypothesis. In any case, however, it is proper to have a critical examination periodically made of the accepted chemical canons; and the example of ozone warns us that a body supposed to be simple may nevertheless exist in very different states, and

that, probably after all, there is only one kind of matter which
assumes different shapes from the association of different proportions of force. Water was at one time supposed to be one of
the elements, because it could not be decomposed, and was subsequently believed to be composed of oxygen and hydrogen
gases, because it was resolved into them by the electric current,
and because those gases by their combination formed water.
There are other explanations, however, conceivable, and it is
right that they should be considered and exhausted.—*Illustrated
London News.*

ON SEA WATER.

IT has been shown during the past year that deep spring
water contains no organic nitrogenous matter, and that the
water of rivers and lakes contains nitrogenous organic matter in
the proportion of about one part of nitrogenous organic matter
to a million of water. The Water of the Sea contains about 100
times as much solid matter as the water of rivers and lakes.
The question has been asked whether the nitrogenous organic
matter increases in anything like that proportion. An examination of sea water collected off the coast of Devonshire (at Teign-
mouth) has been made accordingly, with the object of answering
this query. The result is, that there is about double or treble
as much nitrogenous organic matter in sea water as in average
river water; so that the total solids increase far more rapidly
than the organic matter.

PURITY OF WATER.

M. BELLAMY devotes an article in the *Journal des Connai-
sances Médicales* to the important subject of the Purity of Water,
and proposes the sub-sulphate of alumina as a proper test, which
is preferable to alum. It may be prepared by adding 12 cubic
centimetres of a solution of caustic potash of the strength of 10
per cent. to a solution of 8 grains of alum in 100 of water. A
precipitate is thus formed, which is slowly re-dissolved, and the
solution will keep indefinitely in a limp'd state. This sub-sulphate contains about half as much more potash than alum, which
is a double salt. Of this solution five cubic centimetres are
poured into a litre of the water to be tested. The decomposition
of the salt takes place under the triple influence of the mass of
water, the earthy bicarbonates, and organic matter contained in
it. The latter falls to the bottom of the vessel in the course of
a few hours, being precipitated by the alumina which combines
with it.

THAMES WATER SUPPLY.

DR. WHITMORE, Health Officer of Marylebone, states: "During the present summer, owing to the long-continued drought, the sufficiency of our Metropolis Water-supply has undergone a severe test; but it will afford no small amount of satisfaction, especially to those who derive their supply from either of the Thames companies, to know that while in many large towns in England water has become a scarce commodity, while several of the northern lakes are becoming more and more shallow, owing to the drying up of their tributaries, and while hundreds of homesteads are literally without water, our beautiful river, from which we abstract some 55,000,000 gallons daily, shows no diminution whatever in its depth or its ordinary summer flow; notwithstanding the largely increased quantity of water that has been pumped from the river during the past three months, and to quote from the opinion of an eminent engineer connected with one of the water companies, 'the quantity pumped might have been doubled without any serious diminution of the storeage and flow of the river.' One fact will suffice to show the correctness of this opinion. In the middle of the month of July, it was found that an enormous quantity of water was flowing over the weir at Sunbury, which was estimated at not less than 350,000,000 gallons in the 24 hours; further, it may be stated that the flow over this weir in the summer is quite equal in quantity to that of the summer of the previous year, when it will be remembered the rainfall was very considerable. This may be attributed to the generally improved state of the river banks, and the removal of shoals in the upper reaches of the river; to the repairing also of the weirs and locks, and other works which combine the twofold object of increasing the storeage and purity of the water, all of which are now being vigorously carried out by the Conservators of the Thames, and in furtherance of which the five Thames water companies contribute each £1,000 annually."

LIME AND LEMON JUICE.

The results of the Merchant Shipping Amendment Act, as far as it relates to the supply of Lime and Lemon Juice, are given in a Parliamentary paper recently published. This return, which was asked for at the end of last session by Mr. Alderman Lusk, shows the quantity of juice submitted for inspection from the commencement of the operation of the Act to the 30th ult., the amount passed and rejected, and a summary of the inspector's report in each case. From this it appears that during the past nine months a total of 98,277¼ gallons of lime and lemon juice have been submitted for official inspection in England, of which 42,612 gallons were inspected at Liverpool, 23,810 in London, 11,852 at Leith, 10,160¼ in Glasgow, 5,494 at Sunderland, 2,501 at Cardiff, 422 at Southampton, 381 in Bristol, and 45 at Hull.

This juice was contained in 1,022 casks; 819 casks, containing 76,852½ gallons, having been accepted, and 203 casks, containing 21,425 gallons, rejected. An epitome of the quantities of juice that failed to pass shows that 9,764 gallons were rejected at Liverpool, 2,522 in London, 1,209 at Leith, 6,461 at Glasgow, 1,051 at Sunderland, 148 at Cardiff, and 270 at Southampton. The chief causes of rejection, as described in the inspectors' reports, are severally—1, Deficiency in citric acid; 2, Not up to the standard of the natural product of the fruit obtained by expression; 3, Deficiency in specific gravity; 4, Nauseous taste and disagreeable odour; 5, Adulterated with sulphuric acid; 6, Diluted with water; 7, Dirty; 8, Adulterated with vinegar; and 9, Containing too much pulp. Of the total quantity submitted, 43 casks contained lime, and the rest lemon-juice, so that only a small quantity of the former has as yet been brought into the market for use in the mercantile marine. The largest quantities of juice appeared to have been prepared at Messina and Palermo in Sicily, in York, Sheffield, London, and Stockton-on-Tees. It is probable that lime juice will shortly be imported in great quantity for the purpose of this Act from the West Indies, and especially from the Island of Montserrat, from which a great amount is now sent in a concentrated state for the manufacture of citric acid. An official notice was issued by the Board of Trade in June last by which all vessels bound from any port in the United Kingdom to ports on the eastern coast of North America, situated between the 35th and 60th degrees of north latitude, are exempted from carrying lime or lemon juice; but this exemption does not extend or apply to ships which, being bound to any place within the above limits, are or may be, by the terms of their agreements, also bound to any place out of those limits, nor does it apply to ships bound to any part of the coast or island of Greenland situated within the specified boundaries.

SEA-SICKNESS.

PROFESSOR FORDYCE BARKER, of New York, who has had many opportunities of gathering the experience of others in the treatment of Sea-sickness, lays down the following rules, which have, he says, been thoroughly and successfully tested, usually in all those who suffer most from the neglect to remain exempt from sickness even during a long voyage. They are equally applicable, *mutatis mutandis*, to short voyages. 1. Have every preparation made at least twenty-four hours before starting, so that the system may not be exhausted by overwork and want of sleep. This direction is particularly important for ladies. 2. Eat as hearty a meal as possible before going on board. 3. Go on board sufficiently early to arrange such things as may be wanted for the first day or two, so that they may be easy of access; then undress and go to bed, before the vessel gets under way. The neglect

of this rule by those who are liable to sea-sickness is sure to
be regretted. 4. Eat regularly, and heartily, but without raising
the head, for at least one or two days. In this way the habit of
digestion is kept up, the strength is preserved, while the system
becomes accustomed to the constant change of equilibrium. 5. On
the first night out take some laxative pills, as, for example, two
or three of the compound rhubarb pills. Most persons have a
tendency to become constipated at sea, although diarrhœa occurs in
a certain per-centage. Constipation not only results from sea-
sickness, but in turn aggravates it. The reason has already been
given why cathartics should not be taken before starting. The
effervescing laxatives, like the seidlitz, or the solution of the
citrate of magnesia, taken in the morning on an empty stomach,
are bad in sea-sickness. 6. After having become so far habituated
to the sea as to be able to take your meals at the table and to go
on deck, never think of rising in the morning until you have eaten
something, as a plate of oatmeal porridge or a cup of coffee or
tea, with sea-biscuit or toast. 7. If subsequently, during the
voyage, the sea should become unusually rough, go to bed before
getting sick. It is foolish to dare anything when there is no glory
to be won, and something may be lost.—*British Medical Journal.*

ICED TEA.

THE most delicious and sustaining beverage that can be
drunk in hot weather is good strong tea, cooled down with lumps
of ice. It should be only slightly sweetened, without milk, and
flavoured with a few slices of lemon, which are infused at the
time the tea is first made. A jug of this ready at hand would
suit the complaint of many of our readers to a T, whilst the
thermometer denotes a high temperature.—*Medical Times and
Gazette.*

GREEN DYE.

THE attempts which have been heretofore made to produce a
green dye from aniline have been attended with imperfect success; but M. Keisser, of Lyons, appears at length to have
solved the problem by combining picric acid with the base of
Hofmann's blue. To prepare this new dye one part of Hofmann's blue is dissolved in three parts of alcohol of about
90 per cent. strength. To this is added one part of ethyl,
and the mixture is then heated in a closed vessel for half an
hour. One or two parts of caustic potash are then added; the
mixture is heated for three or four hours, when the alcohol
passes off, and the residual mass is washed with boiling water
and then heated with five or six hundred times its weight of
water, which dissolves the compound, and the solution is filtered
hot. A hot aqueous solution of picric acid is then stirred in

until the liquid acquires a green colour, and, by allowing the
liquid to stand for twenty-four hours, the dye is precipitated,
and may be separated by decantation or filtration. In dyeing a
green colour it is dissolved in alcohol, and is used in precisely
the same way as the other aniline dyes. The colour is a fine
yellowish green.

HAIR DYES.

DR. M'CALL ANDERSON has accidentally discovered what promises to be the most perfect black dye for the hair which has
yet been seen. After having used the bichloride lotion for some
weeks, he changed it for the lotion of hyposulphate of soda;
and the morning after the first application the hair of the part,
which before was bright red, had become nearly black. One or
two more applications rendered it jet black, while neither the
skin nor the clothing were stained. Dr. Anderson saw this
patient a couple of weeks later, and there was not the least deterioration of colour, although, of course, as the hair grows the
new portions will possess the normal tint. He was by occupation a Turkey-red dyer, and was much interested in the discovery,
though rather grieved to find, what medically must be considered
one of its greatest advantages, that it did not dye the linen, and
was therefore unavailable for his purposes."—*British Medical
Journal.*

Some Hair Dyes (says *Cassell's Magazine*) are positive poisons, and poisons, too, of a most virulent character. Of these,
preparations of lead and mercury are the most dangerous, though
they are by no means the only ones that enter into the composition of hair dyes; and what adds to the danger of using them is,
that they are not eliminated from the system in the course of the
circulation, but, on the contrary, they accumulate, and must
eventually be productive of great and serious evils. If people
must use hair dyes, let them carefully avoid such as contain mineral substances; there may or may not be danger in the employment of vegetable extracts, but there is no doubt at all about the
mineral.

INDELIBLE MARKING INK.

DR. JACOBSEN suggests the use of Lightfoot's aniline black
for marking-ink, which, on the score of indelibility, must possess
some advantages over the ordinary nitrate or tartrate of silver
ink. We give the author's receipts for two preparations. He
first prepares a solution, No. 1, by dissolving 8·52 parts of crys-
tallized chloride of copper, 10·65 parts chlorate of soda, and
5·35 parts of chloride of ammonium in 60 parts of distilled
water. Another solution, No. 2, is made by dissolving 20 parts
of hydrochlorate of aniline in 30 parts of water, and adding 20
parts of mucilage (one of gum to two of water), and 10 parts of

glycerine.. To make the ink, which must be prepared as wanted, one part of No. 1 solution is mixed with four parts of No. 2. The mixture will be of a dull greenish colour, and is ready for use at once. It can be applied either with a quill pen or with brush and stencil plate. The colour of the writing will at first be only pale green, but it will darken gradually on exposure to the air, or the black may be produced at once by employing heat. The best way of using heat is to hold the article marked over a vessel of boiling water, but the heat from a lamp or fire will of course effect the purpose. When the articles are washed with hot soap and water, the colour of the writing will be a dark bluish black. It will not be affected by dilute acids nor by alkalis. A strongish acid will turn it green, but an alkali will directly restore the black. A very strong solution of chloride of lime will first turn the colour red and finally discharge it, but only for a time. It will return again in a few days with increased intensity. The colour, in fact, can only be removed by destroying the fabric, which is more than can be said for the silver inks.—*Mechanics' Magazine.*

THE USE OF GAS-TAR.

GAS-TAR and ammoniacal liquor from the gasworks not many years ago formed one of the most repulsive nuisances known to manufacturers. It was either thrown into the river, where it floated in ghastly blue patches, under the name of Blue Billy, or, as at Edinburgh, was conveyed away stealthily at night and emptied into the sea. These offensive products have within these last few years been distilled and transferred into a number of liquids and solids, all of which are more or less valuable. The gas-tar, a material with soiling powers unequalled, and with an odour that is unapproachable, yields benzol, an ethereal body of great solvent powers, which forms the principal constituent of benzine, the most effectual remover of grease stains known, and generally used to renovate kid gloves. Benzol produces with nitric acid nitro-benzol, a body resembling in odour bitter almond scent, which is largely employed in perfuming soap. Could any two products appear more antagonistic to the substance from which they spring? From the same tar we have various mixtures of substances chemically similar to benzol. These are popularly known as "naphtha." One liquid of this kind is the gas substitute of the peripatetic costermonger and cheap jack, besides being the source of illumination of many large factories and yards in which nightwork is done. Another of them, mixed with turpentine, is at once elevated to the dignity of the drawing-room, where it appears in the table-lamp as camphine. Naphtha is also frequently used in dissolving resins, india-rubber, and gutta-percha. Lampblack is made by burning, with slight access of air, the least volatile components of gas-tar. Moreover, if

these be melted and mixed with pebbles, a valuable paving
material is produced, with the appearance of which most of
us are familiar.   Red dyes, but, unfortunately, of only ephemeral beauty, can be made from that once dread enemy to the
gas manufacturer—naphthaline.—*Quarterly Review.*

NEW PIGMENT.

A NEW pigment is stated to have been brought into use in
America, which is likely to have an important influence on the
white lead trade.   In a mine in the State of New Jersey, which
has for 35 years past been worked for lead, a natural chemical
combination has been discovered not heretofore attainable by any
known artificial means, and which is not only suitable as a paint
for ironwork of all kinds, but is specially adapted for the coating
of ships' bottoms, as the particles of copper in the combination
are fatal to animal life.   Messrs. C. and J. Reynolds, of New
York, the Boston White Lead Company, of Boston, and J. S.
Chadwick and Co., of Detroit, are contractors for the entire product of the mine, and the tests applied to it are alleged to have
demonstrated that the material is "superior to any pigment
hitherto made for firmness, body, and durability, and, in fact, in
every essential necessary to form a perfect paint."

NITRATE OF BISMUTH.

IN an essay on the Sub-nitrate of Bismuth, Dr. Monneret enu-
merates the various effects of this valuable medicine.   He was
the first to employ it in nose-bleeding and intestinal hæmorrhage,
in which latter case he administers a tea-spoonful of it in a
table-spoonful of water once an hour.   He has been using it so
in typhus fever for the last five years, and never during that time
has lost a single patient by intestinal hæmorrhage.   The same
salt appears to be a specific for the cure of ozena and otorrhœa.
Sub-nitrate of bismuth, in his opinion, only acts negatively, and
merely as an insulating agent, but it prepares the mucous membrane for the prompt absorption of remedies the action of which
is uncertain.

IMITATION PRECIOUS STONES.

PENDING the discovery of the means of producing artificially
real precious stones—that is, of forming by art stones having the
exact chemical composition and appearance of those formed by
nature—and the arrival of the day predicted by Alphonse Karr,
when a chemist shall present to the Academy of Sciences, a diamond as large as a hen's egg, and apologize for the smallness of
the specimen, the many vain will have to indulge their vanity by
the display of sham jewels.   It is right to say, however, that,

thanks to the researches of Ebelmen, Deville, Troost, and others, small rubies and amethysts have been produced, and even microscopic diamonds have been made. But while we have to wait for larger productions it is well to have the imitations as exact and beautiful as possible, therefore we copy from "Elsner" a receipt for a hard foundation glass, and the proportions of colouring agents necessary to give this glass the appropriate tints of the stones. Elsner takes 43·7 grms. of pure quartz, 22·8 grms. of pure and dry carbonate of soda, 7·6 grms. of borax, 3·4 grms. of nitre, and 11·8 grms. of minium. These ingredients reduced to a fine powder, and well mixed, are brought to perfect fusion in a Hessian crucible over a charcoal fire. To colour this mass, in order to imitate various precious stones, the following must be added:—For sapphires, 0·105 grms. of carbonate of cobalt ; for emeralds, 0·53 grms. of oxide of iron ; for amethysts, 0·205 of carbonate of manganese ; for topaz, 1·59 grms. of oxide of uranium. In all cases the fusion must be perfect, or a clear glass will not be obtained. Cutting greatly improves the appearance of these imitations.—*Mechanics' Magazine.*

ARTIFICIAL DIAMONDS.

FROM time to time we hear of projects for the production of Diamonds by artificial means. In the journal *Les Mondes* we are told that M. Calixte Say (*peut-être nous estropions le nom*) had discovered the true means of fabricating the diamond by vaporizing the iron of a blast-furnace ; and that M. Teasié of Motay proposed to furnish the heat necessary for the operations by the combustion of oxy-hydrogen gas ! We are now told that M. Sais is the author of the process ; and that it consists in forcing through a blast-furnace a current of chlorine, by which the iron in fusion would be converted into a protochloride of iron, which would be volatilized, leaving the carbon intact,—"*Dans ces circonstances, le cristallisation du charbon pourrait s'effectuer !*" Surely, in this present depressed state of the pig-iron trade, our iron-masters might turn their blast-furnaces to account, and by establishing diamond manufactories in the black country, in Cleveland and elsewhere, give a *brilliant* turn to a great native industry.

NEW SAFETY-LAMP.

ACCORDING to the invention of Mr. W. Key, of Bristol, the Lamp is constructed of a metal case, with an orifice 3 in. in diameter, through which the light issues. The glass which fills this orifice is preserved from injury by several crossings of strong wire. The air is admitted through an aperture in the bottom, and the hot air goes out through another at the top of the lamp. Both these apertures, reports the *Mining Journal,* are covered with wire-gauze, protected in such a way as to put a stop to the dangerous practice of miners lighting their pipes through the gauze, which is often done with the lamps now in use. The lamp, made of common sheet-iron, would weigh about 2 lbs., and its price would not be greater than that of the Davy Lamp, while it gives at least twice as much light, at a cost of about 2d. for twelve hours' burning. One difficulty with mining-lanterns has been the breaking of the glass when it is brought into contact with the flame; but with Mr. Key's lamp the light goes out immediately the lantern is held in such a position as to bring the top of the flame under the glass : held in any other position the light remains burning. Another peculiarity of Mr. Key's invention is that, as soon as it is taken into an atmosphere of fire-damp, the flame begins to flicker, more or less, according to the quantity of fire-damp, and eventually goes out. The lamp has been shown to several practical men, who highly approve of it. Altogether, Mr. Key's invention is a very ingenious and useful one, and likely to prove of great service to colliers in their dangerous calling.— *Mechanics' Magazine.*

DOTY'S CONCENTRIC LIGHTHOUSE LAMP.

THE great *desideratum* long sought for, viz., a powerful, brilliant, and reliable light, combined with economy, for coast lights, seems at length to have been attained by Captain H. H. Doty, who has invented and constructed a "first order" Concentric Lamp for burning liquid hydrocarbon oils, the results of which appear to promise what has long been wanted to remedy the irregularity attending the lamps now employed by maritime nations to light up their coasts. It is well known that the safety and success of the commercial marine of every nation depends largely upon, and is increased or diminished by the number and distinctive character of the signal, or danger-lights arranged on its coasts. The demand for good fish-oils being beyond the supply, other and inferior oils have been unavoidably used for lighthouse purposes. The inferior quality and illuminating power of these oils frequently involve the necessity of trimming the lamps during the night, an operation invariably attended with much danger to commerce, and calculated to mislead the navigator and jeopardise life and property to a serious extent. These disadvantages being common to every nation in the world are the cause of many of the marine disasters almost daily brought under our notice. To obviate, as far as possible, the above-mentioned difficulties, and to bring into more general use for coast lights the already well-known liquid hydrocarbon oils, Captain Doty has devoted a large amount of thought and time, and has produced a lamp by which the above-mentioned difficulties will be entirely overcome. It is said that this lamp produces a more powerful light than any now in use, and will burn the full service time with undiminished brilliancy, and without requiring to be trimmed—a saving of more than one-half in the cost of fluid being the result of Captain Doty's improved arrangement of burners. These burners contain only about one pint of oil (without air space), and are supplied, from a reservoir placed outside the lenses of a lighthouse, through a syphon, in such a manner that a constant light and quantity of fluid is always maintained in the burners, and air is guided to the flame by adjustable flanged rings, so that the most complete combustion of the carbon ensues; and the application of Captain Doty's lamp to, and its trial within any first-class lighthouse, can be made without any alteration in the structure or arrangement thereof.

THE OXY-HYDROGEN LIGHT.

OXYGEN is now largely used in the production of the lime-light for photographic enlarging, and the manufacture of this gas may long be carried on with perfect safety, but at last result in an explosion dangerous to life and limb. This arises from the use of an impure sample of black oxide of manganese, containing combustible matter; so that when chlorate of potash is added, and heat is applied to the retort, an explosion is the result. Safety from such accidents may be secured by heating the black oxide of manganese to dull redness in a crucible before using it in the manufacture of oxygen from chlorate of potash. In this roasting operation care must be taken that soot or carbon does not fall into the crucible, or that may result in an explosion. After preparing a quantity of oxygen, it is best to wash out the retort, and to throw the washings on the filter, which will collect the black oxide of manganese. This oxide is not changed by the process of making oxygen, and after having once been used for the purpose, of course it may be employed again with perfect security.

MIXTURE OF GAS AND AIR FOR ILLUMINATING PURPOSES.

IT has been proposed to use the Drummond Light for the illumination of public places in Paris. The manufacture of oxygen for the purpose is indeed the chief difficulty in the way of the application, and the municipal authorities of Paris hesitate to regard the method of M. Tessié du Motay as perfectly satisfactory.

Another inventor comes forward with a scheme for burning a mixture of coal-gas and atmospheric air, so as to produce an intense light. Every one knows that the ordinary flame of a Bunsen's burner is but faintly luminous. The particles of carbon are so rapidly consumed that their momentary incandescence, which, according to the generally received opinion, gives luminosity to the flame, is not permitted. It is necessary, therefore, to introduce some other substance which shall be maintained at a white heat, and so furnish the light. M. Bourbouze employs for the purpose a roll of very fine platinum wire-gauze, upon which he directs a large number of very small jets of the mixed gases. The combustion of these gives a vivid incandescence to the platinum, and furnishes light at a considerably reduced cost. The air, however, has to be supplied by a sort of blow-pipe apparatus, which complicates the matter; but notwithstanding this, the inventor assures us that a great saving is made by the use of his invention.—*Mechanics' Magazine.*

GAS-LIGHTING.

THE illuminating power of a Gas jet, it is found, may be immensely increased by introducing a small cylinder of magnesia into the centre of the flame, somewhat after the manner adopted in the Drummond Light of introducing a small ball of lime. The announcement of the success of this expediment now comes to us from Paris with a very great pretension; but in reality there is little new in it. The species of light now required is an electrical light of moderated intensity, produced in a globe of magnesia or other appropriate material, and the electricity could be distributed to every house by wires in the way in which gas is now distributed in pipes. The electricity should, as in the case of the electric lights for lighthouses, be generated by mechanical power.

MEASUREMENT OF DAYLIGHT.

Mr. R. J. WRIGHT has read before the Royal Society a paper on the subject of an easy manner of measuring the intensity of total Daylight. It has been highly spoken of by our leading philosophers. A white disc, with a black central spot, is made to slide up and down within a metal tube. The tube is mounted vertically, and the observation is made by peering down it, and noting the point, as indicated on a graduated scale, at which the disc and spot can just be discerned in the darkness. The brighter the daylight, the greater the depth of the limit of visibility; the darker the day, the nearer to the tube's edge must the disc be brought. It will be seen that the measurements are differential, and that they may be affected by the sensitiveness of the observer's eye.

EXTENSION OF THE LIME-LIGHT.

ITS principle is exactly the same as that known as the "Drummond Light," but some slight improvements have recently been introduced into its working. Three substances are concerned in the production of the light—viz., two gases, oxygen and hydrogen, and a solid—lime. The mode in which these are used is as follows:—a jet of hydrogen being lighted, a jet of oxygen is turned on so as to mix with it, and the solid incombustible lime being so arranged as to be exposed to the intense heat, it emits a light so pure and so powerful that it is only rivalled by that of the sun.   The method of using and lighting the jet is precisely the same as with a common gas-burner; so that no special knowledge or instruction is requisite for its management.   When the consumption of the lime-light gases does not exceed 1½ ft. per hour, the light produced is equal to four gas-lights, each burning five feet per hour.   Three feet per hour give a light equal to 16 gaslights, each burning five feet per hour.   Six feet per hour give a light equal to 60 gas-lights, so that six feet are equal to 300 ft. of gas.   The two gases may be easily produced, and a very small main would be sufficient to convey them through the streets for the lighting of dwelling-houses—a purpose for which this light is singularly well-fitted, as its products of combustion are quite innoxious, and have no tarnishing effects.   Another advantage is that it exercises no changing influence on colours, every tint being as distinctly observable by its assistance as in the light of the sun.   Various towns in Scotland are adopting the light, and it is expected that the lime-light will gradually supersede the use of gas.—*London Scotsman.*

TEMPERATURE OF FLAMES.

A VALUABLE paper by Bunsen, "On the Temperature of the Flames of Carbonic Oxide and Hydrogen," appears in Poggendorff's *Annalen.*   When a mixture of a combustible gas with oxygen is ignited, the elevation of temperature which occurs can be easily computed by a reference to the heat generated by combustion and the specific heat of the products of combustion.   The rate of propagation of ignition of a mixture of oxygen and hydrogen is 34 metres per second, and of carbonic oxide and hydrogen 1 metre per second.   A mixture of carbonic oxide and oxygen, in the proportions proper for combustion, will, it is reckoned, be heated from 0 deg. to 3,033 deg. centigrade; of hydrogen and oxygen, from 0 deg. to 2,844 deg. centigrade; a mixture of carbonic oxide and air, from 0 deg. to 1,997 deg. centigrade; and a mixture of hydrogen and air from 0 deg. to 2,024 deg. centigrade.

IMPROVED PHOTOGRAPHIC PROCESS.

BY the usual manner of producing photographic impressions, when the invisible impression is formed in the camera, it is removed into a dark room, and at once developed or rendered visible by means of solutions or analogous ponderable agents. Mr. F. B. Gage, of St. Johnsbury, Vermont, U.S., under a recent British patent, employs, on the contrary, what may be termed a partial development in the camera by the aid of diffused light.

The development is afterwards completed in the usual way, taking the precaution to reduce the strength of the developing fluid with water.

Mr. Gage takes a photographic impression in the usual manner, and then places some plain dark dead surface in front of the camera, the sensitive surface still remaining in the camera. He then removes the covering from the lens tube, and exposes the sensitive surface on which the impression has been formed to the light reflected from the dark surface, while the dark surface is kept in gentle motion, so as to prevent the sensitive surface from taking an impression of any wrinkles or other variations on the surface from which the light is reflected. The time of this exposure is varied according to the amount of light reflected and the effect it is desirable to produce. The usual amount of time occupied in this exposure will be from one-fourth to double the time employed in taking the invisible impression; but in some cases it can be extended much beyond this time. For a dark dead surface Mr. Gage usually uses a piece of thick black woollen cloth about 18 in. square, attached by one edge to a stick about 2 ft. long, which is held horizontally and gently moved in front of the camera with the left hand, while the lens tube is uncovered with the right hand. It is not absolutely essential that this dark surface be kept in motion, but it is safer. This exposure of the sensitive surface to the light reflected from a dark dead surface apparently leaves the lightest portions of the impression but little changed, while it effects a much greater change in the darkest portions of the same, and thus harmonizes and properly blends the two, giving to the whole an atmospheric effect never before realized in photographic impressions. It also renders it less difficult to obtain the necessary intensity in negatives.

This invention applies equally well and is operated in the same manner in taking positives or negatives in the camera, and it may be used without further instructions in producing any style of photographic pictures. It is necessary that the dead surface be suitably lighted, and the time of exposure proportioned to the result desired to be produced. The best results are produced when the dead surface is as strongly lighted as possible without sunlight, using a diaphragm to reduce the aperture of the lens to prevent the development being so rapid as to become unmanageable. Mr. Gage has produced excellent results with a silver bath of twenty grains of nitrate of silver to the ounce of water, being about one-half the usual strength in use, the sensitizing of the collodion being proportionally reduced; it will effect a great saving of expense for this reason. The inventor believes his invention also removes the most important obstacle to the production of dry plate impressions by harmonizing the lights and shades, which have heretofore usually been hard and inartistic. Exposing the sensitive surface in the manner described, before the impression is formed, has less tendency to blend the lights and shades than when done afterwards; but it gives a different and peculiar tone to the impression, which in some cases is very desirable, especially in negatives.

When the object to be impressed is strongly lighted, accompanied with deep heavy shadows, it is advisable to illuminate in the manner described the sensitive surface both before and after the impression is formed. This is effected by moving the black cloth before the camera a short time before as well as after, and operating otherwise in the same manner as before described. Some good effect may be produced by admitting transmitted light upon the sensitive surface, or light reflected from yellow and even red and other coloured surfaces, either before or after, or both before and after the photographic impression has been formed. The reflection from a dark dead surface, however, is much to be preferred, and Mr. Gage states that he has found his method of operating perfectly convenient and practical for use.

ACTION OF LIGHT UPON CHLORIDE OF SILVER.

IF in a tube of white glass, from 14 to 15 inches long, you enclose moist Chloride of Silver (freshly precipitated by means of a solution of chlorine in water), and expose it to the direct action of the solar rays, it will be observed, that while the chlorine solution is yellow, the chloride of silver remains white; but after the chlorine solution becomes colourless, the chloride decomposes the water under the action of light. As soon as the chloride of silver blackens at the surface, it should be agitated from time to time, and left exposed for a few days to a direct light, until the whole becomes of a fine black colour. If the tube is now taken into a dark place, the blackness will disappear by degrees, chloride of silver becoming re-formed, and the contents of the tube becoming perfectly white again; and this experiment may be repeated indefinitely. It is an evidence that in their successive reactions the chlorine, oxygen, hydrogen, &c., preserve properties of combination and re-combination. Bromide of silver (and, probably, cyanide) present the same reaction. Iodide of silver only blackens in the sun, after being sensitized by means of pyrogallic acid. It does not blacken visibly without a reducing agent.—*Chemical News.*

MANUFACTURE OF IRON.

A PAPER has been read to the British Association, "On Some Points Affecting the Economical Manufacture of Iron," by Mr. J. Jones. The author estimated the production of pig-iron in Great Britain at 4,500,000 tons per annum, and the make of finished iron at about 3,000,000 tons. He adduced these statistics to show the immense issues involved in the improvements he wished to notice. He then referred to the economical application of fuel in the iron manufacture, more particularly in the finished iron processes, and remarked that the newer blast-furnace plant left little to be accomplished in the economical use of fuel, except in utilizing the waste produces given off in coking the fuel. In puddling, however, great waste of fuel went on, and two modifications of the ordinary puddling-furnace were to be noticed as calculated to save from 20 to 25 per cent. of fuel, and to consume all the smoke usually produced. The Wilson Furnace, in its most recently improved form, consisted of a sloping chamber, into which the fuel was fed at the top; and the volatile matters generally forming smoke were reduced by passing over the incandescent mass of fuel further along the chamber. The air for combustion was delivered into the furnace in a heated condition, and a steam-jet was delivered underneath the grate, by means of which the formation of clinkers was avoided. The Newport Furnace, Middlesborough, had a chamber constructed in the ordinary chimney-stack; and in this were placed a couple of cast-iron pipes, with a partition reaching nearly to the top. These pipes were heated by the waste gases from the puddling-furnace, and through them the air required for combustion was forced by means of a steam-jet, and was thus delivered in front of the grate in a highly-heated condition. These furnaces, of which a considerable number were in operation at the Newport Works, effected a saving of at least 25 per cent. in fuel. The structural modifications would involve comparatively little outlay, and the saving to be effected would recoup that outlay in a single year. The economy represented by applying the new plans to the whole iron trade would amount to about 1,500,000 tons of coal per annum.

The author proceeded to describe the manufacture of iron by the Radcliffe Process, which had been for some time in operation at the Consett Ironworks, Newcastle. The puddled iron, which was usually rolled into rough bars, straightened and weighed, allowed to get cool, then cut up, piled, heated, rolled into blooms, re-heated, and, finally, rolled into finished iron, after a complicated series of operations, was, by the new method, finished off by a continuous and simple process. Five or more puddled balls were put together into a large bloom, under a very heavy steam-hammer, shingled down into a bloom, passed for a short time through a heating furnace, and rolled off into finished iron, not more than half an hour after the iron left the puddling-furnace. Specimens of iron made by the process were exhibited. A great saving in the cost of manufacture was represented by this process in all departments of the manufacture of finished iron; and it was calculated that a saving of 1,500,000 tons of coal alone would result from the general application of this system. Particular stress was laid upon the fact that, in carrying out this process, no extensive or expensive alteration of existing works was required, and a saving of from 3½ to 4 cwt. of puddled iron would be secured upon each ton of finished rails or plates now turned out, the cost

of making malleable iron being reduced to a very considerable
extent. The importance of the whole question, in a national
point of view, was also dwelt upon.

STEEL MANUFACTURE.

MR. FERDINAND KOHN has published a paper "On the Recent
Progress of Steel Manufacture." He stated that at the last meeting of the British Association at Dundee he called attention to a
new process of steel manufacture, viz., on the open hearth of a
Siemens' Furnace by the mutual reaction of pig iron and decarburized iron, or wrought iron, upon each other, which was known
in France as the Martin Process. Within the last year the process has been brought into operation in this country, and he had
now the pleasure of laying before the meeting a few samples of
steel which had been made by the process in the Cleveland district.
The process realizes the old and repeatedly proposed idea of
melting wrought iron in a bath of liquid pig-iron, and thereby
converting the whole mass into steel. The principal elements of
its successful operation, and the points which distinguish it from
all previous abortive attempts were—1st, the high temperature
and the neutral or non-oxidizing flame produced by the regenerative gas furnace of Mr. Siemens; and 2ndly, the method of
charging the decarburized iron into the bath of pig-iron in measured quantities or doses. These doses of wrought iron or steel
are added to the bath in regular intervals, so that each following
charge in melting increases the quantity of liquid mass, and adds
to the dissolving power of the bath, until complete decarburiza-
tion is arrived at. The charge is then completed by adding to
the decarburized mass a certain percentage of pig-iron, or of
the well-known alloys of iron and manganese, and the degree of
hardness or temper of the steel produced depends upon the proportion of this final addition. Having referred to the works in
this country which had begun to work the process, and produced
various tables showing the results of experiments that had been
conducted, he referred to the question of cost, stating that the
materials for producing a ton of steel were estimated to cost
£6 6s., to which must be added wages, repairs of plant, and
royalties to both patentees, which would bring the prime cost
of the Siemens-Martin steel to about £7 10s. per ton, precisely
the same as the prime cost of Bessemer steel. Referring to
the question of how this new process is likely to affect the progress of the Bessemer system, of which it seemed to be a rival,
Mr. Kohn said that, in his opinion, the only influence which it
could have upon the Bessemer steel was to stimulate and assist
it, and widen the sphere of its application. The two processes,
working with different classes of raw material, could never come
into direct rivalry. By working up the waste and offal of the
Bessemer Steelworks, the crop ends of steel rails, and similar

material, tho new process would assist in cheapening the prime
cost of Bessemer steel, in which the waste plays an important
part.

GAS FURNACE.

Mr. C. W. SIEMENS has delivered to the Chemical Society a
lecture "On the Regenerative Gas Furnace, as Applied to the
Production of Cast Steel." Processes were described by means
of which low carbon steel of good quality can be manufactured
from worn railway bars, or produced direct from the ore by dissolving the spongy metal, as soon as formed, in a bath of melted
pig iron. Models of the furnaces were shown, and amongst the
metallic specimens exhibited was a quality of steel containing
2 per cent. of tungsten, the magnetic properties of which were
remarkable, a small horseshoe magnet made from this alloy
being capable of supporting twenty times its own weight from
the armature.

HEATON'S DIRECT STEEL PROCESS.

THE names of Parry, Uchatius, and Martin in earlier, and those
of Bessemer and Siemens in more recent times, are inseparately
connected with the history of Steel-making. Another successful
worker in this direction has lately come before public notice—
Mr. John Heaton, who has brought to a successful issue a direct
method of producing steel, which bids fair to place other processes in the shade. This process has been at work for many
months past in one locality—the Langley Mill Works, in the
Erewash Valley, near Nottingham—upon a manufacturing scale,
and with complete success, both metallurgic and mercantile. The
Heaton Process is a direct chemical reaction, and consists in
applying to the molten crude iron nascent oxygen developed at
the moment of contact between the molten cast iron and such
salts and nitrates as yield oxygen under those conditions. The
idea of decarburizing crude iron by the use of nitrates is, we
believe, to be found in many chemical works, and is, therefore no
novelty. But the fact of its being carried out in practice to a
successful issue is, and therefore reflects the highest credit on
Mr. Heaton. The salt employed by Mr. Heaton is the nitrate of
soda, which is much more plentiful than nitre. It is not decomposed in presence of fluid cast iron with the same intense energy
that nitre is, but still would prove more or less unmanageable as
an agent for the burning out of the silicon, carbon, sulphur,
phosphorus, &c., were it not for the extremely simple but beautifully effective apparatus invented for its application, and which
constitutes, in fact, the essence of Mr. Heaton's patents, of which
he has taken out several in connexion with this process. This
apparatus will be found illustrated at pages 350 and 351 of our
present issue, a detailed description accompanying the engravings.

The process by which the extraordinary results we shall presently describe are produced, is conducted as follows:—Cast iron of any quality is first melted in a common iron-foundry cupola with coke fuel. A known quantity of the liquid iron—usually about a ton—is tapped out into an ordinary crane-ladle, which is swung round to the side of the converter. This latter is a tall cylinder of boiler plate, open at the bottom, between which and the floor a space is left. The converter has a firebrick lining, and terminates in a conical covering, out of which an iron funnel opens to the atmosphere. In the bottom of the converter a number of short cylindrical pots, lined with brick and fire-clay, are adjusted. Into the bottom of one of these pots a given weight of crude nitrate of soda of commerce is put. The surface of the powder is levelled and covered by a thick circular perforated plate of cast iron. One of these pots thus prepared having been adjusted to the bottom of the cylinder, the converter is now ready for use. At one side of the cylinder is a hopper, covered by a loosely-hinged flap of boiler plate. This plate is raised, and the ladleful of liquid cast iron is poured into the converter, and descends upon the top of the cold cast iron perforated plate. The plate does not float up nor become displaced, nor does any action become apparent for some minutes, while the plate is rapidly acquiring heat from the fluid iron above it, and the nitrate getting heated by the contact with it. What follows is so well described by Professor Miller, of King's College, in a Report now before us, that we prefer giving it in the Professor's own words. He says:—"In about two minutes a reaction commenced; at first a moderate quantity of brown nitrous fumes escaped; these were followed by copious blackish, then grey, then whitish fumes, produced by the escape of steam, carrying with it, in suspension, a portion of the flux. After the lapse of five or six minutes deflagration occurred, attended with a roaring noise and a burst of a brilliant yellow flame from the top of the chimney. This lasted for about a minute and a half, and then subsided as rapidly as it commenced. When all had become tranquil, the converter was detached from the chimney, and its contents were emptied upon the iron pavement of the foundry. These consisted of crude steel and of slag. The crude steel was in a pasty state, and the slag fluid; the cast iron perforated plate had become melted up and incorporated with the charge of molten metal. The slag had a glassy, blebby appearance and a black or dark green colour in mass."

The crude steel thus produced from Heaton's converter is broken up, and after the lumps have been squeezed under the shingling hammer, are again heated in a common balling furnace. They are afterwards rolled or forged into bars or masses of any required form. In this condition the material is called by the inventor "steel iron," which, in fact, is a product obtained from the crude steel by taking out the carbon in the re-heating furnace. It is an iron which is nearly free from sulphur and phosphorus, possessing great strength and toughness, and is for structural purposes equal to the renowned wrought iron produced at Lowmoor and Bowling Works. It welds perfectly; it is tough both hot and cold, neither red-short nor cold-short, and forges beautifully at both the test temperatures for iron—a low red, and a clear yellow heat. This steel-iron is in itself a very valuable material, which has been produced ready for market without the intermediate process of piling and balling. From this material Mr. Heaton produces his cast steel in the following manner:—The cakes, after they have been squeezed by the shingling hammer, are broken up, put into ordinary clay melting-pots of the usual size, holding about 60 lb. each. To each 100 lb. of the material, about 2½ lb. or 3 lb. of spiegel-eisen, or its equivalent of oxide of manganese and a little charcoal, are added, and the whole is fused and cast into ingots. It is now excellent cast steel, and when the ingots have been tilted in the usual manner, cast-steel bars are produced fit for any uses to which steel is at present applied. Such is the Heaton Process; its simplicity and directness need neither comment nor praise at our hands.

We have already referred to Professor Müller's Report, which gives the following results of analysis of three samples of metal produced at the Langley Mills under his own observation:—

	Cupola Pig (1).	Crude Steel (7).	Steel Iron (9).
Carbon	2·830	1·800	0·093
Silicon, with a little titanium	2·050	0·200	0·150
Sulphur	0·113	0·018	traces.
Phosphorus	1·455	0·294	0·202
Arsenic	0·041	0·069	0·024
Manganese	0·313	0·190	0·084
Calcium	—	0·319	0·310
Sodium	—	0·144	traces.
Iron (by difference)	92·298	97·038	94·144
	100·000	100·000	100·000

"It will be obvious from a comparison of these results," says the Professor, "that the reaction with the nitrate of soda has removed a large proportion of the carbon, silicon, and phosphorus, as well as most of the sulphur. The quantity of phosphorus (0·298 per cent.) retained by the sample of crude steel from the converter which I analysed is obviously not such as to injure the quality. The steel iron was subjected to many severe tests. It was bent and hammered sharply round, without cracking. It was forged and subjected to a similar trial, both at a cherry-red heat and at a clear yellow heat, without cracking; it also welded satisfactorily." The Professor concludes his report by stating that Heaton's process is based upon correct chemical principles, and that the mode of attaining the result is both simple and rapid.

Besides the Report of Professor Miller, we have before us reports from Mr. Robert Mallet, C.E., and from Mr. David Kirkaldy, upon the general principles of the invention, and the results of the testing of a number of bars of Heaton's steel-iron. Both of these reports are exceedingly satisfactory, and bear valuable practical testimony to the high character of the process and the material produced. Mr. Mallet's conclusions are:—"1st. That Heaton's patent process of conversion by means of nitrate of soda is at all points in perfect accord with metallurgic theory. That it can be conducted upon the great scale with perfect safety, uniformity, and facility, and that it yields products of very high commercial value. 2nd. That in point of manufacturing economy or cost, it can compete with advantage against every other known process for the production of wrought iron and steel from pig-iron. 3rd. Amongst its strong points, however, apart from and over and above any mere economy in the cost of production, are these:—It enables first-class wrought iron and excellent steel to be produced from coarse, low-priced brands of crude pig-irons, rich in phosphorus and sulphur, from which no other known process—not even Bessemer's—enables steel of commercial value to be produced at all, nor wrought iron, except such as is, more or less, either 'cold short' or 'red short.' Thus wrought iron and cast steel of very high qualities have been produced in my presence from Cleveland and Northamptonshire pig-irons, rich in phosphorus and sulphur; and every iron-master, I presume, knows that first-class wrought iron has not previously been produced from pig iron of either of those districts, nor marketable steel from them at all. Heaton's process presents, therefore, an almost measureless future field in extending the manufacture of high-class wrought iron and of excellent steel into the Cleveland and other great iron districts, as yet precluded from the production of such materials by the inferior nature of their raw products. It admits of the steel manufacture also being extended into districts and countries where fuel is so scarce and dear that it is otherwise impossible."

Mr. Kirkaldy's tests go to show that the wrought iron made from Cleveland and Northampton pigs, and tested for tensile resistance, bore a rupturing strain of 23 tons per square inch, and an elongation of nearly one-fourth of the original unit in length. The tilted cast-steel made from the same pig-irons, bore a tensile strain at a rupture of above 42 tons per square inch, with an elongation exceeding one-twelfth of the unit of length. These results show the remarkable quality of the material, and the fitness of the former for artillery, armour-plates, and boilers, and of the latter for rails, shipbuilding, and all other structural uses.— *Mechanics' Magazine.*

Natural History.

ZOOLOGY.

DARWINISM.

A PAPER, by the Rev. F. O. Morris, "On the Difficulties of Darwinism," has been read to the British Association. The difficulties stated by the author, and the way in which they are met by Darwinians, are fully seen in the subjoined discussion.

Mr. Wallace said that the points mentioned by the author really presented no difficulties whatever to the Darwinian theory. He asked, for instance, why female birds did not sing? Mr. Darwin had himself explained the reason; it was the same as that for which the plumage of the female bird was less beautiful than that of the male. In birds, as in all the lower animals, the female chooses the male; and it is the attractions of the latter that lead to the pairing. This applied both to the voice and the plumage. Another "difficulty" raised by the author had reference to the winged beetles of Madeira. Mr. Darwin's theory was that, as Madeira was a single island in the middle of the Atlantic, subject to violent storms of wind, insects from it once blown out to sea could not get back again. Flying insects would thus be at a disadvantage and might become extinct, while those without wings would survive. But there were some beetles in Madeira which could not get on without flying, as they would lose their means of subsistence. It was a remarkable fact, however, that such insects had longer wings than the corresponding animals in Europe, having gradually acquired increased power to enable them to battle against the wind. This Mr. Darwin illustrates by supposing the case of a ship striking against a rock near land. Persons who could swim well would get to the shore; those who could swim imperfectly would probably be drowned in the attempt; and those who could not swim at all would remain on the wreck, and have a good chance of getting ashore the next day by the boats. Thus the advantage would be to those who could swim well and those who could not swim at all, and, in like manner, to insects that could fly exceedingly well and those that could not fly at all. The author referred to the circumstance of apple-trees differing in different years in the quantity of fruit, and said that this did not depend upon the war of apple-trees with each other. Mr. Wallace said we must go back to the crab-apple for the true cause. There was a war in Nature, a struggle for existence, not only between one crab and another, but between crab-trees and every other kind of tree. All these trees produced millions of seeds every year, but not one seed in a thousand became a tree. Why did one become a tree rather than another? The slightest difference in circumstances connected with growth would affect the life or death of a particular seed. Again, the author maintained that cultivated plants and domesticated plants, when allowed to go wild, returned to the original form; and he cited as an illustration the case of the pansy. Mr. Darwin and other distinguished naturalists denied that assertion; and the author should have given proofs of it, if he desired it to be believed. With regard to the moral bearing of the question as to whether the moral and intellectual faculties could be developed by natural selection, that was a subject on which Mr. Darwin had not given an opinion. He (Mr. Wallace) did not believe that Mr. Darwin's theory would entirely explain those mental phenomena.

The Rev. H. B. Tristram said, he himself thought it best to make a compromise between the extremes of Darwinians and the religious party. He thought there was a number of shallow young men who used Darwin's name as a shibboleth, and did not really understand the matter. Mr. Darwin's theory had nothing to do with the soul, nor was there a question as to a Creator, but as to *how* the Creator had created. It was not right that the clergy should be mistrusted by men of science, and blamed by their own cloth too, when they attempted to go into these questions.

Dr. Grierson complained that newspapers and other popular periodicals never presented a correct statement of the Darwinian theory, but invariably caricatured it.

Professor Rolleston said he had thought this matter out for himself, and found he could still keep to the old belief in which he was brought up, whilst accepting the philosophy of Darwin. He agreed with the principle laid down by Archbishop Whately, who said that, if he ever founded a sect, one of its rules should be that no man should ever attempt to prove any proposition in natural science by appealing to the word of God. Natural science people should be left to work out their own conclusions. If they fell into errors, there were plenty of their own brethren ready enough to set them right. If a thing was true, it was true all round, and there was no truth to which it would be contradictory. No doubt if any theory led logically to a conclusion known to be false, the premises must also be false; but it did not appear to him that Mr. Darwin's conclusions were false.

The President of the British Association, Dr. Joseph Hooker, in his inaugural address, thus referred to the progress of the Darwinian theory:—" Ten years have elapsed since the publication of 'The Origin of Species by Natural Selection,' and it is hence not too early now to ask what progress that bold theory has made in scientific estimation. The most widely-circulated of all the journals that give science a prominent place on their title-pages, the *Athenæum*, has very recently told it to every.

M

country where the English language is read, that Mr. Darwin's theory is a thing of the past; that natural selection is rapidly declining in scientific favour; and that, as regards the above two volumes on the variations of animals and plants under domestication, they 'contain nothing more in support of origin by selection than a more detailed re-asseveration of his guesses founded on the so-called variations of pigeons.' Let us examine for ourselves into the truth of these inconsiderate statements.

"Since the 'Origin' appeared, ten years ago, it has passed through four English editions, two American, two German, two French, several Russian, a Dutch, and an Italian; while of the work on Variation, which first left the publisher's house not seven months ago, two English, a German, Russian, American, and Italian edition are already in circulation. So far from natural selection being a thing of the past, it is an accepted doctrine with every philosophical naturalist, including, it will always be understood, a considerable proportion who are not prepared to admit that it accounts for all Mr. Darwin assigns to it. Reviews on 'The Origin of Species' are still pouring in from the Continent, and Agassiz, in one of the addresses which he issued to his *collaborateurs* on their late voyage to the Amazon, directs their attention to this theory as a primary object of the expedition they were then undertaking. I need only add, that of the many eminent naturalists who have accepted it, not one has been known to abandon it; that it gains adherents steadily, and that it is, *par excellence*, an avowed favourite with the rising schools of naturalists;—perhaps, indeed, too much so, for the young are apt to accept such theories as articles of faith, and the creed of the student is also too likely to become the shibboleth of the future professor. The scientific writers who have publicly rejected the theories of continuous revolution or of natural selection, or of both, take their stand on physical grounds, or metaphysical, or both. Of those who rely on the metaphysical, their arguments are usually strongly imbued with prejudice, and even odium, and, as such, are beyond the pale of scientific criticism. Having myself been a student of moral philosophy in a northern University, I entered on my scientific career full of hopes that metaphysics would prove a useful mentor, if not quite a science. I soon, however, found that it availed me nothing, and I long ago arrived at the conclusion, so well put by Agassiz, where he says, 'We trust that the time is not distant when it will be universally understood that the battle of the evidences will have to be fought on the field of physical science, and not on that of the metaphysical.'—('Agassiz on the Contemplation of God in the *Kosmos*,' *Christian Examiner*, 4th series, vol. xv., p. 2.) Many of the metaphysicians' objections have been controverted by that champion of natural selection, Mr. Darwin's true knight, Alfred Wallace, in his papers on 'Protection' (*Westminster Review*) and 'Creation of Law,' &c.

(*Journal of Science*, October, 1867,) in which the doctrines of 'continual interference' and the 'theories of beauty,' kindred subjects, are discussed with admirable sagacity, knowledge, and skill. But of Mr. Wallace and his many contributions to philosophical biology, it is not easy to speak without enthusiasm; for, putting aside their great merits, he, throughout his writings, with a modesty as rare as I believe it to be unconscious, forgets his own unquestioned claims to the honour of having originated, independently of Mr. Darwin, the theories which he so ably defends."*

NATURAL SELECTION.

A PAPER has been read to the Ethnological Society "On the Theory of the Origin of Species by Natural Selection," by the President, Mr. J. Crawfurd. The author reviewed shortly the Darwinian theory of a perpetual sequence of profitable variation in every species of plants and animals, proceeding to show that in authentic history, however remote, there is no trace of such variation; but that the mummies of the ibis and kestrel hawk, and drawings of the ox, ass, dog, and goose, which existed in ancient Egypt, show them as identical with the animals living at this day. The arguments of the Darwin school are chiefly derived from the variations to be met with in animals and plants; but these seldom occur in the wild state, but only after subjection to the control of man. We did not find the disposition towards variation in all species, the ass and the camel being notable instances, and because whenever under man's influence it does take place, it results in a weakening in the animal of those qualities which render it most fit to maintain "the struggle for life." After a return to the wild state, the bird or animal loses those qualities it had acquired, and merges into the common stock. This, if the theory of progressive and profitable development were correct, it should not do, but should impart its own properties to its fellows. The same thing was seen in plants,—the rose and pine-apple, for instance, which, by cultivation, gained qualities agreeable to man, but lost the power of reproduction, and were thus weakened in "the struggle for life."

ORIGIN OF MYTHS.

THE popular notion of Myths is, that they are free and unrestricted growths of fancy, and that the study of such baseless, unsubstantial fabrics of the imagination can lead to no precise or scientific results. But wider knowledge must dissipate this idea by showing that myths are intellectual developments to be traced to definite causes, like other products of the human mind.

---

* The *Athenæum* has replied to these statements.

Thus the myth that, on a certain hill there was a battle of giants and monsters, will be probably interpreted by the fact that great fossil bones are really found on the spot. Again, the story of the presence of a race of men with tails in a particular district is apt to indicate the real existence of a tribe of aborigines or outcasts, like the Miautsze of China or the Cagots of France. There are two "philosophic myths," invented again and again in the infancy of science to account for strictly physical phenomena. The Polynesian myth of Mafine, the subterranean god who causes the earthquake by shifting from shoulder to shoulder the earth which he carries, and many other similar myths, come under the common heading of myths of an earth-bearer, found in various regions to account for the occurrence of earthquakes. The myth of the Guaranis of Brazil, that a jaguar and a huge dog pursue the sun and moon, and devour them, which causes eclipses, is an instance from the wide-spread group of eclipse-myths of a similar kind. On this and other evidence the writer argued for the possibility of discovery in the phenomena of civilization, as in vegetable and animal structure, the presence of distinct laws, and attributed the now backward state of the science of culture to the non-adoption of the systematic methods of classification familiar to the naturalist.—*Mr. E. B. Taylor, Proc. British Association.*

A VERY RARE DISEASE.

THERE exists on record in medicine up to the present day only one case of enormous overgrowth of all the bones of the body with wasting of the muscles. It was observed by Saucerotte. The second has been recently observed by Professor Friedrich, and is described at length in Virchow's *Archiv.* bd. 1868. Saucerotto's case was that of a man, aged 39, the weight of whose body rose, in the space of four years, from 119 lb. to 178 lb. by the increase of his bones, the muscles all the while wasting. The head of this man was monstrous, the eyes being pushed forward to the level of the forehead, and the lower jaw having attained an enormous thickness. Attacks of insensibility and difficulty of breathing characterized his complaint. M. Friedrich's patient, aged 26, is one of six children, of whom all are healthy, except the youngest, who has a minor form of the same affection. At 18 he began to find his feet and ankles growing large and heavy; then the hands and fingers. The lower and upper limbs are of enormous size (the measurements are given), due entirely to the growth of the bones, the muscles being wasted. Standing, walking, and lying are all difficult. He finds comfort from cold baths. His internal organs appear healthy, and there is no disturbance of the brain. Medicines have been ineffectual, but latterly the disease has seemed to be stationary.
—*British Medical Journal.*

AUSTRALIAN ZOOLOGY.

THE DUKE of EDINBURGH has brought, in the *Galatea*, a varied collection of colonial birds and animals. In Tasmania, he procured a very fine wombat, which was presented to him by Lady Dry, wife of the Chief Secretary of that colony; this wombat was so tame and docile that it soon became a general favourite with all on board the *Galatea*. From South Australia, Victoria, and Tasmania, his Royal Highness obtained a large collection of beautiful parrots and other birds, and this collection received some valuable additions in Sydney. His Royal Highness received as a present, from a gentleman in the Hunter River district, a pair of very fine emus; and on the Saturday prior to the sailing of the ship, he was presented with a pair of large and very tame kangaroos—one reared and presented by the Colonial Secretary, Mr. Parkes, and the other by Mr. J. T. Ryan, M.L.A. His Royal Highness also received from Mr. Parkes the mongouste, which killed the snakes at the museum on the occasion of the Royal visit to that establishment. This little animal was as docile and as playful as a kitten. The same gentleman also presented to his Royal Highness a pair of native cats, which, having been taken from their mother when very young, were reared and rendered as tame and tractable as ordinary domestic cats.—*Hobart Town Mercury.*

WALRUS.

DR. MURIE has given to the Zoological Society an account of the morbid appearances observed in the Walrus lately living in the Society's Gardens, the death of which appeared to have resulted from extensive ulcerations in the stomach, caused by the presence of numerous entozoa. These notes were accompanied by a description, by Dr. Baird, of the entozoon in question, which was regarded as a new species, and proposed to be called *Ascaris bicolor.*

ELEPHANTS.

IN a review of Sir S. W. Baker's work "On the Nile Tributaries of Abyssinia," in the *Athenæum*, the traveller is taken to task for having omitted to notice that Bruce had preceded him in giving an account of the sword-hunters of the Hamram Arabs, of whom he indeed makes a feature in his book. A correspondent of that journal points out that it is still more interesting, as attesting the handing down of customs from generation to generation, to know that the manner of hamstringing Elephants, as practised by those people, was an art as perfectly understood by the ancient barbarians (see Strabo, lib. xvi., p. 772, and Diod. Sic., lib. iii., p. 161) as by Bruce's and Baker's "agageers," or "ele-

phant-hunters." A relish for the flesh of elephants and hippo-
potami, so signally illustrated in Baker's work, appears to have
been a characteristic of the dwellers in a region so favoured by
large game from time immemorial ; for we are told by Agathar-
cides that Ptolemy would have redeemed the life of the elephant
at any price, as he wanted elephants for his army ; but he met
with a refusal from the native hunters, who declared they would
not forego the luxury of their repast for all the wealth of Egypt.
It is recorded in the celebrated inscription of Adule, that Ptolemy
invaded Asia with his land and sea forces, and with elephants
from the country of the Troglodytes and Ethiopians. The latter
may refer to the region in question, situated between the Settito
and the Khor el Gash, and which probably constituted one of the
chief hunting-grounds of the Ptolemies. The port for the em-
barkation of elephants was, it is well-known, Ptolemais Theron,
built by Eumedes, and which Ptolemy, the geographer, tells us
was in the latitude of Meroo, that is to say, in 16 deg. 24 min.
N. lat. ; or, in his tables, 16 deg. 25 min.

<hr>

GOLDEN EAGLE.

A SPLENDID specimen of the Golden Eagle has been captured
in the little isle of Herm, one of the islets opposite Guernsey.
Major Fielden, to whom the shooting in the island belongs, had
found sensible marks of its presence in the destruction of his
game, and at last, on the 16 ult., his keeper was fortunate enough
to secure the precious depredator. This eagle weighs 7 lb. 12 oz.,
measuring 3 ft. from beak to tail, and 7 ft. 2 in. from wing to
wing.

<hr>

THE GREAT BUSTARD.

MR. H. STEVENSON has read to the British Association a paper
"On the Extinction of the Great Bustard in Norfolk and Suffolk."
After referring to some very early allusions to the existence of
the bustard in this country, and to the gradual diminution and
extinction of the species in the different English counties, the
author said that Norfolk was the last county to reckon the bustard amongst its resident species. The two latest "droves" had
their head-quarters in the open country round Swaffham, and in
that near Thetford. The Swaffham drove formerly consisted of
twenty-seven birds, but the number subsequently decreased to
seventeen, sixteen, and eleven, and, finally, dwindled down to five
and two. All accounts agreed in stating that the last remaining
birds were hens. One great cause of the extinction of the bird
was the introduction of improved agricultural implements, which
destroyed the eggs. The precise time of extinction could not be
determined with accuracy. The last-known specimens were seen
about the year 1838; but it had been stated that some of the

birds had lingered on till 1813 or 1845. The other drove, near
Thetford, consisted of thirty or forty birds; but the number
gradually declined to twenty-four, eighteen, fifteen, nine, seven,
six, five, and two, the last survivors being hens only. Some persons supposed that the bird could be taken by dogs, but this
was not confirmed by the testimony of trustworthy eye-witnesses.
After referring to the local distribution of the bustard in the
county, and to the appearance of occasional immigrants from the
Continent, Mr. Stevenson thus concluded his paper: "Having
served its purpose in its day in the great scheme of Nature, the
great bustard has passed for ever out of our local Fauna. Better
thus to have perished a few years earlier than to have met with a
no less certain and more melancholy end. Had it still existed in
1808, some reigning belle, some leading votary of fashion, would
inevitably have decreed that bustards' plumes should be 'the
thing' for the season. Then, indeed, its fate would have been
sealed at once, and the last British bustard would have been *cut
up for hats!*"

___

THE CUCKOO.

THE Rev. Mr. SMITH has contributed to the *Zoologist* a resumé
of the researches of Dr. Baldamus respecting the natural history
of the Cuckoo, by which it appears that the cuckoo never con-
structs a nest, but lays her eggs upon the ground, and conveys
each egg in her beak to the nest of some other bird, where it is
hatched. The cuckoo, before laying, looks round to select the
nests in which her eggs are to be deposited, and she never places
more than one egg in one nest. She continues to watch these
nests during the process of incubation, and it is the parent bird,
and not the young cuckoo, which removes the other young birds
from the nest. The eggs of the cuckoo are small, relatively with
the size of the bird, and they have been observed to assimilate
in the markings to the eggs of the foster-mother, whence it has
been inferred that the cuckoo had the power of determining the
markings on its eggs. But Dr. Baldamus says that this is not
so, but that the eggs of different cuckoos have different markings,
and that each selects the kind of nest in which similar eggs are
found. The parent cuckoos migrate before the young ones are
strong enough to accompany them; but the young ones follow
by unerring instinct.

___

SALMON CULTIVATION IN ENGLAND.

MR. FRANK BUCKLAND states that, in his official capacity as
Inspector of Salmon Fisheries, he has lately visited many of the
rivers of England and Wales. The supply of Salmon has been
much increased owing to the protection of the Legislature. Still
there remains much to be done. He complains bitterly of the

impediments thrown in the way of parent salmon ascending to the spawning-grounds by weirs. He instanced Diglis Weir, which was the "hall door" of the great Severn; Chester Weir, that blocked the Dee; Tadcaster Weir, on the Wharfe, &c. Besides these large weirs there were many other mill-weirs on rivers that would otherwise be highly productive. The study of sal-mon-ladders was of the greatest importance. He exhibited seve-ral models of ladders which might be applied to weirs at a reasonable cost, and without interfering with the water supply. The question of pollutions was a very serious one, not only for the fish, but also for the public health. He instanced the Dovey, the Tees, the South Tyne, &c., which were suffering under "bush" from the lead mines; and he deprecated the habit of allowing chloride of lime to run into the rivers. Paper-makers were the great culprits in this matter. The law of pollutions should be made much stronger. He earnestly requested the at-tention of the D Section to the question of "close time" for salmon, as the evidence went to show that the Welsh, Cornish, and Devonshire rivers were "later rivers" than the Severn, Dee, Tay, Wye, &c. Mr. Buckland then expounded his theory as to the cause of the failure of oysters for the last six years. The cause of the success this year, he considered, was warm weather and tranquil water. He had published, in *Land and Water*, temperatures taken daily at five different oyster-fisheries. The results, he thought, confirmed his theory. He had obtained a heavy fall of spat at his experimental fishery at Reculvers, near Herne Bay. There had also been a fall of spat in the rivers Crouch, Roach, and on the grounds of the Herne Bay Oyster Company, but he believed that the Colne and the Black-water had not been so favoured. He called the attention of the public to his "Museum of Economic Fish Culture," at the Horti-cultural Gardens, Kensington. He had hatched and sent away to different rivers nearly 40,000 salmon and trout last year; and in his collection would be found models, coloured casts of fish-nets, and other implements connected with the improvement of British fisheries.

THE PARR CONTROVERSY.

BRIEFLY stated, the controversy is, or at least was, whether a little fish known in Scotland as the Parr, and in England as the semlet, was the young of the true salmon (*Salmo Salar*). "It is," said one body of the disputants. "It is not," rejoined another. And in that state the discussion remained for a long period. But some clever persons, who took an interest in the economy of the salmon fisheries, being resolved that the parr question should not rest on such an unsatisfactory basis, deter-mined to see and observe for themselves, and began a series of what were in reality, although the phrase was then unknown, piscicultural experiments. The experimenters are both dead, and they have gone to their graves unrewarded, although it is quite certain they did royal service to the cause of the salmon fisheries. The names were Shaw, of Drumlanrig, forester to the Duke of Buccleuch, and Andrew Young, of Invershin, who was at the time of experimenting in the employment of the Duke of Sutherland, at Dunrobin Castle. They gathered the eggs of the salmon, and kept them till they came to life, and grew into parrs, detaining them till they were seized with the migratory instinct, when they were found to have changed into what in Scotland are called smolts, having a totally different appearance from the parr, being scaled fish, ready to encounter the salt water, an element in which we know the parr cannot live. The most curious circumstance attending their experiments was the independent conclusions arrived at by the two men; one found that the parr changed into smolts, and became scaled fish at a period of twelve months from the time of their being hatched; whilst the other asserted that these fish did not become parr till they were two years old! Thus the parr question remained till the Stormontfield experiments began; nothing being settled but that the parr ultimately became young salmon, and even that was very grudgingly admitted by many of the controversialists. A curious turn was given to the controversy by the establishment of artificial breeding-ponds at Stormontfield on the river Tay. It was there found that neither Shaw nor Young was right, but that a moiety of the parr became smolts at the end of twelve months, whilst the other half of any given brood of salmon did not change till the fish were two years of age! How is that? will be asked. Well, we cannot tell; nobody can tell; it is one of those curiosities of fish 'growth which nobody can understand.—*Gentleman's Magazine.*

THE WHITEBAIT AND THE HERRING.

Dr. A. Günther, of the British Museum, has given to the Zoological Society a *resumé* of his researches into the distinctions between the different fish of the Herring family. The British species of this important group are the herring, the sprat, the pilchard (which is identical with the sardines of the French coast), and the two species of shad. These species are readily distinguished from one another by the numbers of their vertebræ and that of their scales, the relative position of the fins and that of the teeth. One of the most important results arrived at by this eminent ichthyologist is the absolute identity of the Whitebait and Herring. In the last volume of the *Catalogue of Fishes in the British Museum*, Dr. Günther describes the whitebait as a purely nominal species, introduced into science in deference to the opinion of fishermen and gourmands, and states that every example of whitebait examined by himself were young herrings.

The late Mr. Yarrell, who has been followed by most naturalists, regarded whitebait as a distinct fish, but the circumstances that it has the same number of vertebræ (56) as the mature herring, the same number of lateral scales, and an identical arrangement of fins and teeth, a combination of characters found in no other fish, prove conclusively that it is the fry or young of the herring ; moreover, an adult whitebait in roe has never been discovered. With regard to the effect on the supply of herrings occasioned by the destruction of the young fry, it is probable that the number of eggs deposited by the mature herring is so large and disproportionate to the number of fish that attain maturity that the capture of a portion of the fry could have no appreciable result in diminishing the multitude of mature fish.

SEA-GULLS AND THE HERRING FISHERY.

ONE of the principal sources of the prosperity of the Isle of Man is the Herring Fishery ; and one of the principal means of ascertaining the whereabouts of the fish is the presence of flocks of Sea-gulls, which hover over the shoals of fish. During the last few years, however, so many of these birds have been destroyed (in order to supply the demand for plumes for ladies' hats), that great fears were entertained in the island that the bird would become extinct, and that thus the fishermen would be deprived of one of the best indications of the presence of fish. In order to avert this calamity, the Legislature of the Isle of Man recently passed an act for the preservation of sea-gulls. The penalties under this measure are very severe. The first prosecution under the Act has lately been heard before the High Bailiff of Douglas. The defendant was Mr. John Gold, a noted dealer in sea-gulls' plumes. It appears that a few weeks previously, the police at Ramsay, a town in the north of the island, telegraphed to the police at Douglas to the effect that three hampers, supposed to contain sea-gulls, had been forwarded per coach, to Douglas, directed to Mr. Gold. Accordingly, on the arrival of the coach in Douglas, three policemen met it, examined the hampers, and found they did contain sea-gulls. The defence was that Mr. Gold did not know who had sent the birds, that he did not order them, or know anything about them. The Court held the charge proved, and fined the defendant.

THE GOURAMI FISH.

WE call the attention of our readers and the public to the offer, by Mr. Henry Lee, F.L.S., of a prize or reward of £20 to the person who shall first take to him, at the office of Land and Water, four or more living specimens, male and female, of the Mauritius fish Gourami. The Gourami is so highly esteemed as an article of food that its flavour is by some thought even more delicious than that of the salmon or turbot, and it would, without doubt, be a great and useful addition to the food-fish of Great Britain. Even before the year 1850 various attempts were made to introduce it to foreign countries. Some of these were successful, others the contrary; but there is every reason to believe that, by careful attention during transit, the Gourami may, without any great difficulty, be safely brought to England. It is very hardy, and will live for a long time in rather impure water. It is thought that specimens of this fish, about four inches long, are those most likely to bear the voyage well. Dr. Meller, Director of the Royal Botanical Gardens, Mauritius, has kindly promised to supply as many as he may think desirable to anyone who will seriously undertake the care of them.

PEARL FISHERY.

A CORRESPONDENT of the *Illustrated Australian News* directs attention to a discovery of considerable importance—the existence of an extensive Pearl Fishery on the north-west coast of Western Australia. He describes the fishing ground as stretching along the coast no less than 1,000 miles. "There had been upwards of 60 tons of pearls obtained up to December, when circumstances obliged me to leave," he writes, "and these were purchased on the spot at the rate of £100 per ton. The banks at Perth will advance £100 per ton, not including the inside pearls, which are valued from £1 to £20 each.

WALKING FISHES.

DR. SHORTT is expected shortly to arrive here from India, bringing with him about a dozen and a half of the Walking Fishes of India, Murral and Korava, many of them intended as a present to the Zoological Society's Gardens from Dr. Day. The largest species, known as Ophiocephalus striatus, grow to upwards of 3 ft. in length, and, if they succeed in England, will make a capital addition to our lakes and canals. The smaller variety, Ophiocephalus guchua, will perhaps be more interesting than useful, as they only grow to about 1 ft. in length. Pains have been taken to accustom them by degrees to confinement before shipping them in tin boxes. Dr. Day is said to have come to the conclusion that they breathe air direct from the atmosphere, as well as air in solution in the water in which they live.

THE JOHN DORY.

ASK most people what the derivation of John Dory (the fish) is, and they will tell you it is Jean-Doré, the French Golden John. Now, this is obviously wrong, when if you ask a fishmonger in Paris for a Jean-Doré, he does not know what you mean. The true derivation, then, is this, the name of the fish in Spain is "Janitore," so named after St. Peter, who is the janitor or porter of heaven; it is the fish which he pulled up with the tribute-money. The fish also bears the thumb-mark in its head. So easily—please pronounce it in Spanish, Janitore—Jean Dory! John Dory!—*Cornhill Magazine.*

THE TUATERA, OR NAVARA LIZARD. BY W. TEGETMEIER.

THE Zoological Gardens have recently received a living specimen of a very interesting lizard from New Zealand. The animal was first described by Dr. Dieffenbach in his *Travels in New Zealand*, where he states:—I had been apprised of the existence of a large lizard, which the natives called Tuatera or Navara, and of which they are much afraid. But, although looking for it at the places where it was said to be found, and offering great rewards for a specimen, it was only a few days before my departure from New Zealand I obtained one, which had been caught in a small rocky islet called Karewa, which is about two miles from the coast in the Bay of Plenty. From all that I could gather about this tuatera, it appears that it was formerly common in the islands, lived in holes, often in sandhills near the seashore, and the natives killed it for food. Owing to this latter cause, and no doubt also to the introduction of pigs, it is now very scarce; and many even of the older residents have never seen it. The specimen from which the description is taken I had alive, and kept some time in captivity; it was extremely sluggish, and could be handled without any attempt at resistance or biting.

This specimen was presented to the British Museum by Dr. Dieffenbach, and is in a most perfect state of preservation. Subsequently a few other specimens reached England; but no museum out of this country was fortunate enough to possess an example; and it is, I believe, owing to the interest of Dr. A. Gunther that this specimen has been received. This lizard possesses the strange interest that attaches itself to those animals that are slowly, but not the less surely, being improved out of existence by human agency. It is, in fact, following the moa, the dodo, and the solitaire. Dr. Gunther, in his monograph on its singular anatomy, which has been published in the "Transactions of the Royal Society," remarks:—Evidently restricted in its distribution, exposed to easy capture by its sluggish habits, esteemed as food by the natives, pursued by pigs, it is one of the rarest objects in zoological and anatomical collections, and may one day be enumerated among the forms of animal life which have become extinct within the memory of man. It is in the anatomical structure rather than the outward form that this lizard is of so great a degree of interest to zoologists. Its dentition is most peculiar. The teeth appear as if they were prominences of the jaw-bones, the edges of which are polished like the teeth, and perform their functions when the latter are ground down by use. The front teeth are four in each jaw in the young animal; but as the lizard advances in age those of the upper jaw coalesce into two pairs, and appear very like the chisel-shaped incisors of a rat or other rodent. The food of the tuatera appears to consist of young birds, as the remains of these animals have been found in its intestines. But most probably, as these lizards are of very sluggish habits, they would prey only on the young of such birds as feed on the ground. The wing-cases of beetles have also been discovered in their intestines. The general skeleton of the lizard offers many peculiarities; the vertebræ are concave on both aspects, like those of a fish, and from each rib proceeds a process which laps over and rests upon the one behind—an arrangement which is identical with that found in birds. The tuatera is remarkable for a very highly developed arrangement of abdominal ribs, distinct from those attached to the spine. These are nearly twice as numerous as the true spinal ribs, and correspond in number with the scaly plates on the under surface of the body. Dr. Günther suggests that these ribs assist in locomotion, as the tuatera lives in holes in sandhills on the seashore, and has short limbs and feeble claws, that are unfitted for dragging the heavy body. He states:—" I do not for a moment entertain the idea that an individual with the limbs disabled could glide from the spot where it lies, nor am I convinced that the action of the abdominal apparatus is constantly superadded to that of the limbs; but in the case of a lizard living in the rocks and sandhills of the seashore, the occasions must be frequent when the feebleness of its claws is assisted by its ventral plates. If the supposition should be confirmed that the tuatera lives in holes where the free action of the limbs is naturally more or less impeded, the abdominal ribs would be of material service." The scientific name *Hatteria punctata* has been given to this animal, which forms the type of a distinct group, placed between the lizards and the crocodiles. The skeleton of this animal, with its singular fish-like vertebræ and abdominal ribs on the one hand, and highly developed bony skull and bird-like ribs on the other, offers a strange combination of very low and high vertebrate organization. " And this," writes Dr. Günther "is more significant, as the animal occurs in a part of the globe remarkable for the low and scanty development of reptilian life. The New Zealand of the present period is inhabited by only a few (about nine) small species of generally distributed geckos and skinks, and a single species of frog; and it is not probable that this small list will be considerably increased by future researches. With more confidence we may look forward to discoveries of remains of extinct forms." The tuatera now in the Zoological Gardens is a young animal, some of the adult specimens reaching two feet and upwards in length. —*The Field.*

VENOM OF TOADS.

THE Toad, formerly considered as a creature to be feared, does in reality possess a venom capable of killing certain animals and injuring man. The *British Medical Journal* says that this poison is not, as is generally thought, secreted by the month; it is a sort of epidermic cutaneous secretion, which acts powerfully if the skin be abraded at the time of contact. Dogs which bite toads soon give voice to howls of pain. On examination it is found that the palate and tongue are swollen, and a viscous mucus is exuded. Smaller animals coming under the influence of the venom undergo true narcotic poisoning, soon followed by convulsions and death. Experiments made by MM. Gratiolet, Cloez, and Vulpian, show that the matter exuding from the parotid region of the toad becomes poisonous when introduced into the tissues. A tortoise of the species *Testudo Mauritanica* lamed in the hind foot, was completely paralysed at the end of fifteen days; and the paralysis lasted during several months. Some savages in South America use the acid fluid of the cutaneous glands of the toads instead of the curara. The venom exists in somewhat large quantity on the toad's back. Treated with ether, it dissolves, leaving a residuum, the evaporated solution exhibits oleaginous granules. The residuum contains a toxic power sufficiently strong, even after complete desiccation, to kill a small bird.

THE POISON OF THE COBRA.

AN extensive series of experiments has been conducted in India by Dr. Shortt to test the efficacy of a large number of reputed antidotes to the Poison of the Cobra. These antidotes have been brought forward by claimants for the prize of 1,750 rupees offered by the Maharajah of Travancore for the discovery of a specific, but not one of them has been found to possess any counteracting influence over the rapid and deadly action of the poison. In consequence of communications which he has received from various parts of Europe, Dr. Shortt has made arrangements with Dr. Tilbury Fox to superintend any experiments which may be made in England with the view of discovering any real antidote which may be in the possession of any one in Europe. Dr. Fox has been supplied with an ample stock of active poison, both in the liquid and crystalline form, for the purpose. Dr. Shortt will finally test the virtues of any specific which may possibly be discovered.

A PROCESS FOR KILLING PULMONIFEROUS LAND MOLLUSKS FOR ANATOMICAL PURPOSES.

THE great contractility of the Pulmoniferous Mollusca at the instant of death, presents an obstacle to anatomical researches.

It becomes difficult to recognise the true position of the organs, and it is frequently noticed that the delicate parts, such as the dart sac for example, are ruptured and become entangled in the liver, and other glands.   On placing such mollusks, however, in a bottle of water, closed, and having the air excluded, we observe that they extend all their external organs, such as the tentacles and foot; the jaw is also extruded, and, if a little tobacco be added, the lingual ribbon turns itself outwards, and can be easily recognised.   The limaces, enclosed with a morsel of cigar, die very quickly, with their jaw and lingual ribbon exserted.   The clausiliæ, however, withdraw themselves before death.   To pre-vent this happening, close the aperture of the cell with wax or cotton, and pierce near the lip a hole sufficiently large to admit of the passage of all the parts of the animal contained in the shell, and too small to allow of its retracting when the tissues are extended with water.   In this way Mörch has obtained a preparation of *Clausilia laminata*, in which can be seen the jaw protruded, and the tentacles extended.   He recommends this process for the examination of the jaws of the Pulmonifera, which is thus ren-dered extremely easy.—*Journal de Conchyliologie; Scientific Opinion.*

SHETLAND SPONGES.

THERE has been read to the British Association, a Report on Shetland Sponges, with Description of a remarkable new genus, by the Rev. A. M. Norman.   More than eighty species of sponges have been met with, and thirty-three of these have not as yet been found elsewhere.   Their description, from the specimens procured by the Dredging Committee, are contained in Dr. Bowerbank's work.   Specimens of all these species were exhi-bited to the Section, and, in addition, several new and as yet undescribed species.   Among these are three unusually fine forms : one, a fan-shaped *Isodictya*, exceeding in size all other species of the genus except *I. infundibuliformis*, and of most elegant structure, to which Dr. Bowerbank proposes to give the name *I. lacinisæ* : the second, which is another of the largest of British Spongiadæ, is the type of a new genus, characterised by having the skeleton without fibre, composed of an irregular net-work of polyspiculous fagot-like bundles, the spicules of which are compactly cemented together at the middle, but are radiating at the terminations, which will be named by Dr. Bowerbank *Raphioderma coacervata ;* the third is also the type of a new and very remarkable genus, for which the name of *Oceanapia* is proposed.   Fragments of this sponge had been frequently dredged in previous years ; and portions of the bulbous base are the *Isodictya robusta* of Bowerbank ; while the cloacæ are his *Desmacidon Jeffreysii.*   During the present year's dredging, the committee have at length procured perfect specimens, which prove to be very remarkable both in form and structure.   The

form and size are that of a full grown turnip, the rind being of
usual sponge structure, though unusually close, firm and compact. From the crown of the turnip there arise more or less
numerous cloacæ—sometimes simple, sometimes branched, and
from three to seven inches high. From the opposite side (whence
the roots of the turnip would spring) there proceed somewhat
similar processes to those of the crown, but of much smaller size.
These appear to be used for purposes of attachment, as, in one
instance, they tightly grasp a stone. The whole interior of the
sponge is, as it were, a cup filled with sarcode, of which the
largest example must have contained fully half a pint; internal
tubes, continuous with the external cloacæ, of delicate structure,
proceed downwards from their base into the centre of the sponge,
and are there bathed in the mass of sarcode. The distribution
and arrangement of the fibres and spiculæ are very varied in the
different parts of the sponge; but the whole is built up by rather
short, stout needle-formed spiculæ, which are acute at both ends.
The only other form is an excessively minute semicircular spicule,
which is abundant on the dermal membrane.

DREDGING IN THE SHETLAND ISLANDS.

THERE has been read to the British Association, the last Report on Dredging among the Shetland Islands, by Mr. J. G.
Jeffreys. In spite of the weather, which was this year unusually
cold and boisterous, some further results were obtained. A
fine species of Pleurotoma (P. carinata, Phillippi) was added to
the British Fauna, having been first discovered as a Sicilian
fossil and since recorded as inhabiting the coasts of Upper Norway. Several of the rarer species, peculiar to the Zetlandic seas,
also occurred. Other departments of marine zoology would be
reported on by Messrs. Norman and Waller and Drs. Günther
and M'Intosh. Mr. Jeffreys then compared the mollusks of our
North Sea with those from the Mediterranean and Adriatic,
which he had carefully investigated, as well by his own dredgings
in the Gulf of Spezzia as by the examination of nearly all the
public and private collections. Although the littoral species of
the northern and southern parts of the European seas exhibit a
considerable difference, there is a remarkable identity between
these which inhabit deeper water. Out of 317 Zetlandic species
no less than 244 are found living south of the Bay of Biscay,
203 being found north of the British seas. This concordance
partly arises from different names having been applied to the
same species by British and foreign writers on the subject. A
summary of the results from all the dredgings by the author in
Shetland was given under several heads, including the comparative size of specimens of the same species from the northern and
southern parts of the European seas, the colour of shells from
deep water, the geographical and bathymetrical distribution of

species, the identity of certain fossil and recent shells, the devolution of species, and the course of the Gulf-stream with respect to the oceanic molluscs.

The Rev. A. M. Norman spoke of the great value of Mr. Jeffrey's dredging researches. The crustacea, polyzoa, and sponges had fallen to his care, and he believed that most important additions to our knowledge were made through Mr. Jeffrey's constant excursions.

THE BORING OF LIMESTONE BY ANNELIDS.

MR. E. RAY LANKESTER observes that, in the discussions concerning the Boring of Molluscs, no reference has been made to the Boring of Annelids—indeed, they seemed to be quite unknown. He brings forward two cases, one by a worm called Lencodore, the other by a Sabella. Lencodore is very abundant on some shores, where boulders and pebbles may be found wormeaten, and riddled by them. Only stones composed of carbonate of lime are bored by them. On coasts, where such stones are rare, they are selected, and all others are left. The worms are quite soft, and armed only with horny bristles. How, then, do they bore? Mr. Lankester maintains that it is by the carbonic acid and other acid excretions of their bodies, aided by the mechanical action of the bristles. The selection of a material soluble in these acids is most noticeable, since the softest chalk and the hardest limestone are bored with the same facility. This can only be by chemical action. If, then, we have a case of chemical boring in these worms, is it not probable that many molluscs are similarly assisted in their excavations? Mr. Lankester does not deny the mechanical action in the pholas and other shells, but maintained that in many cases the co-operation of acid excreta was probable. The truth is to be found in a theory which combines the chemical and the mechanical view.

ORGANISMS IN THE ATLANTIC.

THERE has been read to the British Association a paper, " On some Organisms which live at the Bottom of the North Atlantic, in depths of 6,000 to 15,000 Feet," by Professor Huxley. In the year 1857 Professor Huxley examined and reported upon specimens of mud obtained from the bottom of the North Atlantic, and his observations appeared in a Report, published in 1858, on the so-called Telegraph Plateau, under the title, " Deep-Sea Soundings in the North Atlantic Ocean." He there described certain small oval particles of carbonate of lime, as coccoliths. Dr. Wallich, in 1860 to 1861, described, in papers on Life from the Deep-Sea Soundings, in addition to the coccoliths, bodies which he called coccospheres, and which he considered to be the source of the former. Mr. Sorby has made the interest-

ing discovery of coccoliths in the chalk, and has also examined
the deep-sea soundings, and thrown some light on their history.
But now, Professor Huxley has re-examined his specimens and
others with a much higher power of the microscope than he previously used, namely, a one-twelfth of Ross; and he has been
able to throw considerable light on the history of both coccoliths
and coccospheres. He shows that, first, there are two kinds of
coccoliths, both in the Atlantic mud and in the chalk—the one
having a more complex structure than the other kind; secondly,
that both these are developed in a gelatinous granular substance;
thirdly, that they are not derived from the coccospheres, as Dr.
Wallich supposed. The slimy material which is drawn up in the
deep-sea soundings owes its stickiness in great part to the presence of agglomerations of small protoplasmic masses of soft
gelatinous matter, in which the coccoliths and coccospheres
appear. Professor Huxley considers that the coccoliths have the
same sort of relation to the living gelatinous matter around
them which the crystalline spiculæ of the sea-jellies or radiolaria
have to their surrounding jelly. The gelatinous matter of the
coccoliths contains granules aggregated in groups, about one-
thousandth of an inch in length, or less, and each of these
Professor Huxley would regard as a distinct being. The largest
coccoliths are the sixteen-hundredth of an inch in length.
Professor Huxley described the whole structure of coccospheres,
coccoliths, and jelly-like matter with great minuteness. He considered that we had here evidence of a wide-spread and new form
of life at this great depth in the Atlantic sea-bed, but it was impossible to say whether these jelly-like organisms with their
coccoliths were animal or vegetable. They reminded him of the
Urschleim of the Germans, in their wide-spread occurrence, forming a living paste on the floor of the Atlantic, such as some persons supposed more complex organisms to be developed from.
Each mass of granules was probably a distinct being, with its
own processes or pseudopodia when living, but when dead they
run together and form the slimy mass.

AGRICULTURAL ANTS OF TEXAS.

In the *Science Gossip* is an interesting paper, by Dr. Gideon
Lincecum, on the Agricultural Ants of Texas. There is, it appears, in that country a species of Ant remarkable not merely for
the social and industrial instincts for which the ant is elsewhere
famous, but which is distinguished by the habit of sowing and
reaping a certain species of grain, which, when deprived of its
husk, is stowed away for use in suitable granaries. When these
ants have established a colony, they construct a pavement of
several feet diameter round their habitation, by paving the surface of the earth with small pieces of grit, in the way we pave a
street. This pavement is kept scrupulously clean, and around it

a species of grain-bearing grass (*Aristida stricta*) is sown and carefully weeded. The seeds of this plant, under the microscope, resemble the rice of commerce, and the seeds are sown in the autumn, so that the autumnal rains may bring the ant-rice up. When the crop is ripe, they cut off the grain and carry it, husks and all, to the interior of their dwelling. The husks are then removed, and are brought out and carried clear of the pavement, and always thrown out on the leeward side. In long-continued droughts, the ants make their way into the wells, and many of them fall into the water, where they cling together and collect in lumps. If one of these ants is thrown into water it will be drowned in about fifteen minutes; but out of the wells lumps of ants are drawn up in the buckets, which must have been in the water for many days, and yet the ants forming the lump are still found to be alive. It is consequently concluded that the balls of ants, by concerted action, put themselves into slow rotation, the effect of which is to give sufficient breathing-time to each constituent member of the mass to keep it alive.

SILKWORMS' EGGS.

An Italian correspondent of the *Times* writes as follows:— "You are aware that of late years there has been a very large importation of Silkworms' Eggs from Japan to Italy, with a view to repairing the ravages of disease among the worms in this country. To a considerable extent the plan has proved successful, although the descendants of the Japanese by no means invariably escape— at least, in the second and third generation—the mysterious affliction which has long played such havoc with one of the most important and lucrative of Italian productions. Latterly, complaints have been heard even of the eggs imported direct from Japan. To all appearance they were of good quality; but it seemed they were not proof against the general epidemic. A circular that has been addressed to the presidents of agricultural committees, by the Minister of Agriculture and Commerce, accounts otherwise for the deterioration. A system of fraud has been detected. The Minister declares it to have come to his knowledge that a great number of cases containing empty papers of the kind in which the Japanese eggs are usually sent, have been received by a Milan house. These papers are duly stamped and inscribed in Japanese; they bear the Custom House mark usually put on them at Yeddo on their way to the Yokohama market. The Minister declares it beyond a doubt that these papers were intended to receive Italian eggs, which would be sold to the public as genuine Japanese—thus discrediting the only eggs on which is now based a hope of reviving sericulture in this country. The discovery appears to have been due to the refusal of certain Japanese houses to submit their papers of eggs to the inspection and stamp of the Italian agents and consuls in Japan.

The Minister promises great vigilance to defeat these frauds, for, he adds, he has reason to believe that other persons are on the alert to follow the example of the Milan house above alluded to. According to an official publication, in 1863 the total value of the cocoons produced in the provinces which now compose the kingdom of Italy, was upwards of four millions sterling. To obtain this there had been imported 56,129 kilogrammes of eggs, at a cost of twenty-four millions of francs, or nearly a million sterling."

BRITISH WASPS AND NESTS.

MR. JOHN HOGG has exhibited two admirably-photographed plates of two kinds of Wasps' Nests, which he had collected at Norton, in the south part of the county of Durham, from the years 1831 to 1856, both inclusive. Plate 1 represents the inner portion of the rather large nest of *Vespa arborea*, built on the under side of a branch of the larch-tree, which he discovered in a neighbouring wood. A few years before, he captured, in the same vicinity, two neuters of a new wasp, which, being sent to Professor Westwood and Mr. F. Smith, the latter entomologist gave the name of *arborea* to that new species, because of " its habit of constructing its nest in trees." (See *Zoologist*, p. 171, June, 1843.) The structure of the outside of that nest is strong, and rather coarse; the numerous cells were empty, owing to the lateness of the season (October) when discovered; but these are regularly distributed and well formed. Other naturalists coincided in the great probability of that nest being the fabric of the tree-wasp, *V. arborea*. Plate 2 exhibits four very delicate and beautifully built nests of the *V. Britannica* of Leach, or, according to other entomologists, *V. Norvegica*. They are of a grey colour, and composed of a fragile paper-like substance, but varying in size. Affixed alongside the nests is a wasp taken out of each one respectively, and they are all of that identical species, which is small, of a dark colour, and rough, with black hairs. The facial lines and marks are also the same, and quite distinct. Plate 3 shows another nest of the same social wasp, which, of a larger size, was taken this summer in a neighbouring garden. Mr. Hogg then observed upon the great use of photography in accurately illustrating natural objects, and so easily preserving the true representation of any rare plant or animal.

THE TSALTSAI, OR ABYSSINIAN SPEAR-FLY.

MR. SAMUEL SHARPE, in a communication to the *Athenæum*, remarks :—" In all sciences we are much interested in avoiding a variety of names for the same object, and in acting upon the rule that the name by which it is first known shall not, without good reason, be put aside for a new name. The above-mentioned

Fly, the dreadful scourge of Abyssinia, was first brought to this country by the traveller Bruce, who called it, as he had there heard it called, the Tzaltzala-fly. It has since been brought here by Dr. Livingstone from South Africa, and it is called in his book the Tsetse-fly. Bruce had very properly conjectured it was the fly mentioned, but without its name, in Isaiah vii. 18, 'The Lord shall hiss, or whistle, for the fly that is in the uttermost parts of the rivers of Egypt.' But it is to Dr. Margoliouth that we are indebted for the remark that it is twice mentioned by name in the Hebrew Scriptures. In Deut. xxviii. 42, we read, 'All thy trees and fruit of thy land shall the Tsaltsal consume,' or, in the Authorized Version, the locust. But it would seem that the writer was not well acquainted with its habits, as it does not destroy vegetables. The next passage is yet more important, because it had hitherto baffled the commentators. Isaiah, in chap. xviii. 1, addresses Abyssinia as 'The land of the winged Tsaltsal, which is beyond the rivers of Ethiopia.' Here, then, we have Bruce's name for this fly supported by the Hebrew writers; and the two together should make the naturalists give up the new name lately introduced by Dr. Livingstone. In Job xli. 7, Tsaltsal is a spear, or harpoon, with which fish are killed; and hence the formidable little spikes attached to the fly's mouth may have given to it its name. In order to distinguish the insect from the piece of metal, Isaiah calls it 'the winged Tsaltsal,' or the spear-fly."

RARE INSECTS.

THE following have been described to the Entomological Society:—Mr. Stainton called attention to the history and figure of a small Lepidopterous Insect, published, in 1750, in the *Mémoires de l'Académie Royale des Sciences de Paris*. The habits and transformations were described with great particularity from the observations of M. Godchen de Riville, made in the island of Malta; and though the insect was quite unknown, except from M. de Riville's description, it was clear that it belonged to the genus Antispila. The larva was apodus, and fed upon the leaves of the vine. Mr. M'Lachlan mentioned that *Anax Mediteraneus*, an African dragon-fly, once captured in the island of Sardinia, but which had been rejected from the list of European Libellulidæ, had last year occurred in swarms at Turin, and in other parts of Italy. Mr. F. Smith exhibited a larva, found by Mr. O. Janson by digging in a sand-bank, which was believed to be that of a Xantholinus, attached to the underside of which were four parasitic pupæ, probably of a Proctotrupes. Mr. F. Smith also exhibited specimens of a Longicorn beetle, *Cerosterna gladiator*, and of a large Acheta, from India, which had caused great damage to young plantations of Casuarina, along the Madras Railway. Dr. Cleghorn said that the Acheta appeared suddenly,

in September last, after some rain at the end of the hot season.
During the night the larvæ emerged from the sand and crawled
up the young trees, generally biting off the leading shoots. He
employed little boys to burrow in the sand, to extract them from
the tortuous passages which they made therein, and by this
means destroyed bushels of them. The Cerosterna was also very
mischievous; but its attacks were principally directed to the
bark. Mr. F. Smith exhibited eight species of larvæ, all of
which were described as " borers," and as being very destructive
to trees in India; amongst them was the now notorious "coffee-
borer," *Xylotrechus quadripes* of Chevrolat. Captain Taylor,
who has been long resident in Coorg, gave his personal experiences of the coffee-borer, and reported that the evil was now
on the decrease.

VALUABLE INSECTICIDE.

It is stated that Petroleum Oil possesses the highest efficacy as
a destroyer of all kinds of insects injurious to plants or animals,
and the less purified, and consequently the cheaper, it is the better. Thirty parts should be mixed with 1,000 of water, and applied where required. The *Medical Times* states that vermin of
houses may be destroyed by introducing into the holes or cracks
a few drops of petroleum.

A HINT FOR TRAVELLERS.

A WELL-KNOWN German traveller, F. Jager, in his *Sketches of
Travels in Singapore, Malacca, Java* (Berlin, 1866), describes
the powder of the *Pyrethrum roseum* as a specific against all
noxious insects, including the troublesome mosquitos and those
which attack collections. He says, " A tincture prepared by
macerating one part of the *Pyrethrum roseum* in four parts of
dilute alcohol, and, when diluted with ten times its bulk of
water, applied to any part of the body, gives perfect security
against all vermin. I often passed the night in my boat on the
ill-reputed rivers of Siam without any other cover, even without
the netting, and experienced not the slightest inconvenience. The
' buzzing,' at other times so great a disturber of sleep, becomes a
harmless tune, and, in the feeling of security, a real cradle-song.
In the chase, moistening the beard and hands protects the hunter
against flies for at least twelve hours, even in spite of the largely
increased transpiration due to the climate. Especially interesting
is its action on that plague of all tropical countries, the countless
ants. Before the windows, and surrounding the whole house,
where I lived at Albay, on Luzon, was fastened a board, six inches
in width, on which long caravans of ants were constantly moving
in all directions, making it appear an almost uniformly black surface. A track of the powder, several inches in width, strewed

across the board, or some tincture sprinkled over it, proved an insurmountable barrier to these processions. The first who halted before it were pushed on by the crowds behind them; but, immediately on passing over, showed symptoms of narcosis, and died in a minute or two, and within a short time the rest left the house altogether."—*British Medical Journal.*

BOTANY.

A TROPICAL AIR-PLANT.   BY CHARLES WRIGHT.

A WONDERFUL tree—if tree it can be called—grows throughout the West India Islands, in South America as far south as Brazil, and perhaps in Florida. It is not remarkable for its beauty, nor for its great size, but for its irresistible power of destroying other trees.

It is an epiphyte (*Clusia rosea*, Linn.), perhaps a true parasite. Whether it ever germinates in the ground I know not; nor do I know why it should not, if it can sprout from a woodpecker's hole in a palm. Certain it is, that of hundreds which I have seen, I never saw a young plant attached to the soil. It grows on many kinds of trees, and at almost any height above the earth. In some situations it grows feebly. On a palm it never or rarely attains to any considerable size; whether there is an incompatibility between the two growths, or whether, as is commonly the case on these trees, it germinates at too great a height. On the spreading branch of a tree it thrives better, but seems there to be not in its proper place. In any case, its main development is downward. When on a branch remote from the trunk, the descending axis—root or trunk, whichever it may be—is like a cord, increasing to the size of a rope, or a hawser, or growing even larger; rarely branching, but sometimes, near the ground, sending off stays. The ascending axis makes little more than a bush, while the root may be thirty or forty feet long. In one respect this is like a true root—it branches irregularly; while, on the ascending trunk, the leaves and branches are in pairs.

In order to attain its full development, it seems necessary that it should germinate at a point from which the descending axis shall pass in proximity to the trunk of the tree; and it has seemed that, if this point be very high, it is a circumstance unfavourable to its rapid growth.

Supposing, then, our plant to start under favourable auspices, not very high above the ground, and from a hole or a fissure in an erect trunk, the ascending stem presents nothing of special interest; but the root, passing down near the foster-tree, is most singularly affected by it. It would seem as if possessed of a most grateful affection for that which gives it support; so much so, as to multiply arms with which to embrace it. It sends off, from time to time, at irregular distances, from one side to the other, slender, almost thread-like branches, which pass horizon-

tally around the tree, till they meet on the opposite side and
unite; or, it may be, if two should *not* meet, they would pass
entirely round it and unite again with the main root. On this
point, I either made no careful observations, or my memory is at
fault. Gradually the foster-tree is embraced by a succession of
these cords. But, by the same regular growth, these cords
spread upward and downward, till they become hoops. And
these hoops often send off branches from one to another; and
these in their turn widen, till the tree is inclosed in a living
cylinder or a cylindrical network of bands, having immense
strength; and as these seem to increase only laterally, the
growth of the tree is checked, and its destruction is inevitable,
sooner or later, according to its less or greater power of endur-
ance.

A tree on which the Copey has woven a pretty complete
net cannot long retain its vitality. Its circulation is stopped,
and it dies. But this *seems* not to check the growth of the
destroyer, so long as the trunk remains erect. But when they
both fall, the parasite cannot long survive. It would seem
that it required either elevation or an erect position for its
existence. I can recall to mind but one instance of a Copey
growing from the ground, and it is probable that in this
case the place whence it started was low, and it had time to
reach the soil and fasten its roots there before the death and
decay of its foster-parent. Copey is, probably, the aboriginal
or Carib name of the plant, which, like many others, has been
retained. Scotch lawyer, or Scotch attorney, by which name it
is known in Jamaica, is not altogether flattering to legal gentlemen of Caledonian extraction.—*The American Naturalist.*[*]

GIGANTIC TREES OF NEW ZEALAND, AND THE CEDARS OF LEBANON.

A paper has been read to the British Association, "On the
Wellingtonia gigantea, with Remarks on its Form and Rate of
Growth as compared with the *Cedrus Libani*," by Mr. J. Hogg.
This paper commenced with an historical sketch of Mr. White-
head's discovery of this colossal North-West American tree in
1850, in the Calaveras Grove, in California. The situation of this
grove, at an altitude of above 4.000 feet above the sea-level, is in
lat. 38° N. and long. 120° 10′ W. The author, having described
the finding of more of these stupendous evergreen conifers by
Messrs. Dowd and Lewis in 1852, mentioned the names and
dimensions of some of the largest trees in the Calaveras Grove.
Of these, the "father of the forest" measured 110 feet in circumference at the ground, and was about 435 feet in height. Other
trees of vast size had been likewise detected growing in Crane
Flat, in Calaveras County, on one of the tributaries of the Big

* Quoted in *Scientific Opinion*, a new periodical compiled from valuable
sources.

Creek; on an upper branch of the Fresno river; and in the Mariposa Grove, between the Big Creek and the Merced. Again, in 1857, Mr. Clark discovered two smaller groves of the same species; and subsequently, an eighth and more extensive *habitat* was noticed about twelve miles east of the Fresno Grove. Mr. Hogg gave the names of the most remarkable of these "mammoth trees," with their estimated sizes, and said, that instead of there being *two groves* only, as was at the first reported, there existed eight or nine in all, and that the whole of these localities comprised more than 1,200 trees. He then pointed out the generic differences of the Wellingtonia, and remarked that that name had been inappropriately assigned to it by the late Dr. Lindley in 1853, instead of the more proper American one, Washingtonia. That botanist, supposing that this conifer did not grow "above two inches in diameter in twenty years"—being equivalent to twenty-four lines in twenty years—concluded that its age was 3,000 years. Four years afterwards the author made another estimate of its longevity, and taking *three lines* in diameter for its *mean annual* growth—the tree being of quick growth—reduced its age to 1,344 years. The mode of computation of the age of this tree, by calculating its concentric zones, cannot strictly be relied on, because this, like some other forest-trees, may experience *two* growths in the year; at least, such is the opinion of several botanists. Mr. Hogg then compared the annual increase in height of a young Wellingtonia, eight years old, with a cedar of ten years, and the result was in favour of the former tree by more than two years. Also the cedar of Lebanon does not produce cones until it is near thirty years of age; but the Wellingtonia was proved to do so, accompanied with male flowers, when only six years old. Whilst the mode of growth of the young cedar is by parallel and horizontal branches, that of the young Wellingtonia is upright and pyramidal; but as the latter advances in age its lower branches fall off for one-third or more of its entire height. As the longevity of the Wellingtonia in the primeval forest of America has been estimated at 3,000 years, so that of the oldest cedar in its native *habitat* of the Lebanon has been given at 2,500 years. Mr. Hogg having illustrated his paper by copies of photographs of a patriarchal tree in Lebanon, and of several immense Wellingtonias, remarked that the growth of large trees, or the formation of the concentric zones in the stems, required further investigation.[*]

THE BAOBAB TREE.

Mr. J. J. Monteir has exhibited at the Royal Institution specimens of the inner bark of the gigantic Baobab Tree, or *Adansonia digitata*, bleached and unbleached, and also specimens

---

[*] See an exhaustive paper on the *Wellingtonia gigantia*, in *Natable Things of Our Own Time*, by the Editor of the present volume, pp. 89—103.

of very strong paper made from it, an application which he has
lately discovered. As this tree grows abundantly all over the
continent of Africa, and has hitherto never been applied to any
useful purpose, its utilisation in the paper manufacture promises
to be of very great importance both to British commerce and
negro civilisation, by creating a new field for native industry and
trade in the place of the traffic in slaves. The inner bark is
merely stripped off the tree, beaten, and dried in the sun. The
trunk sometimes attains a circumference of 100 feet.

TIMBER-TREES OF INDIA.

DR. CLEGHORN observes, When the British Association met at
Edinburgh, in 1850, a committee was appointed to consider "the
probable effects, in an economical and physical point of view, of
the destruction of Tropical Forests." Their Report was presented
in 1851 at Ipswich (See Vol. Brit. Assoc.) Attention was thus
directed in India to the importance of preserving every influence
which tends to maintain an equilibrium of temperature and hu-
midity, of preventing the waste of valuable material, and the
special application to their various uses of the indigenous Timbers
of the country. A few days later, forest establishments were
sanctioned in British Burmah and the Madras Presidency. In
1864, the Government laid the foundation of an improved general
system of forest administration for the whole Indian empire,
having for its object the conservation of state forests, and the
development of this source of national wealth. The executive
arrangements are left to the local administrations, the general
principles being laid down; the most important of which is, that
all superior Government forests are reserved and made inalien-
able, their boundaries marked out to distinguish them from waste
lands available for the public. Act 7, of 1864, was passed, de-
fining the nature of forest rules and penalties, and this Act has
been adopted by most of the local governments. Valuation surveys have been made to obtain trustworthy data as to the geo-
graphic distribution of the more valuable trees, the rate of growth,
and the normal yield of the forests. In British Sikkim and the
Dooars of Bootan there are large tracts of Sal (*Vatica robusta*)
not yet surveyed. The produce of these forests is required for the
extension of the East Bengal Railway, and for the doubling of
the East India line now in progress. In the Darjeeling district
the higher slopes above 6,000 feet have been reserved, and plantations, both of temperate and sub-tropical trees, have been
formed. In the Terai several thousand mahogany trees have
been raised. The recent surveys have added much to our knowledge of the forest resources of these districts. In Kumaon and
Gurhwal the area surveyed was about 400,000 acres; a large
part of this is covered with *Pinus longifolia*, bearing an average
of fifteen trees per acre. The Himalayan box is plentiful in

certain localities, and has come into use in the schools of art for wood engraving. The Goruckpore forests cover 120,000 acres, and consist mainly of Sal (*Vatica robusta*), with an average of twenty-five well-grown trees to the acre. The northern limit of indigenous teak is in Bundlekhund. It has been planted in the Punjab, but in that dry climate it is poor and stunted. The management of the forests of the North-West Provinces is second in importance only to that of Burmah. From the survey in Oudh, it appears that more than half of the government forests consist of Sal: the other reserved woods are sissoo, toon and ebony. Considerable sums have been expended in clearing the Sal trees from the destructive twiners. In the Punjab, the forests growing on the banks of the five rivers have been formed into as many ranges under skilled officers, and timber operations have been conducted, with more or less success, in the intramontane districts. Long leases of the Deodar forests, in the territories of the Rajahs of Chamba and Bussahir, have been negotiated. Wood is the only fuel at present available in quantity for locomotive purposes. The requirements of the railway alone is estimated at 50,000 tons annually, and the yield of the old shikargahs, or fuel-reserves being inadequate, skilled management has been brought to aid in the increased production of fuel. Selected tracts have been trenched before planting, and cattle and camels strictly excluded. The suitability of some Australian trees to the arid plains of the Punjab is remarkable, and several species of acacia, casuarina, and eucalyptus have been tried with apparent success. The northern limit of Sal is on the bank of the Beas river in the Kangra valley; but here it is small and stunted. The character of the forests in the Central Provinces, Mysore, Hydrabad, British Burmah, Madras, and Bombay Presidencies, were noticed in detail, and the result of the recent surveys shown upon a large map. The importance of continuing the forest surveys and of demarcating the reserved tracts was urged, and the want of a *Flora sylvatica* of India insisted upon. —*Proc. British Association.*

CINCHONA IN JAMAICA.

THE cultivation of the Cinchona in Jamaica is beginning to occupy the attention which its importance demands. Mr. Thompson, in his circular, has confirmed what has been said on the natural capabilities of Jamaica for the production of this invaluable plant, to which a Correspondent adds:—"The subject at this moment is of immense importance, from the fact that the supplies of bark from South America are gradually diminishing. It would be sheer waste of time and capital to attempt to cultivate the varied species of the cinchona in the island, for some would not yield a sufficiency of quinine to repay expenses of cultivation. Those, then, which contain larger amounts of

the vegetable alkaloid should be introduced. And they are the
cinchona cansaya and *cinchona rubra.* In these quinine exists in
excess. Mr. Thompson's estimate of the annual yield is rather
low. From what has fallen under the writer's observation, he
is inclined to anticipate a larger yield. But this will, of course,
depend on the fertility and height of the locality selected. The
febrifugal properties of the cinchona bark are increased in proportion to the degree of moisture the foliage derives from the
mists surrounding elevated mountains. To select localities
favourable for this purpose the botanist's attention should be
directed to such districts as Stoneyhill and Newcastle ; the mean
heights of which range from 1,800 to 2,500 feet above the level
of the sea, and are only about nine to ten miles distant from
Kingston. These districts are physically fertile and humid.
From a small sample of bark (of the *cinchona rubra*) cultivated
in the Botanical Gardens at Bath, Jamaica, a large amount of
pure quinine was obtained by the writer, with a very small proportion of the allied alkaloid-cinchonine. It is to be hoped that
the gentlemen who are now so advantageously conducting the
political affairs of the island will give every support in their
power in furtherance of the cultivation of one of the most useful
plants in our *materia medica.*"

PLANTS WITHOUT ROOTS.

M. DUCHARTRE describes certain plants which vegetate without
roots. In South America, people will suspend such plants from
a balcony by a thread, without their being in contact with anything else, and yet they will grow and blossom in this strange
position. In our hothouses we see them, simply stuck upon a
piece of wood or cork, thrive beautifully without any roots. The
question therefore is, How do they live? M. Duchartre, to discover the secret, has instituted a series of experiments on a plant
of this family, the *Tillandsia dianthoidea.* Two tolerably equal
sprigs of it, without a vestige of roots, were tied separately to
two slices of dry cork cut from the same piece. They were then
freely suspended in the air, and weighed from time to time. One
of these plants, A, never got any water at all; the cork of the
other, B, on the contrary, was moistened every second or third
day. After the experiment had been continued for 103 days,
A had lost one-third of its weight, but had nevertheless produced
blossoms and a small root. B, on the other hand, was extremely
vigorous, and had increased one-eighth in weight. M. Duchartre
thence concludes that these plants require water for their nourishment, but that they do not for all that absorb the moisture of
the atmosphere.—*Land and Water.*

VEGETABLE HAIR.

UNDER the above name we have an invention, by M. Werner Stanfen, of 16, Rue Auber, Paris, and which has recently been patented in England. The invention consists in the manufacture of a species of Vegetable Hair from the fibrous material which grows through and proceeds from the bark situated near the foot of the palm known as the *Levistonia chinensis*, Roxb., or *Latania chicensis*, Jacq. The fibrous material and adherent bark, as imported in the rough state, being first disintegrated by an opening machine, is boiled in an alkaline lye composed of from 5 lb. to 10 lb. of soda or potash, dissolved in 100 gallons of water. This operation, which occupies from half-an-hour to two hours, according to the strength of the lye, is continued until the gummy, resinous, and ligneous matters adhering to the fibres are completely removed. The material thus cleansed is exposed to the action of a mordant preparatory to its removal to the dyeing vat, charged with the required colour, to which is added from 1 lb. to 4 lb. of oil soap for every 100 lb. weight of fibre. The dyeing process being completed, the mass is dried, either in the open air or artificially, and is afterwards submitted to the action of ordinary opening and combing machinery, by which the filaments are glazed and divided to the required degree of fineness. The fibrous material thus obtained is to be applied to the different purposes for which horsehair, bristles, and other kinds of hair have hitherto been employed as articles of commerce. When intended as a substitute for bristles (as, for instance, in the manufacture of brushes), the coarser fibres are selected and left straight, but when intended for stuffing and similar purposes in lieu of horsehair, they are curled in the same manner as the latter, after which they are steeped in water till softened, and finally baked at a temperature of about 140 deg. Fah.—*Mechanics' Magazine*.

NEW PIGMENT.

PROFESSOR CHURCH has extracted from the brilliant red feathers of the bird known as the " plantain-eater," a new animal Pigment which he calls " Turacine." The feathers may be washed nearly white in water containing a little soda of ammonia, the water assuming a pink hue, and by evaporation the pigment is obtained, and is found to be a species of cruorine, in which the iron usually present is replaced by copper. The source of the copper has been traced to the plantains, on which the birds subsist.

THE ALMOND AND THE PEACH.

PROFESSOR K. KOCH has read to the British Association a paper " On the Specific Identity of the Almond and the Peach."

The author stated that he had travelled over the mountains of the Caucasus, Armenia, some parts of Persia and Asia Minor, during four years, for the purpose of studying the origin of our fruit-trees. Although he could not assert that he had found them perfectly wild, or run wild, he nevertheless had collected much interesting material. He believes that our pears and apples, cherries, most prunes, also peaches and apricots, are not natives of Europe. Only certain bad varieties of prunes have their origin from the *Prunus insititia*, the tree which grows in a wild condition in the woods of Europe. After discussing the wild stock of our cherries and pears, Dr. Koch stated that apricots do not grow wild in Oriental countries, but may, perhaps, come from China and Japan, as also the peaches. In the east of Persia, however, a peach-shrub grows, which is intermediate between the almond and the peach-trees. For some time naturalists and gardeners have asserted that there is no difference between almond and peach trees; that the latter is merely a variety in which the dry peel of the almond has become fleshy, and where, at the same time, the stone has acquired a rough surface. Botanists say also that the petioles of the almond-tree have at the superior end small glands, which are absent in the peach. But the nectarine, which is but a smoothed peach, exhibits these same glands. The flowers are not readily distinguishable of peach and almond. On the shores of the Rhine a double-flowered variety grows, as to which it is not certainly known whether it is peach or almond. In England and France, also, there is a plant which is well-known as the peach-almond, and which is a constant variety. This plant occasionally produces a branch bearing good peaches, but, as a rule, its fruit is intermediate in character. The property of atavism seems to prove the derivation of the peach from the almond; for occasionally a sound peach-tree will produce a branch bearing almond-like fruit.

REMARKABLE MEXICAN PLANT.

A REMARKABLE Plant has been in flower at the Royal Botanic Gardens, of the same genus as the American aloe. The popular belief in England is that it is the nature of these plants to bloom but once in a century, but the fact seems to be that this peculiarity is attributable to them only in a cold climate like our own, and that in their native Mexico they reach the period of bloom in about fifteen years. Having flowered, they rapidly pine away. The flower at the Royal Botanic Gardens is called the *Fourcroya longæva*, and was raised from seed received from Mexico about twenty-eight years ago. In its native country it produces flower-spikes more than 40 ft. long.

FIELD'S TEA FLOAT.

MESSRS. FIELD, of Grey Street, Newcastle-upon-Tyne, have invented an improved method of making tea, the apparatus for which purpose consists of a cylindrical sieve, surrounded by an air-tight chamber which forms the lid, and causes the whole to float when placed in water. The tea is put in the sieve and placed in the tea-pot, the required quantity of water being added. The result of practical trials is a saving of 25 per cent. of the tea used. The *rationale* of this useful invention is as follows:— The hottest particles in a body of water are always those which are uppermost. Now we know that to boil the tea itself is to utterly spoil it, as the active principle (theine) becomes dissipated, much of the virtue of the beverage is lost, and it becomes unpalatable. On the other hand, the successful brewing of tea depends upon the herb being exposed for a sufficient time in water of a temperature as near the boiling point as possible.— *Abridged from the Mechanics' Magazine.*

TREES AND THEIR TRADITIONS.

MR. MENZIES, in his *History of Windsor Great Park and Windsor Forest,* gives these curious facts:— On Trees struck by lightning, oaks are, of all forest trees, the most dangerous. If they have a large spreading head, they are shivered into shreds; if they have a long tapering stem, they are less dangerous, for the stem acts as a conductor, and the lightning ploughs a furrow in its side. He has seen a beech struck once; an ash, an elm, and a cedar each once; but never any other trees except oaks: and whereas the beech, the ash, the elm, and the cedar were struck in open spaces, where they stood alone, the oaks were selected by the lightning, and struck in the midst of thick wood. Another observation discredits an ancient tradition. He has never seen the mistletoe growing on the oaks at Windsor; but on the poplar, the thorn, and the maple; and chiefly on the lime. The only oak in England on which he has seen the mistletoe growing is one seven miles from Godalming. Some of his friends have tried to plant the mistletoe on the oak; but have never succeeded. A third note is on the relations of birds to trees. A fir forest is always a silent one. The reason is, that birds almost always choose deciduous trees to sing in, and corroborates his view by the testimony of an officer attached to the Baltic expedition in 1856, who describes the silence of the great pine forests in Norway, Sweden, and Russia as very remarkable. The last observation has to do with incisions on the barks of trees. On one of the old beeches, near the obelisk, in the park, there is still clearly legible a name, with the date of 1743; and near the iron gates of Virginia Water there is one with the still earlier date of 1707. If these are genuine incisions of those years, they prove that when the cut is made quite through the bark, the form of the letter is preserved longer than one could suppose.

Geology and Mineralogy.

GEOLOGICAL TIME.

A PAPER has been contributed to the *Philosophical Magazine*, by Mr. James Croll, "On Geological Time and the Probable Date of the Glacial Period." It is well known to geologists that there have been great changes in the temperature of the earth at different epochs, and that after great heat had prevailed for a long period, great cold succeeded, of which the evidences are to be found in the traces of glaciers in countries now warm or temperate, and which glaciers have carried forward great stones, and grooved and scratched the underlying rocks, in precisely the same way as existing glaciers are found to do. Indeed, there is reason to believe that there have been several successive periods of heat and cold; and the cause of these changes of temperature has been an interesting subject of speculation. One cause to which these changes have been imputed is an alteration in the eccentricity of the earth's orbit; and in 1782, Lagrange, in a memoir communicated to the Berlin Academy, determined the superior limits of the eccentricity. But it was subsequently shown that, although the sun, when he is over the southern hemisphere, is nearer the earth than when over the northern, yet that, in consequence of the increased velocity of the earth when nearer the sun, it does not receive more heat in the same time, and that the diminution of the heat consequent on the increased velocity is an exact compensation to the increased heat consequent on the increased nearness. Although, however, this is so, it nevertheless can be shown that the glacial epochs may result from the variations in the eccentricity of the earth's orbit, as, when the eccentricity is at its superior limit, the effect will be to lower very much the temperature of the hemisphere whose winters occur in aphelion, and to raise to a nearly equal extent the temperature of the hemisphere whose winters occur in perihelion. Mr. Croll has computed the variations of eccentricity for a million of years back, and he finds that there are two periods of great duration, during which the eccentricity was great. The one of these periods extended from about 980,000 to 720,000 years ago, and the other extended from about 240,000 to 80,000 years ago. Mr. Croll believes the period of the boulder clay to be the latter epoch, and the Upper Miocene period to be the former epoch. By referring to the drainage area and deposit brought down by the Mississippi, it appears that the continent of America is lowered one foot by denudation in about 1,388 years, at which rate it would be lowered about 500 ft. in 700,000 years. But there is warrant for such a conclusion in the observed geological phenomena. When we arrive at the Miocene forma-

tions, the marine shells are found to differ almost wholly from
those now existing, whereas in the glacier or boulder clay formation only about five per cent. of them differ from forms now
existing. If we suppose the glacial period to have occurred a
million years ago, then, if we suppose the transformation of
species to be pretty uniform, about twenty millions of years
must have elapsed since the time of the Lower Miocene period,
and, by a similar mode of computation, we should infer that
sixty millions of years must have elapsed since the Eocene period,
one hundred and sixty millions of years since the Carboniferous
period, and two hundred and forty millions of years since the
beginning of the Cambrian period. But there are good grounds
for believing that our globe cannot be two hundred and forty
millions of years old; one of which grounds is, that the sun could
not have lasted for that time without being burnt out. It is found
that the heat which falls on a square foot of the earth's surface
exposed to the vertical rays of the sun is equivalent to 83·4 foot-
pounds per second; or about 7,000-horse power per second is
radiated from every square foot of the sun's surface. If the sun
consisted of coal, it would be burnt up in 5,000 years. But a
pound weight falling to the sun from an infinite distance would
produce 6,000 times the heat generated by a pound of coal. If
we suppose the sun to have been originally a nebulous mass
filling the whole solar system and an indefinite space beyond it,
the total amount of heat produced by the gravitation of the particles into a condensed globe would suffice to maintain the sun's
heat for over twenty millions of years, even supposing the particles to be quite cold. But if we suppose them to be very hot,
the heat generated would suffice for over three times this period.
Such a heat could be generated by the collision of two great
globes like the sun.

———

ORGANIC CONTENTS OF MINERAL VEINS.

Mr. CHAS. MOORE for some years has been paying attention to
the curious fact that most, if not all, our Mineral Veins contain
organic remains; and that he had found 115 species alone at the
bottom of a lead mine on the Mendips, including not only fresh-
water but land shells, and that it could be proved by their presence that the minerals belonged to the lias and not to the older
rocks in the veins of which they were found. During the past
year he examined 134 different samples from the mines of the
north of England, in 80 of which he had obtained specimens. A
more lengthened examination was needed before he could arrive
at precise conclusions as to the age of some of the Yorkshire and
Cumberland mines; but as he had found some seeds—Flamingites
gracilis in one of them—which had hitherto only been found in
the coal measures, he concluded in this instance they must be at
least as young as that period, as in the case of the Mendip mine.

o

He had under these peculiar circumstances found several genera of fresh-water shells, including valvata, in considerable abundance, from which he concluded a land area could not be far distant from the vein during their formation. These were mixed with a marine fauna, including several genera of fishes, viz., petalodus, eteuoptychius, squaloraia, hybodus, and acrodus, together with numerous species of brachiopoda, entomostraca, &c., and for the first time a beautiful nummuline-like species of foraminifera, together with the genera Nodosaria, Cristellaria, Dentalina and Ilotalina. The author concluded by saying the operation for the discovery of these remains was very difficult, owing to the minute examination needed, and to the generally intractable character of the vein-stuff yielding them.—*Proceedings of the British Association.*

ENCROACHMENT OF THE SEA.

An American paper notices the wearing away of the coast of New Jersey by the action of the Sea. It appears that the dimensions of many farms have been seriously affected, and the men are living who used to plough lands which now cannot be found. It is stated that the Seven-Mile Beach, opposite Scavile, has worn away 100 yards in the last twenty years. Dennis Creek is said to have lost more than a mile of its length by the wearing away of the marsh at its mouth in the last seventy years. The tide is found to be rising to higher points upon the land than formerly, and the salt grass is killing out the fresh grass and timber. Numbers of farmers along the sea-shore of Cape May can point to pieces of land which were covered with timber when they came into possession of the land, but are now covered with marsh, and the timber has been killed out. Where the marsh abuts upon the upland, fallen timber is often found buried, and the stumps of trees are seen standing with their roots in the ground where they originally grew. Large numbers of stumps of pine, cedar, and other durable woods are seen standing in the waters. In digging a ditch through a tide pond, magnolia and huckleberry roots were found under the mud. Then, after 4 ft. more of mud, large pine stumps were found, while cedar snags were found 4 ft. and 5 ft. under the pine. They were standing with 4 ft. and 5 ft. of water above them at low water. Other facts and cases are cited showing the sinking of the coast of this State below the ocean. The whole amount of this subsidence is supposed to be 17 ft. or more, and it is calculated that it proceeds at the rate of 2 ft. in a century, or about a quarter of an inch a year. This may seem slow, but when it is recollected that the major portion of the southern part of the State has but little elevation above the level of the ocean, it will be perceived that great changes may occur as the subsidence proceeds.

A DEEP SEA DREDGING IN THE GULF STREAM.

At the meeting in August of the United States National Academy of Sciences, a paper was read by Count Purtales, who has recently been employed by the Coast Survey to Dredge the bottom of the ocean along the course of the Gulf Stream, in parallel lines, crossing the current, the lines being somewhere about ten miles apart. In starting south-easterly from Florida, he found the bottom for four or five miles made up of the common coral sand of that neighbourhood, with very scanty traces of life. The next area, from 90 to 300 fathoms, and the first part of the way forming a plateau, is a rocky floor made of very hard limestone, derived from living shells. Life was abundant, consisting of lamp-shells, starfishes, crustaceans, and molluscs generally. There were also many bones of the manatoo, a dolphin-like animal, found in shallow water. The third area was the regular and common ocean bottom from 250 to 800 fathoms, covered by the chalky remains of foraminifera. —those minute animals found several years since on the telegraphic plateau in the North Atlantic. He also exhibited a map of the bottom of the ocean off the coast, and found, first, extending from the north of Florida to Montauk Point, near Block Island, Rhode Island, a bottom of siliceous sand, perhaps 100 miles wide. Outside of it was a calcareous bottom, occupying the whole area south of Georgia. Between the two of the Carolinas, is a limited deposit of green sand, containing the foraminifera. A letter was read from Professor Agassiz, warmly eulogising Pourtales' papers, and saying that he had solicited the honour of publishing the maps and other results in one of the volumes issued at the Museum of Comparative Zoology in Cambridge, Massachusetts. It opens, he said, an entirely new chapter in natural history. It disclosed what had never been before known, the various fauna at the bottom of the ocean. Among the animals obtained were some that had been extinct since the cretaceous and tertiary periods.—*New York Times.*

DEPTHS OF THE OCEAN.

Facts are multiplying in disproof of the late Professor Edward Forbes's theory that the Depths of Ocean would be found devoid of life and colour. More than fifty years ago, General Sabine witnessed the bringing up of a living star-fish, of large size, from a depth of 800 fathoms in Baffin's Bay, and since then, other evidence, similar in kind, has been brought forward by Dr. Wallich and other investigators. The latest facts bearing on the question, gathered by Dr. Carpenter and Dr. Wyville Thomson, have been laid before the Royal Society, when Mr. Gwyn Jeffreys stated, in confirmation, that shells brought up from his deepest dredgings off the Shetlands, were as brightly coloured at those found in shallow waters. It appears, indeed, that so far from being lifeless, the deep-sea bottom teems with animal life and with creatures of a very remarkable kind, some of which connect long-past geological periods with the geological action going on in our own time. The high importance of this fact will be manifest to all who have studied the question from a scientific point of view. And now fresh corroboration comes from abroad. The soundings and dredgings under the United States' Coast Survey were resumed last year; and though, through interruption by yellow fever, the season was but a short one, the fact was clearly ascertained that, in the sea between Key West and Havana, "animal life exists at great depths in as great diversity, and as great an abundance, as in shallow water." And in the exploration in the last year the dredge brought up, from a depth of 517 fathoms, "a very handsome Mopsea, a crab, an Ophiurian, and some annelids."—*Athenæum.*

STATISTICS OF COAL.

An interesting Blue-book has been issued, containing Reports from Her Majesty's Secretaries of Embassy and Legation respecting the production of Coal in different countries. According to these Reports, the production of coal in Belgium, in 1866, from 286 mines, was 12,774,662 tons; the quantity exported in the year was 3,938,768 tons, nearly all of which was sent to France. With reference to the exhaustion of the coal mines, a subject to which public attention has been directed in Belgium, it appears that in Hainault alone, of a coal-producing surface of 54,173 hectares, only 23,423 hectares had been explored in 1860. It is estimated that there were about 4,700 millions of tons yet to be worked at an easily workable depth, and the exhaustion of the Hainault coal-fields above a depth of 1,000 metres would not take place before the expiration of a century and a half. In Brazil, large coal-fields have been discovered in the province of St. Catherine's. In China, coal has been discovered at Ponghou, the chief island of the Pescadores. It is reported that no coal useful for steam purposes had yet been found; a judicious miner, however, could alone settle the question as to the extent of these mines and the quality of the coal. At Iwanai, in the Island of Yeddo, in Japan, coal-mines had been discovered. An experiment was made with some of the coal picked out from the surface of the seams, in the galley-fire of Her Majesty's ship *Salamis*; 79 lb. of coal yielded 17·27 per cent. of ash, 1·5 per cent. of clinker, an average volume of smoke, and a strong durable flame. Another coal-field was found at Yeddo, in the immediate vicinity of the port of Hiogo. The natives had been working it for the last ten years, but not continuously. Prussia, as is well known, is rich in mineral fuel, especially in very good coals. The quantity of coal to be obtained by the working of the coal-pit of the river Saar would suffice for the supply of 3,000 years, at the rate of 2,500,000 metrical tons per annum. The coal-pits of the River Ruhr extend over ten miles in length on the Lower Rhine, a Prussian mile being equal to 24,001 Prussian feet, nearly 4⅓rds English miles. There were 65 strata of coal more than 20 inches deep, the united thickness of which gives a pure coal 210 feet. It has been estimated that the produce of those pits will last more than 5,000 years, at the rate of 1,000,000 metrical tonnen per annum. In 1865 there were 409 pits at work in Prussia, producing 371,812,299 centners of coal, value £4,954,986; they gave employment to 89,192 persons. Of the 409 pits in work, 393 were in possession of the companies and private persons, and 16 belonged to the State. Of the coal sold, 28 per cent. went to the interior, 22 per cent. to the States of the Zollverein, 45 per cent. to France, and 5 per cent. to Switzerland. Hanover possessed 33 coal-pits. The more considerable fields of brown coal were in the provinces of Saxony and Brandenburg. In 1865, there were 511 of these pits at work, producing £710,437.

An Appendix to the Consular Reports shows that in Tasmania, workings have been opened in the Douglas River coal-field. Coal of good quality for steam purposes has been discovered on the east coast of South Brani Island, at Adventure Bay; and a bituminous coal of fair quality has been discovered near Hamilton. Coal deposits are reported in Trinidad; the finest quality was found at Point Noir; it burnt rapidly with much flame and little smoke. A Report by Mr. Oldham, Superintendent of the Geological Survey of India, shows that the British territories cannot be considered as either largely or widely supplied with coal. Extensive fields existed, but they were not distributed generally over the districts of the Indian Empire. Up to the present time little more than surface workings had been carried on. Specimens of coal from 74 localities showed that the average composition per cent. was, fixed carbon 52·2, volatile matter 31·9, and ash 16·5, against an average composition of five English specimens, of fixed carbon 68·1, volatile matter 29·2, and ash 2·7. He states that the very best coal of the Indian fields only touches the average of English coals, and that Indian coals are not capable of more than two-thirds, in most cases not more than one-half, the duty of English coals. These results of the quality of Indian coals would show the groundless nature of the hopes which have been expressed that the coal-fields of India, Borneo, Australia, and New Zealand, would not only contribute large supplies, but would also serve to coal the ocean steamers trading between Europe and those far distant regions. As far as Indian coal was concerned, Mr. Oldham feared it would never supplant the better fuel now obtainable elsewhere for ocean voyages.— *Times.*

HÆMATITE IRON ORE IN SCOTLAND.

THOUGH it has long been known or suspected that Ironstone was to be met with in the volcanic rocks that form the greater part of the Garleton Hills, it is only within the last twelve months that its existence in rich abundance has been conclusively demonstrated. The discovery promises to be most important. Mr. Aitchison, the new tenant of West Garleton, within the boundary of which farm the hill in question is situated, while getting the soil turned over, was struck with the frequency with which the plough brought fragments of ironstone to the surface, as if broken off from larger masses beneath. Knowing something of the value of the ore, he recommended the proprietor of the land, Sir T. B. Hepburn, of Smeaton, to make a systematic examination of the hill to test the character of the underlying rocks. Mr. Robb, the local geologist, on being consulted, strongly urged that the ground should be at once examined. This was done, and after breaking ground without success at several points, the search was at length rewarded by the workmen coming on a vein or veins of Hæmatite Iron, of the richest and most valuable character. The vein, where at present struck, crosses the hill in a slanting direction from the south-east to the north-west, and has been traced nearly along the whole breadth of the hill, a distance of some four or five hundred yards. In some places the ore is met with in small boulders, interspersed with the ordinary trap rock of the hill, and in others it is met with in a solid mass of as yet unknown depth. This occurs on the crest of the slope, the iron being within five or six inches of the surface, and penetrating downwards in a solid bed. The workmen proceeded to uncover this section of the vein to the depth of seven feet, without any appearance of its termination; and as veins of a similar character are in all probability to be met with in immediate proximity, there cannot fail, at the most moderate calculation, to be many thousand tons of the ore. Hæmatite iron, singular to say, has nowhere else been found in Scotland, the well-known " blackband " being the only description of the ore hitherto met north of the Tweed.—*Haddingtonshire Courier.*

CARRARA MARBLE AND ALABASTER.

THE quarries of Carrara produce, besides the statuary Marble for which the district is famous, another description of white marble, and a bluish marble, called Bardiglio. Mr. Herries reports that in the course of three years, 1863-65, there were exported from Carrara 126,928 tons of marble, the average annual value of which was upwards of a million francs. The export and other duties levied by the State amounted to about 80,000f. per annum. 2,238 persons, or about one-seventh of the whole population of the district, were employed in working the marble.

There is a greater variety of marble in the adjoining district of Massa, but the elevated position of the quarries, and the want of transport, prevent them from being completely worked. Oriental Alabaster is found in the Siena country. In the neighbourhood of Voltena there are 29 quarries, which yield annually 680,000 kilogrammes of alabaster of different colours and qualities. The alabaster statuettes, vases, and ornaments of all sorts that are sent from Italy to all parts of the world are made principally at Voltena.—*Mechanics' Magazine.*

DIAMONDS AT THE CAPE.

PROFESSOR TENNANT has made to the British Association a communication " On the Recent Discovery of Diamonds in the Cape Colony." This gem, he stated, had been recently found somewhat abundantly in the above district ; and he exhibited the casts of some weighing nine carats, worth £300. Some agate, chalcedony, and other precious stones found in the same deposit had been sent him, but he would have preferred some of the sand and mud in which they were deposited. One diamond, found very recently, weighed as much as fifteen and a half carats. He was of opinion that before long we should have a large collection of diamonds from the Cape Colony, adding that, although we had heard a great deal of diamonds being found in Australia, those stones were not worth now so many pence as pounds had been asked for them.

KENT'S CAVERN.

MR. W. PENGELLY has made the Fourth Report to the British Association " On the Exploration of Kent's Cavern, Devonshire, the Lecture Hall and South West Chamber." The rich cave-earth of unknown depth was completely sealed up with the stalagmitic floor, which, in its turn, was covered with a layer of black mould ; the objects found in the overlying black mould were less numerous than in similar accumulations of former Reports. Amongst them were several pieces of pottery, a spindle-wheel, a roughly cut piece of red sandstone, portion of a bone comb, small red earthenware pan, marine shells, small piece of smelted copper, the entire lower jaw and almost complete skull of a badger, part of a human upper jaw with eight teeth, four of which remained in their sockets, and the internal cast of a fossil shell. All these articles were described in detail in the Report, illustrated with the specimen themselves, all ranged by Mr. Pengelly in drawers which severally represented the black earth, stalagmite, and the red cave-earth, in which was intermingled blocks of stalagmite, breccia, &c. He then enumerated the articles found in the second floor, or stalagmite, which included a fine molar of the rhinoceros, the pre-molar of a hyæna, two or three molars of a bear, a large part of the humerus, probably of the bear, &c. Since the time of the rhinoceros, the increase of thickness of stalagmitic floor had been barely sufficient to cover these relics. A few relics of charred wood were found in the same floor. Passing to the cave-earth, report said that it was of the ordinary typical character. A description was then given of the fossils in the cave-earth, and of those entombed in the breccia mingled therewith. While split bones bearing teeth-marks had been found in the red cave-earth, no split bones had been found in rocky breccia. The bones scored with teeth had probably been gnawed by the hyæna, and the split bone that was found was an evidence of the presence of man. A long explanation was given respecting the breccia, which was rolled by some means into the caverns, and the Report stated that it was probably older than red cave-earth, and though it had not given up any charred wood or longitudinally split bones, yet the com-mittee thought it would be premature to draw at present any inference from this negative fact. But while the labours of the past twelve months had not added anything to the knowledge of the early antiquity of man, yet up to this time no comparatively modern objects had been found below its place, and no ancient one had been found within the modern niche. The lower floor of the stalagmite had kept the two apart. Probably ancient cave-men made use of unpolished flint implements, of which spe-cimens had been found in the red cave-earth, to split the bones of animals, employed fire in the preparation of food, and selected stones for crushers or hammers.

Mr. Pengelly stated that bones in four different states of con-dition were found in the above cavern—namely, entire, crushed, fractured, and split. The crushed bones were due to blocks of breccia and stalactite falling upon them; to hyænas he attributed the fracturing of others, in corroboration of which he produced bones fractured by hyænas in the Zoological Gardens, that pre-sented exactly the same appearance; whereas he concluded that man split up the bones to obtain marrow, and what recent bones had been split up by the like agency had a precisely similar aspect. This paper gave rise to a discussion upon the Report.

Mr. J. Evans was of opinion that, taking the evidence given by the bones themselves, they must have been split and gnawed by animals. With regard to what had been suggested by Mr. Pengelly, of man's existence being synchronous with that of the hyæna and cave-bear, he thought it hardly likely that, even in so remote a period, such a "happy family" could have lived together as was inferred. Mr. Evans gave a description of the flint flakes and tools found in these caverns, and referred to the molecular changes which took place in block flints so as to make them actually soft. He has found flints in the Reading and Woolwich series, which he could cut with a knife—a fact of great importance to geological chemists.

CAVE FAUNA OF ENGLAND AND THE CONTINENT.

THE late Dr. Falconer, the Palæontologist, was necessarily led in the course of his researches to the study of what is termed the Cave Fauna of England and the Continent, upon which numerous communications and valuable scattered notes were either published by himself or have been found among his papers. He was the first to establish as members of this fauna *Rhinoceros hemitæchus* and *Elephas antiquus*; and among the first to recognise the validity of the evidence upon which the contemporaneousness of man with several extinct forms of animals has been based and has since been firmly established. He was the originator, and one of the most active promoters, of the exploration of the Brixham Cavern. He also made extensive explorations and discoveries tending to the same end in Sicily, where, in the Grotta di Maccagnone, he found stone implements of ancient type, commingled with the remains of hyæna, hippopotamus, and other animals long extinct, at any rate in Europe. On his outward route, also, on this visit to the South, he visited the extraordinary collection of flint implements which had been brought together with such indomitable zeal and energy by M. Boucher des Perthes, from the ancient river gravels of the Valley of the Somme, and was perhaps the first eminent authority to recognize the importance in the primeval history of man of M. des Perthes' discoveries, the value of which has since been universally recognized, and which have deservedly crowned him with lasting fame.

Mr. Falconer must thus be reckoned among the foremost, if not the foremost, of those who have within the last few years given a new impulse to the study of the antiquity of the human race, a subject which before 1859 had scarcely attracted attention, or, as in the case of the Kent's Hole Cavern in the time of Dr. Buckland, was even shunned as an unholy thing.

Dr. Falconer took the deepest interest in the important researches of Captain Frederick Brome in the ossiferous caves and fissures in Gibraltar, and with the view of making himself fully acquainted with the site of these extraordinary explorations, he proceeded, in company with Mr. Busk, and on the invitation of the Governor, Sir W. Codrington, to Gibraltar in the autumn of 1864. A brief account of the caverns and their contents was given to the Governor in a letter since published in the *Journal of the Geological Society.—Review, in the Times.*

THE IGUANODON.

A PAPER has been read to the British Association "On the Skull and Bones of the Iguanodon," by the Rev. W. Fox. The paper showed from these bones, which had lately been discovered, more especially the skull of a small individual, that there were several species of this genus of saurians. The Iguanodon, which was perhaps as well understood as any of the extinct reptiles, had always created much interest among naturalists. The jaw-bone of a young animal of this genus had been discovered by the author in the Wealden strata of the Isle of Wight. Only a few months ago he discovered the skull-bone of a new and small species. The author detailed at some length the anatomical structure of the teeth and bones; and, in conclusion, expressed his belief that the skull in question belonged to a new species, and that future researches would disclose more individuals.

Professor Huxley, who thanked Mr. Fox for his important contribution, explained that all fossil saurians hitherto discovered were defective of the fore-paw and end of the snout. By the aid of the black-board he showed what an important addition had been made to our knowledge of the structure of the skull of this Dinosaurian, more especially in its showing almost as clearly the structure of the skull as would that of the modern lizard show to which reptile it belonged.

THE SEWALIK FOSSILS.

ALTHOUGH Dr. Falconer was the first to determine the tertiary age of the Sewalik-hills, which had previously been referred to the "New Red Sandstone" period, he was not led to this determination from the study of any animal remains, for none had then been found in the strata. Nevertheless, he expressed the confident opinion "that the remains of mastodon and other large extinct mammalia would be found." How amply this prediction has been fulfilled is well shown in the vast collection of Sewalik fossils in the British Museum, presented by Captain (now Sir) Proby T. Cautley and Dr. Falconer, by whose laborious exertions they have been displayed, and in great part figured and described, in the *Fauna Antiqua Sivalensis*, which forms the contents of the first volume. Among these, and not the least remarkable, is the gigantic *Colossochelys Atlas* —a monstrous tortoise, upon which, as suggested by Dr. Falconer, it is not improbable that the Hindoo myth of the world being supported on the back of a tortoise might have been founded. There seems, at any rate, reason to believe that this now extinct form may in the Miocene period have co-existed with man, seeing, as Dr. Falconer has shown, that several existing species of tortoise are found associated with it in the fossil state. —*Review, in the Times.*

NEW FOSSILS FROM SWEDEN.

PROFESSOR OTTO TORELL has exhibited to the British Association a series of slabs marked by the impressions of various land plants, known to geologists as *chondrites*. He had these Fossil plants from a formation *much older than any from which fossils have hitherto been obtained.* The rocks from which they were derived were of an age similar to those of the Longmynd rocks in Wales. The undersides of many of the slabs were pitted with the marking of rain drops, and the conclusion which Professor Torrell came to was, that the character of the plants and the meteorological markings of them indicated that they had been deposited under shallow water conditions. This he corroborated by showing that a bed of shingle or conglomerate was associated with them, which he judged to have been part of an old sea-beach. The same slabs were marked by the trails of marine worms that had crawled over them. The casts of some of the worms were distinctly to be seen on the surface of the slabs. The slabs were handed round and scrutinized by many geologists present.

Sir Charles Lyell described the facts stated by Mr. Torrell as one of the greatest discoveries that he remembered to have been made in the progress of geology in his time. Mr. Torrell had long been known as an explorer of the northern regions. He had undertaken four voyages, and had spent four summers in Spitzbergen and other parts of the northern regions. No one was more thoroughly acquainted with the Fauna of the seas of those countries, or had thrown more light on the geology of the glacial period than Mr. Torrell. But his knowledge was not confined to modern deposits; they would see by his paper that he had been exploring the most ancient rocks that had furnished us with fossil Flora. This discovery of the earliest land plants yet brought to light had been made in a district where the geology was so clear that the merest tyro might read it off. It was impossible that any mistake could be made as to the chronological order and succession of beds. Older than rocks which exist in the Alps, they presented themselves in the form of inclined and inverted strata; they were really at the basis of the Cambrian system, or the primordial rocks of Barrow, and these rocks were older than the oldest rocks of Murchison's Silurian formation. Sir C. Lyell proceeded to explain the great antiquity of the rocks in which these fossil plants had been found.

FOSSIL PLANTS IN GREENLAND AND SPITZBERGEN.

SIR C. LYELL, at the late Meeting of the British Association, called attention to the Fossil Plants collected by Mr. Whymper in Greenland. Mr. Whymper had undertaken to make the collection at the instance of the British Association and the Royal Society, whose curiosity had been excited in consequence of the remarkable proofs that had been furnished, by plants previously found in Greenland and Spitzbergen, of a much higher temperature having existed in Miocene times throughout the north polar region than at the present time. Mr. Whymper had faithfully and successfully executed that commission, and had brought

home a rich collection of plants, which was now in the hands of
Professor Heer. There were, besides, proofs of a much colder
climate than now exists having intervened since this ancient
warmer period and the present day. This was during the glacial
period, when Norfolk and Suffolk were visited by icebergs,
covered in part by land ice, brought down by northern regions,
where glaciers were much more extensive than they are now.
The shells of this period now found fossil in Norfolk and Suffolk
were decidedly of an Arctic character. Mr. Whymper had not
brought home many new species of fossil plants, but he had
brought home fruits of species of which previously only the
leaves were known, thus enabling Professor Heer to confirm his
guesses as to the true character of the plants. Important and
satisfactory conclusions had thus been arrived at with regard to
several of the fossil species: for example, the magnolia, which
can now be affirmed to have grown there in Miocene times. The
existence of an oak with leaves 6 in. long, of the vine, plane, and
other trees implied great heat in summer, while the number of
dicotyledons showed that there could have been no great cold in
winter. With reference to the causes of these great changes of
climate, he adhered still to the opinion that the principal cause
was the altered distribution of land and sea, and the consequent
alteration in the direction of the marine currents between the
equatorial and the polar regions.

Professor Rolleston said that an examination which he had
made of sixteen skulls brought home by Mr. Whymper proved
that the Greenlanders belonged to the Esquimaux race, and had
no affinity with the Norsemen. They were the skulls of a savage
people, and remarkable for the tendency to obliteration of the
sutures, as in the skulls of carnivorous animals. The form
showed that there was no ethnological connexion between the
Greenlander and the Red Indian.

FOSSIL OYSTERS.

Mr. WHITTLE, of Chorley, has sunk a new shaft down the
Arley seam of coal, between the Addington and Horwich stations,
on the Lancashire and Yorkshire Railway, about a couple of
miles from the foot of Rivington Pike. Two seams of coal being
passed, at a depth of 130 yards the sinkers cut through a
bed of Fossil Oysters 2 ft. 4 in. in thickness. How far the bed
extends it is impossible to say. The oysters are petrified into
one solid mass as hard as flint, are all perfect in form, and small
in size, rather less perhaps than the London natives. The conclusion which immediately suggests itself is that the sea must at
some very remote period have washed the foot of the Rivington
range of hills, two miles distant. The whole of West Lancashire
is alluvial land, and at one time was covered with a forest of
oak, there being abundant proof of this in the fact that trees are

frequently found embedded in the moss, and also in the bed of
the Ribble. The skull and antlers of the gigantic Irish elk, found
not long ago in this river, also point to the fact that animals of
the mammoth tribe must have roamed through the forest which
covered the country at a period since the oysters were embedded,
and the upper coal strata formed.

FOSSIL ECHIDNA.

An extract has been read to the Zoological Society, from Mr.
G. Krefft, Sydney, stating that, amongst other fossil remains
which he was now arranging for the Australian Museum, he had
discovered a portion of the humerus of an extinct species of
Echidna, from the Darling Downs, indicating the former existence
of a gigantic form of this Monotreme in Australia.

THE SUBSIDENCE OF HAWAII.

The progress of the Subsidence of Hawaii, after the great
submarine eruption which took place some months previously,
has been gradual, the depression of the southern and western
shores amounting to several feet ; while that of the northern and
eastern shores has not exceeded a few inches. It is important
to attend to this peculiarity. Had the change of the sea-level
been uniform in all parts of the island, we might have been
doubtful as to the seat and nature of the disturbance. The
gradual elevation of a widely extended portion of the bed of the
Pacific Ocean would, in that case, have accounted as satisfactorily for the diminution in the height of Hawaii above the sea
level as the sinking of the island itself. But the inequality in
the motion of subsidence points at once to the real character of
the disturbance, which doubtless only affects Hawaii and its
immediate neighbourhood. We do not think, however, that the
sinking of the island has been a gradual process. Far more
probably it took place simultaneously with the outburst of submarine fire which caused so much mischief last April. The
manner in which the Honolulu correspondent of the *New
York Tribune* refers to the discovery of the change does not certainly convey the notion that any slow process of change has
been detected. " It is believed," he writes, " that Hawaii is
slowly sinking into the ocean. Ever since June last unusually
high tides have prevailed along the southern and eastern shores,
and it is now evident that the island has sunk," &c. It appears
to us that the inhabitants do not fear that any further sinking
will take place, unless—which seems unlikely—there should presently be other submarine explosions.

Very interesting also is the account of the waves which spread
across to Hawaii from Peru, after the great earthquake of
August 13-16 last. " The first of these waves was observed at

Honolulu on the night of the 13th of August, and at almost the same moment it was noticed 200 miles to the south-east. They were more powerful at those parts of Hawaii which are nearest to the South American coast. They rolled in at the rate of three or four per hour for four days. They were not like the sweep of the furious breakers that lashed the shores during the April eruptions, but appeared to be the effects of some gigantic oscillation across the Pacific." There can be no doubt whatever that these waves were due to the great earthquake which recently desolated the coast of Peru and Ecuador. The waves must have been transmitted across the whole distance which separates Hawaii from Peru in the course of a few hours, or probably at the rate of about 1,000 miles per hour. This does not exactly agree with the statement that the wave was noticed almost at the same moment at two places 200 miles apart (and lying nearly in the direction of the waves' progress); but probably there is some mistake in this account. The difference of time must have been nearly a quarter of an hour.

It would be important if we could determine the exact rate at which sea-waves traverse the Pacific Ocean, since it has been shown that a relation exists between the depth of an ocean and the rate at which waves of given dimensions traverse it. We say of given dimensions, but it is not necessary that all the dimensions of a wave should be known in order that its ascertained velocity should suffice for the determination of the depth of the ocean across which it has been propagated. All that is necessary is that the breadth of the wave should be known—that element being measured from summit to summit of successive rollers. The following Table is derived from one calculated by Mr. Airy, the Astronomer-Royal. It presents the relation between the breadth of sea-waves, their velocity, and the depth of the sea they are traversing :

Depth of water in feet.	Breadth of the wave in feet.				
	1,000	10,000	100,000	1,000,000	10,000,000
	Corresponding velocity of wave per hour in miles.				
1	3·88	3·80	3·80	3·80	3·86
10	12·21	12·22	12·22	12·22	12·22
100	38·40	34·64	38·66	34·66	38·68
1,000	48·77	115·11	122·18	122·27	122·27
10,000	48·77	154·25	364·92	386·40	346·84
100,000	48·77	154·25	487·79	1151·11	1223·70

A little consideration of this table will show that the velocity alone affords an important indication of the depth of the ocean traversed by a wave ; because in many instances the change due to the breadth of the wave is slight. Thus, when the depth of a sea is 1,000 ft., waves of 100,000, of 1,000,000, and of 10,000,000 ft. in breadth, all travel at the rate of a fraction over 122 miles per hour.

An earthquake which occurred in the year 1854 supplied an opportunity of applying Mr. Airy's formula. A shock of earthquake had been felt on board the Russian frigate *Diana* at 9.45 a.m. on the 23rd of December in that year. She lay at anchor in the harbour of Simoda, near Jeddo. At 10 o'clock a large wave was seen rolling into the harbour, while the water on the beach was rapidly sinking. "As seen from the frigate," says the narrative, "the town appeared to be sinking." Fifteen minutes later another wave rolled in, and until 2.30 p.m. similar waves continued to come and go, the frigate being thrown on her beam-ends no less than five times. A few hours afterwards several distinctly marked waves, of extraordinary volume, arrived at San Francisco. No doubt can exist that these waves and those which destroyed Simoda had the same origin. In all probability the seat of disturbance lay very near the shores of Japan. Now, it was calculated by Captain Maury that the San Francisco wave had a breadth of 256 miles and a velocity of 438 miles per hour; and that the San Diego wave had a breadth of 221 miles and a velocity of 427 miles per hour. It would follow from these premises that, according to Airy's formula, the average depth of the North Pacific between Japan and California is, by the path of the San Francisco wave, 2,149 fathoms, by that of the San Diego wave, 2,034 fathoms. Either result indicates an average depth of about two-and-a-half miles.

The rate at which the great waves of disturbance produced by the earthquake of Peru traversed the portion of the Pacific Ocean which lies between South America and the island of Hawaii, would indicate that the ocean is here much deeper than it is farther north. A similar result follows from the evidence we have respecting the rapid passage of the great wave produced by the submarine eruption of April 20th, from Hawaii to the shores of Mexico and Oregon. In that case a distance of 5,000 miles was traversed in little more than five hours. Probably the average depth of the part of the Pacific which lies between Polynesia and South America is little less than four or five miles.—*Daily News*

VOLCANIC DISTURBANCE AT SANTORIN.

At the Geological Society, M. Lenormand has referred to the discovery investigated by the French Commission during the late Volcanic Disturbance at Santorin. Santorin forms a group representing the *debris* of a great volcanic cone, and as it yields largely the hydraulic cement called puzzalona, extensive quarries, or excavations, have taken place, more particularly of late years, in consequence of the supplies required for the Suez Canal. These deposits belong to the period before the first eruption of Santorian took place, and are therefore pre-historic. It being announced to the French Commission that remains of buildings had been discovered, on proceeding to the spot its members found numerous walls of buildings with their contents, which had been buried like Herculaneum and Pompeii, and so preserved. The inhabitants were found to have possessed houses with walls of masonry, and the roofs of which had been supported with wild olive-wood. This is remarkable, as, in consequence of the volcanic eruptions, the olive can no longer grow on the surface of the island. The utensils and food of the people were found, stone mortars and coarse pottery, much resembling those of the Lake-village Period, but with knives and other implements of obsidian of the most exquisite finish, and of all of which specimens are now deposited in the Imperial Museum. A more remarkable feature was the presence of finer pottery, evidently imported, and which had been made with the wheel, and some gold beads, not of molten metal, but apparently hammered into shape with stone tools. M. Lenormand considers the pottery was imported in return for Santorian oil, the produce of the island; but it is doubtful whether the wild olive would be very productive. It is as likely the pottery was derived from the return cargoes of fishermen. M. Lenormand considers it remarkable that no metal tools were imported. The gold can hardly be considered the produce of the island, but of the main land, from the regions afterwards known by the Gold-sands of the Pactolus. The real interest of the discovery depends not so much on obtaining further evidence of the continuance of rude populations, as in the parallel fact of the carrying back, and particularly in Asia Minor, of the epoch of comparative civilization denoted by the improved pottery and the gold ornaments.

SUBMARINE EARTHQUAKES.

THE bark *Euphrosyne*, Captain Christie, arrived at St. Helena, and the Captain reports that on the 9th of October, when in lat. 26 36 S., long. 52 32 E., he experienced strong gales, with thunder and lightning, and a tremendous sea. The barometer fluctuated greatly. This weather continued for some days, and then moderated, but at midnight on the 8th—9th of November, in lat. 16 40 S., long. 4 W., the sky suddenly became densely overcast, a noise like distant cannonading ensued, the sea became greatly agitated, and the compass became erratic, meteors became frequent, and the ship shook from stem to stern, as if in convulsion. Then large fish leapt out of the water, and the rumbling increased. The agitations continued till sunrise, and then the weather became settled again. Captain Christie thinks that he must have been passing over some great submarine convulsion.

EARTHQUAKES IN ENGLAND.

SHORTLY after half-past 10 on Friday night, October 30, a distinct shock of Earthquake was felt in many places in the West of England, and in South Wales. The unusual sensation lasted for about a couple of seconds, and is presumed to have been occasioned by what is termed an earth-wave. Five years ago a similar shock occurred, which was felt throughout a considerable part of the West of England.

The night on which the shock occurred was cloudy but light, the moon being nearly at the full. Taking the various accounts that have appeared from observers of the phenomenon, we find its indications variously described. Some describe it as a trembling of the earth, as if a laden waggon was passing along the street; others imagined some animal had got in their bed-room and was scampering about the floor. The motion of articles of furniture, the jingling of glasses, and the oscillation of beds, were very common. Most of the reports agree that these indications were accompanied by a noise like that of a high wind, or of a vehicle or train passing. Reports from towns and villages westward of Worcester describe the shock in very similar terms. In Leamington, three shocks followed each other in rapid succession. The shocks were noticed by Dr. O'Callaghan, the President of the local Philosophical Society, who, during a residence in the West Indies, had on more than one occasion felt the shock of earthquakes. On Friday night, Dr. O'Callaghan called attention to the occurrence, and found that the shocks had been observed by some of his domestics. At Exeter, many persons noticed the earthquake. Three distinct shocks were felt by the inmates of a house in York-buildings. The occupant of a house, who has long resided in the West Indies, recognized the shaking as that of an earthquake. The inmates of the Exeter Eye Infirmary felt the building shake, and experienced a peculiar sensation.

Reports from Merthyr state the shock to have been rather severe, the movement appearing to be from south to north and continuing for three or four seconds. A few minutes before the occurrence there was that strange calm prevalent which has been noticed during the time of a complete eclipse. Great alarm was felt by the inhabitants. At Twynrodyn, and other high quarters, the people rushed out of their houses, some attributing the shock to explosions underneath in some coal or mine pits; but Merthyr is not undermined, and no mine or coal stratum exists there; this fact was conclusive that the disturbance proceeded from a greater and mysterious cause. We will conclude with a report from a meteorologist of some years' standing, who writes from Port Talbot, South Wales. He says:—"The night was wet, with a drizzling rain, and a strong S.W. wind blowing. Barometer stood at 29° 50'. Soon after 10 P.M. the wind abated; then followed a lull, with an overcast sky. At 10.35 P.M., as I

P

was sitting with a friend in my drawing-room, on the ground-floor, suddenly we experienced a dead, heavy thump, as if from the fall of a bale of goods close outside the window, followed, after the interval of a second, by a tremulous motion through the room, which shook the floor beneath our feet, and rattled some ornaments on a side-table, the shock passing from N.W. to S.E., and occupying about five or six seconds of time. The effect upon me and my friend was unmistakable. We at once exclaimed, 'An earthquake!' My house stands alone on a hill, with gardeners' cottages near. My servants, who were in their bedrooms, ran out to know what was the matter, and those down stairs were equally at a loss to know the cause of the noise. I find that persons who were out of doors at the time did not feel the shock."

The following account of Earthquakes that have occurred in England has appeared in the *Spectator*:—" In 974, Wendover tells us that a great earthquake shook all England. In 1081 one occurred which was attended ' with heavy bellowing.' In 1089 there was ' a mickle earth-stirring all over England,' says the *Saxon Chronicle*, and the annalist notices that the harvest was especially backward. In 1110, says Florence of Worcester, ' there was a very great earthquake at Shrewsbury. The river Trent was dried up at Nottingham from morning to the third hour of the day, so that men walked dryshod through its channel.' In 1133 there was a great earthquake in many parts of England. In 1165, says Matthew Paris, ' there was an earthquake in Ely, Norfolk, and Suffolk, so that it threw down men who were standing, and rang the bells.' The same writer records another in 1187, when many buildings were thrown down; and another in 1247, which he speaks of as especially violent, on the banks of the Thames, where it shook down many buildings. One feature of it was that some days afterwards the sea became preternaturally calm, as if the tides had ceased, and remained so for three months. Next year the western parts of England were the great sufferers. In the diocese of Bath, wide rents opened in the walls, and a cupola on the tower of Wells Cathedral was dashed down upon the roof. At St. David's, great damage was done to the Cathedral. Two years later a shock was felt in Buckinghamshire, which caused more panic than injury, the accompanying sound being like thunder underground. It was noticed that the birds were driven wild with fear. In 1275, says Matthew of Westminster, there was a general earthquake, ' by the violence of which the Church of St. Michael-of-the-Hill, outside Glastonbury, fell down, levelled with the soil.' Many other English churches suffered in a less degree. From this time earthquakes seem to have been less common; but in 1382 there was one which shook down some churches in Kent, and which a poet of the time has described rather vividly. Three years later there were two shocks but they seem to have been very slight, as Walsingham only interprets the first to mean an expedition against Scotland, and

the second a vain excitement in the political world. I have carefully confined myself to historical notices. But there are legends
which ascribe the destruction of whole cities or armies to convulsions of this kind. Camden records that the town of Kenchester
was destroyed by an earthquake. The *Chronicle of Evesham*
says the same of Alcester, but as the visitation in this case was a
special judgment on the smiths of the town, who drowned St.
Egwin's preaching with the noise of their hammers, there is
reason to hope that it was a purely local infliction. Reginald of
Durham says that at Mungedene-hill, near Norham-on-Tweed,
the earth opened and swallowed up many thousand Scots who
were then ravaging St. Cuthbert's lands. These traditions may,
perhaps, be taken to show that the popular fancy in England recognized earthquakes as an occasion of violent change. Of course
I do not pretend that my list is exhaustive even for the 12th and
13th centuries, nor have I touched upon such geological theories
as Mr. Geikie's, that there has been a great upheaval of Scotland
about the Antoine Wall, or a great recession of the sea since
Roman times."

EARTHQUAKE IN SAN FRANCISCO.

THE San Francisco *Alta California* of October 21, has the
following account of this Earthquake:—"At 7.51 o'clock this
morning the most severe earthquake which has occurred since the
occupation of California by the Americans shook our city. This
is the first earthquake that has ever caused loss of life in San
Francisco, and the amount of damage caused is unquestionably
greater than that caused by the shock of October 8, 1865. The
morning was moderately warm, and a dense fog covered the
town. There was not a trace of a breeze perceptible. The first
indication of the approach of the earthquake was a slight
rumbling sound, as of something rolling along the side-walk,
coming apparently from the direction of the ocean. Whether
this proceeded from beneath the surface of the earth, or from
the agitation of loose bodies on the surface, is uncertain. The
most general opinion appears to be that it was from the latter.
The shock commenced in the form of slow horizontal movements,
not perpendicular, as in the great earthquake of 1865. The
oscillations continued from 10 or 15 seconds, growing more rapid
and more violent for six for seven seconds, then increasing in
force and rapidity for four or five seconds, then suddenly ceasing.
The motion was so great that water was thrown over both sides
of a pail not more than two-thirds full, which was on the ground
on the summit of Russian-hill, where the shock was felt with less
force than in any other part of the city. At 8.42½ there was a
very slight shock, just perceptible. At 10.23 A.M. a third shock,

quite a sharp one, was felt, and a panic was created in the principal streets, crowds rushing from every building. At 11 A.M precisely, a fourth and very slight shock was felt. The fog cleared away and the sun shone out in a cloudless sky, while a slight breeze sprang up at that hour. The great shock of 1868 produced a wholly different effect on buildings from that of 1865. In October, 1865, glass was broken and shivered in atoms in all the lower part of the city by the perpendicular oscillation, while comparatively few walls were shaken down or badly injured. The earthquake of to-day broke very little glass in any part of the city, but the damage by the falling of cornices, awnings, and walls was immense. The shock was principally felt on ' made ground ' and the flats, where the foundation is known to be unreliable at all times. On the eastern shore of the bay, and, in fact, all the way around it, everything built on the flats has suffered severely. It is a noticeable and gratifying fact that not a single building constructed as it should be in a city liable to earthquakes like San Francisco has suffered to any extent at all. When the great shock culminated there was a stampede from every building in the city. Hundreds of horses on the streets took fright and ran away. As the minor 'tapering off' shocks were felt the excitement increased instead of diminishing, as it should have done—seeing all our experience goes to show that the worst invariably comes first in San Francisco earthquakes."

GREAT EARTHQUAKE IN SOUTH AMERICA.

A TERRIBLE Earthquake has visited the western seaboard of South America, destroying eight cities, the most important on that coast; and making its terrible influence felt northward from Arica, the central point, to Callao, a distance of about 650 miles; and southward to Cobija, in Bolivia, a distance of 280 miles or more.

According to advices from Central America, published in the New York papers, Arica, Arequipa, Ilay, Iquique, Pasco, Juan-Cavolica, Ibarra, and numerous other towns of Peru and Ecuador, were totally destroyed by a succession of earthquakes, which lasted from the 13th to the 16th of August. Most of the inhabitants of the Peruvian seaport town of Arica, and of Arequipa, chief town of the Peruvian littoral province of Arequipa, escaped with their lives.

The havoc, it is estimated, must at any rate have stretched over a distance of 1,200 miles, from Ibarra, a town of Ecuador, fifty miles to the north-north-west of Quito, the capital of that Republic, and within less than a degree of the equatorial line, down to Iquique, a seaport and island in the southern provinces of Peru, in the 20th degree of south latitude.

Arica has often been devastated; Arequipa, which is built on the slope of a volcano, has a similar history; Lima and Callao,

which have now escaped, but which lie near the line of the
convulsion, were once destroyed together, when 18,000 persons
were buried in their ruins. It is inevitable that we should con-
template such a catastrophe through the medium of our own
familiar ideas, and Humboldt has told us what it is for a
European to feel his faith shaken in that immobility of the
earth which is part of our experience and almost of our religion.
But to the natives of Peru or Ecuador the phenomenon which
seems so appalling to us must appear less terrible.

The shock appears to have been chiefly felt on that Peruvian
coast which has been correctly described as the Home of Earthquakes.   Iquique, in lat. S. 20° ; Arica, 18 ; Islay, the
port of Arica; Tacna, 18°; Arequipa, 15°; Pisco, 13° (a
little seaport town, not, we take it, the inland city of Pasco,
nor Pasto, as has been suggested), are pretty clearly specified.   Huancavelica, also mentioned, is in lat. 12°, and at
some distance inland. All these are, comparatively speaking,
within the same region. But the remaining intelligible names
are those of Tacunga and Ibarra, cities of the Republic of
Ecuador, nearly under the equinoctial line, and therefore about
a thousand miles distant from the nearest of the Peruvian places
above specified. It is a little strange that the same earthquake
should have destroyed towns so distant without any reported
effect on the vast and peopled tract between ; still more strange
that the intelligence of each separate visitation of a calamity
affecting some 1,500 miles of country, and occurring between
Aug. 13 and 18 (in a region, too, almost destitute of communications), should have converged so as to reach New York less
than a month afterwards, by Sept. 12th.

The following is a later and more correct account :—In Peru
an earthquake has swept Iquique into the sea. The ports of
Mejillones, Pisagua, Arica, Ilo, and Chala have shared its fate
Arequipa, the second city of the Republic, is levelled to the ground,
and Cerro de Pasco is demolished. In the table-land of Ecuador,
which is a sort of monster bubble blown up by Cotopaxi and its
sister cones, not less than 20,000 persons have been killed.
Ibarra and other towns in the province of Imbabura are in ruins.
Quito still stands, but threatens at every moment to become a
heap of ruins, and the towns adjoining it have all but wholly
disappeared. Where Cotocachi was is now a lake, and there and
at Ibarra almost the entire population has perished. But it is
only too clear that one of the richest regions in South America
has been prostrated utterly for the time, and laid desolate.

The characteristic of the present earthquake is that it had
almost as many centres as there were cities to destroy, and not, as
is usual, one particular focus. But otherwise it followed its
ordinary routine—a subterranean groaning, an upheaving and
yawning of the earth, an ebbing of the sea, where the town was
on the coast, and its return in one enormous tower of water to

merge and bear all back with it. English engineers and merchants draw the scene with the poetry of an intense reality. One
writing from Iquique describes how the sea first gave forth a
moaning, then shrank back hundreds of yards into the bay, then
reared itself in a solid wall fifty feet high, and there was then
but one roar and crash, and now "the sea was on us, and at one
sweep dashed what was Iquique on to the Pampa." Another,
the British Vice-Consul, tells that the earth opened, belching
out dust with a terrible and overpowering stench, and the air was
darkened as at midnight; that he and his were flying to the hills
over ground which trembled under their feet, when suddenly
"a great cry went up to heaven such as few men have heard—
'The sea is retiring!'"—that the sea drew every ship along with
it out of the bay, snapping anchors and chains like packthread;
then came in, "with an awful rush, carrying all before it in its
terrible majesty, bringing the whole of the shipping with it,
sometimes turning in circles, as if striving to elude their fate;"
and in a few minutes all was over, the city drowned, and every
vessel either stranded—one, he writes, a mile in shore— or bottom
upwards. The walls of houses were "blown out as if jerked"
at the inhabitants, and numbers perished under the ruins. None
had heart to try to disinter the dead bodies; but there is the cold
comfort that the survivors "do not think any are buried
alive." In the province of Imbabura the few left uninjured had
fled from a pestilence threatened by the stench from the exposed
dead. In Callao a fire broke out the night after the earthquake,
and destroyed 2,000,000 dols. worth of property. The open
country itself was insecure. Chasms appeared in the desert, the
mountains opened and shut, and great rocks fell down. On the
third day the ground was still rocking and trembling, and there
had been in all no fewer than a hundred secondary shocks; but
the real alarm and ruin were concentrated into less than ten
minutes.

VESUVIUS.

VESUVIUS was in Eruption during the whole of February, and
exhibited a most magnificent spectacle. For four months, too, it
continuously poured forth lava in spite of the predictions of the
learned that the end was drawing near; so that our professors
were under the necessity of distinguishing three phases of the
eruption—the first extending from the night of the 12th or 13th
of November to the 15th of January; the second from the 15th
of January to the 11th of February, during which interval it exhibited great varieties with less activity; the third from the 11th
of February to March 1st, a period of diminution, according to
Palmieri, though appearances and daily reports refuted the hypothesis. About the beginning of this period little lava flowed, and
the impression was that the eruption was nearly over; but on the

12th greater activity, which continued till the 15th of February, was observed; the detonations were loud, generally in the morning and evening; some stones were thrown out, and then comparative silence ensued. During these two or three days several shocks of earthquake were registered, and the apparatus of variation was disturbed. A few sublimates were collected on the summit, but not sufficient to mark the end of the eruption. On the 17th yet greater activity was perceptible, and its periodicity was confirmed; twice a day Vesuvius put forth all its energies, interesting equally the scientific and curious. The guides of Vesuvius, who reside in Resina, tell that in their town great shocks were felt, sufficient to make their doors and windows tremble. Columns of stones, they add, were shot into the air to the height of 300 metres, something little short of 900 ft.; the lava, too, progressed considerably, forming at the foot of the cone five different streams, which poured over like cascades.

On the 18th the great cone began again to roar with considerable violence, and two shocks of earthquake were felt distinctly in Resina; the very summit of the cone fell in, forming by the obstruction of the material three craters, which threw out large quantities of stones, and offered, as it were, a magnificent display of fireworks. The entire cone trembled, too, as if shaken by an earthquake, for four or five seconds; while later in the day the lava forced an opening in the direction of the Piano delle Ginestre, and flowed onwards in a stream of full 10 metres in width. From this time to the end of the month the mountain continued to thunder, and to throw out masses of red-hot lava, which, dividing into many streams, presents at a distance a spectacle of great beauty. Frequent though slight shocks were felt at Resina, and, indeed, in other places in the immediate neighbourhood, but they created no alarm, as the people were so accustomed to their return. On the night, however, of the 27th, there was one unusually strong. In March, about the 11th or 12th of the month, a mouth was opened towards the north on the very walls of the crater, stones in abundance were projected out, and a fresh stream of lava was poured forth. At the same time there were three small openings in the top of the cone, discharging columns of stones into the air to the height of upwards of 1,000 feet. All the force of the eruption became concentrated on the back of the mountain, in full view of Pompeii. At this commenced more brilliant displays, exceeding anything witnessed since November, 1857. The mountain appeared to be bursting with subterranean energy, the entire summit was on fire, the blackened walls became red with heat, and whilst the lava surged, not flowed, over, heavy stones, which the eye could detect and measure, were thrown up upwards of 1,500 feet. It was one of the grandest spectacles ever witnessed. Palmieri's report states: "Vesuvius has broken its monotony. For two days the lava has diminished in quantity; but the activity of the cone has in-

creased to such an extent as to surpass even that which was
observed in the early days of the eruption. Strong and continuous detonations, globes of smoke, often of an ashy character, and
copious bodies of projectiles, discharged to the height of 450
metres, are the phenomena which distinguish this new phase of
the eruption. The tube in face of the mountain, perhaps from
its being obstructed, receives only a small quantity of lava; that
which descends from the north side of the cone issues through a
small mouth, and has not arrived, therefore, at the Atrio del
Cavallo. The instruments have resumed all their activity; for
the entire soil trembles without ceasing. . . . In the Atrio
del Cavallo the thunders of the eruption are re-echoed with remarkable distinctness by the vertical rocks of Monte Somma,
sounding like the discharge of two batteries of artillery between
contending enemies. All this appears like a new effort for a fresh
discharge of lava."

At Naples, the thunders of the mountain were heard far above
the din of this most noisy city, interrupting us amidst our private
conversation. This by day! But who can describe in terms sufficiently eloquent the wondrous spectacle which Vesuvius presented by night, as seen from the Hotel de Rumie, in Santa
Lucia! Palmieri, as has been quoted, predicted a copious
discharge of lava; and so it happened. The mouth on the eastern
side of the mountain discharged its stream with greater vigour, the
instruments were violently agitated, but the detonations were less
loud and frequent. The mountain had found relief. Towards
the Atrio del Cavallo another large stream of lava flowed down.
It surged over from the summit. You could see it rising wave
over wave, and then pouring down. Already it has run full 400
metres, or about 1,400 feet. Nothing could exceed the beauty of
the spectacle! There were no scoriæ on the surface, so that the
spectator gazed on a pure stream of living fire, which flowed on
silently and tranquilly, yet with irresistible power, sometimes
swelling in its course as it received fresh contributions at its
source, and falling off at the sides in, as it were, so many cascades. The fine ashes were ejected and carried to a great distance.
Resina was covered with them; so was Naples in the line of their
progress; so were the houses at Posilippo, full ten miles distant
from Vesuvius Visitors came in at night powdered all over, unconscious of their disordered toilette. It is calculated that on the night
of the 12th and 13th, the black ashes covered a superficies of the
mountain of about six kilometres! "Admitting," says one,
who desires to make a nice estimate of the quantity thrown out,
" that the ashes have fallen in a like abundance to that which has
been observed on many points, where it has been collected in the
proportion of 14 kilogrammes to the cubic metre,—that is, 14
millions of kilogrammes for the square kilometre,—about 84
millions of kilogrammes must have fallen on the superficies spoken
of above." From the 13th to the 14th, the ashes were carried by

the force of the wind as far as Sicily and beyond Gaeta, the wind being furious and variable at the time. They have been collected also, and one may say of course, in the island of Capri, Procida, Ischia, and Ponza.

With regard to the first object proposed, Prof. Phillips, who was accompanied by Mr. Lee, of Caerleon, and our friend Cozzolino, the Vesuvius guide, discovered a peculiarity in the current of 1794, that which flowed through Torre del Greco even into the sea. It was of a much harder and more solid character than the other currents, attesting its fluidity, a quality which does not belong to the great proportion of the lava that is ejected. Somma, it was found, was inaccessible except by making a long detour round the very base of the mountain and ascending deep gorges and precipitous rocks, involving the sacrifice of much time and labour. The desire was to ascertain whether there was anything in the appearance of the mountain to corroborate the theory of Sir Charles Lyell, that volcanoes are formed not by the upheaving of the crust of the earth, but by the gradual accumulation of material thrown out. This wish, however, it was found impossible to carry out for the reasons above stated. Cozzolino communicated two facts connected with Somma which were received with interest—that there was a bed of tufa on it full of fossil plants, and that there was ten years ago a bed of the same formation with fossil marine shells. As to the probable depth of the volcanic motive power, the general opinion was that of Mr. Mallet, who fixes it at from six to eight miles below the surface of the earth. Dr. Phillips inspected with much attention the Lago d'Agnano, the Monte Nuovo, and Astroni, all three of which he declares to be for the most part of tufa formation; the depth of the lowest parts of these was ascertained to be about 20 feet above the level of the sea, and from these coincidences he claims for them a common character. On visiting the Geological Museum in the University he was greatly delighted to find many specimens of tufa with fossil marine shells imbedded, thus confirming an opinion long entertained by him, that though the elements of tufa are of volcanic formation, the peculiar structure was subaqueous.

On Saturday, November 15, was the great eruption. A column of fire rose continually to a great height, obedient to an impulse which seemed to be given every two or three seconds. The light fell for some miles across the bay, and the waves by their undulations seemed to increase its intensity, giving it the appearance at times of a path inlaid with millions of flashing diamonds, at others of a path of solid fire. On Sunday night the side of the mountain was covered with fire, while the clouds which obscured the summit prevented us from seeing what was going on.

On Monday, the clouds had cleared away, and an observer records, "I witnessed, not a column, but a huge body of black smoke, rising, I should say, upwards of 2,000 feet in the air; it was not a rigid column, but through the glass appeared to be formed of innumerable circlets, rolling one over the other and mingling in their ascent. How grand it was! What an idea of power it gave! Unfortunately for the perfect beauty of the spectacle, a south-west wind carried it inland, and we shall doubtless hear of whole districts being covered with that impalpable powder which generally insures a good harvest in the following season. The course of the lava, which flowed most abundantly, was marked by a white smoke, which rose all the way from the summit down to a point long past the Atrio del Cavallo."

Professor Tyndall and Sir John Lubbo k examined the phenomena of the eruption, and found the country all round Naples very smoking and hot, showing the existence of extensive subterranean fires, but they gained no information of scientific value. On different occasions they ascended the mountain from different sides, and in one instance, when a hurricane of wind favoured them, they went further than the guide would lead them, and had a look down the fiery tube of the crater itself. The wind was so strong, that on the way Sir John Lubbock was blown down flat on his face. They also explored some hot subterraneen galleries in the side of the mountain, and visited the Grotto del Cano, the well-known cavern, where the floor is covered several feet deep with carbonic acid gas. The heavy invisible gas, in fact, runs out of the cavern in a great stream, and will in the open air put out torches when they are held near the ground. He repeated some of the commoner experiments with the carbonic acid gas, by collecting some in his hat, and carrying it away a little distance from the cave, where it was poured over lighted matches, and put out the flame. A little dog is kept near the cave to be half suffocated by immersion in the gas when visitors arrive; and Professor Tyndall protested against the cruelty of the experiment, which, he says, serves no useful purpose, and ought to be stopped.

GEOLOGICAL APPLICATIONS OF PHOTOGRAPHY.

DURING the recent visit of the British Association to Norwich, we had the pleasure of examining some very beautiful and unique Photographic Negatives. Each negative was of solid translucent stone, of the very hardest character, and its details were not produced by man, for they consisted of sections of the organs of extinct animals which once lived and breathed, but whose internal structure has been marvellously preserved in the fossil state, during the lapse of untold ages. Mr. James Thomson, of Glasgow, is the geologist who first carried out the excellent idea of sawing thin plates out of those few fossils which are translucent, then of polishing the plates, and printing from them by photography, thus gaining the power of indefinitely and truthfully multiplying pictures showing the internal structure of extinct animals.

Mr. Thomson has confined his own work to animals of the Pollop or coral variety. Corals belong to the same family as the anemone, and differ only from the anemone in the fact that they secrete solid matter. This solid matter is secreted at the base of the Pollop, and the animals themselves vary much in size, from the little "coral insect," making the beautiful branch-like bunches of coral so well known, to individuals as big as the common anemone, who each form only their own single large secretion, and do not by union build up large tree-like structures. These large fossil corals, as found singly by the geologist, are about the length of a common hen's egg, but not so broad; they are thicker at one end than the other, and somewhat resemble a curved pear in shape. When found they are often imbedded in a thick superficial crust, so that none but a good zoophyte geologist would be likely to know the real contents of the stony mass.

These larger fossil corals are those which have been photographically printed, and not the smaller ones, which require the aid of a microscope to be examined. Many Pollops now existing find their fossilized representatives in geological strata; but those collected by Mr. Thomson are extinct species, whose remains are found in the lowest parts of the Scottish coal-field. They are found imbedded either in shale or limestone, and on the sea-coast, near Dunbar, they are very plentiful at a level below high-water mark. In many instances, however, they are so much crushed that it is difficult to ascertain the specific character of their internal structure.

The first part of the process, after finding a coral, is to have it cut transversely by a lapidary, and as they are excessively hard, in consequence of the great quantity of silica they contain, the workman is obliged to use diamond-dust in the operation. Next, they are cut longitudinally. These two operations must be superintended by the geologist, otherwise the important portions of the structure of the animal will be missed, as the slightest touch will take away the interesting parts. The pieces are next fixed to a piece of thick plate-glass, by means of a mixture of beeswax and resin in equal proportions by weight, which mixture is warmed over a spirit-lamp when required for use. After the piece of coral is attached to the glass, it is ground down to the required thickness with a mixture of fine emery-dust and putty-powder upon the leaden plate of the lapidary.

The finishing polishing operations are performed by the geologist himself, who rubs them down upon a hone, kept wet by means of water. The manipulations in this part of the process must be very delicate, and Mr. Thomson has, in difficult cases, been sometimes occupied for ten hours in finishing a single specimen. Sometimes, also, after working many hours, a few touches too much will destroy much of the value of the speci-

men. After they are finished, they are detached from the thick
piece of glass by the aid of gentle heat, and a number of them
are fixed in rows upon a sheet of patent plate-glass, the longitudinal sections being placed in one row, the transverse sections in
a row beneath, and so on alternately. Nothing has been found
to answer better to cement them to the glass than common gum
arabic. Any semi-transparent substances known to the geologist may be treated in the same way, such, for instance, as thin
sections of teeth, bones, and agates.

As regards corals, Mr. Thomson has made upwards of two
thousand sections, and has undertaken to prepare a duplicate set
for the British Museum. He has been working at the subject
for seven years, and wishes that others would in the same
manner make sections of the extinct corals of England, Wales,
and Ireland. One of the corals thus sectioned, the C. Fungitis
of Ure, was first figured by the Rev. David Ure in his history of
Rutherglen and Eastkilbride, in 1793, and there has been a
great deal of controversy about it ever since 1815, naturalists
not being agreed to what genus it ought to be assigned. This
fossil is the property of the Royal Society of Edinburgh, who
sent it to Mr. Thomson to be sectioned. At the British
Association at Norwich, a small grant of £25 was made to
Mr. Thomson, to aid him in carrying on his work with the
Scottish corals.

Photographic prints from these hard-stone negatives have
been taken by Mr. Robinson, of Glasgow, upon albumenized
paper, in the ordinary manner. We have pointed out to
Mr. Thomson the perishable character of the pictures thus
taken from his valuable little originals, and he contemplates
getting other copies printed in carbon. The most beautiful
prints he can obtain from them will be of course photographic
transparencies upon glass. In these pictures, to secure perma-
nency, it is desirable not to use bichloride of mercury in toning,
to well wash the prints after fixing, and to cement with hardened
Canada balsam a second sheet of glass over the negative, so as
completely to protect the film from the action of the atmosphere.
Mr. Thomson has already had transparencies and micro-photo-
graphs taken in the copying camera from his stone negatives.—*Mechanics' Magazine*.

THE BRONZE AGE.

SIR JOHN LUBBOCK, in *Pre-historic Times* speaks in the
following terms of the use of Bronze in the Iron Age :—" The
Iron Age, in which that metal had superseded bronze for arms,
axes, knives, &c.; bronze, however, still being in common use for
ornaments, and frequently used for the handles of swords and
other arms, but never for the blades." " I shall endeavour in the
present one (chapter) to show that, as regards Europe, the bronze

arms and implements characterize a particular period, and belong
to a time anterior to the discovery, or at least the common use,
of iron." He, however, states that, "as regards other civilized
countries, China and Japan for instance, we as yet know nothing
of their pre-historic archæology. Now, without entering on the
vexed question of whether or not there ever was a Bronze Age in
any part of the world distinguished by the sole use of that metal,
it is a fact that in those two countries to the present day, in the
midst of an Iron Age, bronze is in constant use for cutting instruments, either alone or in combination with steel. The principal seat of the manufacture is in the Canton province, where
every schoolboy may be seen with a clasp-knife made of a sort of
bronze; case, spring, and blade being all made of this material.
To form the cutting edge of these clasp-knives, a thin piece of
steel is let into the bronze blade; but knives made entirely
of bronze, and occasionally ornamented and riveted with copper,
are not uncommon; I have met with them as far north as Shanghai. In Japan, I have seen similar implements. But, though
the use of bronze in these countries has thus survived to the
present day, there is abundant evidence that at a former date it
was much more prevalent. Thus up to the Han dynasty, about
the Christian era, the ordinary coins of the country were made
of brass or bronze, in imitation of knives and swords; showing,
apparently, that in the earliest ages, when the use of some medium
of exchange was found essential, the weapons in common use
presented themselves as the readiest currency. The word in use
by the Chinese for their copper, or rather bronze, currency (the
alloy being properly a mixture of copper, zinc, and tin), which is
the only actual coin in circulation, is T'sien, a precisely similar
sound to the verb "tocut"; the phonetic in the written character
in both cases representing two spears. Nor is historical evidence
of the prevalent use of an alloy of copper for weapons of war at
an ancient date wanting. Thus Woo, the founder of the Chow
dynasty, B.C. 1121, reviewed his army on the plain of Muh; in
his left hand he is represented as carrying a weapon of yellow
metal. Although Dr. Legge supposes this means ornamented
with gold, the simpler interpretation seems the best. About the
same time, amongst the precious articles displayed at the funeral
of King Ching, we find red knives and cloths ornamented with
foo, explained in the " Urh-ya," a book of Confucian date, as de-
noting figures of axes, from the wooden handle being black when
" compared with the glittering head and edge"—a comparison
which seems unlikely to have suggested itself were the axes
formed of iron or steel. In The Tribute of Yu, however,—a
book to which a high antiquity cannot be denied, however we
may differ about its authenticity,—we have a glimpse at a still
earlier stage of civilization; but it is strange that here, as well as
at the present time, no material seems to have been in exclusive
use. Amongst the articles of tribute from the several provinces,

we find constant mention of stone arrow-heads and other implements, of the three grades of metals supposed, with good reason,
to be gold, silver, and copper, and, in one place, of iron and steel.
I have once or twice seen in China socketed bronze weapons, like
the celts of Europe, stated to be very ancient, but have only succeeded in obtaining one as yet. I have seen no stone axes,
though possibly the present sceptre of official authority derives
its traditional shape from the Stone Age. Thos. W. Kingsmill,
Shanghai, China.

SEA-BOTTOM SOUNDINGS DURING THE NORTH GERMAN POLAR EXPEDITION.

EHRENBERG has communicated to the Academy of Sciences,
Berlin, of which he is a member, a short notice of the specimens
brought up from the Sea-bottom by Soundings during the North-
German Polar Expedition of last season. The specimens are
thirty-nine in number, collected from lat. 73° to 80° N.—an
area extending from the Bear Islands and beyond Spitzbergen to
the coast of Greenland. Six of them were taken, it appears, between 80° and 81°, and in long. 13°, 14°, 15°, and 16° east from
Greenwich. As regards depth, thirty-two of the specimens were
brought up from less than 100 fathoms, four from 135 to 170
fathoms, two from 240 to 250 fathoms, and one from 300
fathoms. This latter was in lat. 76° 36' N. and long. 15° 52' E.
These depths, though not great, have, as Professor Ehrenberg
remarks, the advantage of certainty, which cannot always be
claimed for soundings at 1,000 or 2,000 fathoms. But after
Ehrenberg has had them under his microscope, we shall not have
long to wait for explicit information on these points; and further
light will be thrown on the question, which, in his opinion, is the
most important of all, namely, Whether the six classes of microscopic creatures already described in *Microgéologie* are found un-
mixed, or mixed with other hitherto unknown forms, within the
Polar Circle? To obtain conclusive proofs of the relations of
organic life in its minutest forms throughout the globe would be
worth all the cost and labour bestowed in obtaining them. Moreover, according to the nature of the specimens brought up,
whether fine or coarse, slimy or powdery, will, as is thought, be
the evidence of streams, swirls, or quietness in the depths of the
ocean. Should Mr. Petermann and his friends attempt another
expedition in the coming summer, it is to be hoped they will rely
more on the dredge than on the sounding-lead for specimens from
the bottom. Taken in connection with the results obtained by
the expedition under Drs. Carpenter and Wyville Thomson (an
interesting Report of which has been printed in the *Proceedings
of the Royal Society*), a higher value attaches to the specimens
brought home by the German explorers, and expectations of
Ehrenborg's description can hardly fail to be lively. We quote
this interesting abstract from the *Athenæum*.

Astronomy and Meteorology.

THE TOTAL ECLIPSE OF THE SUN.

THE following account of the phenomena attending the Total Eclipse of the Sun, which occurred on the 18th of August last, has been prepared from the official report of Mr. J. Pope-Hennessy, Governor of Labuan, and from the observations of Captain Reed and the officers of Her Majesty's surveying vessel *Rifleman*.

The observatory spot was Barram Point, on the north-west coast of Borneo, in lat. 4° 37′ 15″ N., long. 113° 58′ 28″ E., where a small tent was erected on the north side of the river, close to the casuarina trees, which show as the extreme of the point when approaching it from the north-eastward. The telescopes were of the kind ordinarily used on board ships. They were suspended to tripods, made by lashing three boat-hook staves together, and afforded very fair means of observing with accuracy. No special instruments or instructions had been furnished to the *Rifleman*.

The observers were Governor Pope-Hennessy, Captain Reed, Navigating-Lieutenants Ray and Ellis, and Mr. Doorly, midshipman. Equal altitudes were obtained for determining the exact mean-time at place, and the data of the various observations reduced to the exact mean-time. Dr. O'Connor landed to note the physiological phenomena, and Mr. Wright, midshipman, to watch the magnetic needle. Mr. Petley, and the other officers left on board the *Rifleman*, had charge of the barometrical and thermometrical observations, and they were also directed to watch the variations, if any, in the magnetic needle.

At 11.50 four solar spots were visible, lying nearly in a line parallel to the plane of the horizon; those spots are subsequently referred to as Nos. 1, 2, 3, and 4.

No. 1 spot (that farthest to the left, or eastward) was much the largest; it was surrounded by a distinctly visible penumbra.

No. 2 spot was small, bold, and clearly defined.

No. 3 was a sort of double spot, surrounded by a penumbra.

No. 4 was a small sharply defined spot, similar to No. 2 in size and shape, but encircled with a bright luminous space, which was not observed round any of the other spots.

At 11 h. 56 m. 7 sec. the first contact of the moon took place with the lower and left-hand quarter of the sun; and at 1 h. 23 m. 13·6 s. the total obscuration occurred. At 1 h. 29 m. 25·8 s. the sun's limb reappeared, and at 2 h. 48 m. 31·7 s. the separation of the sun and moon's limbs happened.

During the six minutes and twelve seconds (nearly) of total eclipse not the slightest change of any kind could be observed in the magnetic needle, nor did it move or vibrate in any way on the reappearance of the solar spots. No movement of the needle, either, could be detected on board the ship.

The general phenomena of the Eclipse are thus described by Mr. Pope-Hennessy :—" I confine myself to copying from the rough notes I took at Barram Point, and from the note-books of Captain Reed and his officers, also taken on the spot. I have not time to arrange the materials before me in anything like scientific order, and the absence of any works of reference renders me still less able to do justice to the facts we collected. We were very fortunate in the weather. The day was bright and clear—not a cloud near the sun. A few round white clouds that lay on the horizon hardly moved. There was a slight breeze from W.S.W. The sea was breaking heavily on the shore, and it had a slight brownish blueish tinge all over, except where the white breakers approached the land. The grove of casuarina trees behind us had the same deep green colour which they always exhibit on a fine day in the tropics. A few swallows were skimming about high in the air. We also noticed some dragon-flies, butterflies, and a good many specimens of a large heavy fly like a drone-bee. When we left the ship, at 10 o'clock, the barometer was 30 ·00 ; the mean of two thermometers in the shade was 85° ; the dry thermometer exposed to the sun was 61° ; and the wet thermometer exposed to the sun was 83 ·5. During the progress of the eclipse, the barometer fell steadily from 29°·96 to 29°·81. The mean of the two thermometers in the shade was 85°, without any change whatever, from 10 o'clock till the close of the eclipse. At the close of the eclipse, 2 h. 43 m., 31·7 s. it rose to 86°. The dry bulb thermometer, hung in the sunlight, fell from 96° to 85° as the moon was covering the sun, and rose from 85° to 96' as the sun was reappearing. The wet-bulb thermometer fell from 83°·5 to 83° at the total eclipse, and rose to 89° at the termination of the whole eclipse. Ten minutes before the total eclipse there seemed to be a luminous crescent reflected upon the dark body of the moon. In another minute a long beam of light, pale and quite straight, the rays diverging at a small angle, shot out from the westerly corner of the sun's crescent. At the same time Mr. Ellis noticed a corresponding dark band, or shadow, shooting down from the east corner of the crescent. At this time the sea assumed a darker aspect, and a well-defined green band was seen distinctly around the horizon. The temperature had fallen, and the wind had slightly freshened. The darkness then came on with great rapidity. The sensation was as if a thunderstorm was about to break, and one was startled on looking up to see not a single cloud overhead. The birds after flying very low disappeared altogether. The dragon-flies and butterflies disappeared, and the large drone-like flies all collected on the ceiling of the tent, and remained at rest. The crickets and cicadæ in the jungle began to sound, and some birds, not visible, also began to twitter in the jungle. The sea grew darker, and immediately before the total obscuration the horizon could not be seen. The line of round white clouds that lay near the horizon changed their colour and aspect with great rapidity. As the obscuration occurred, they all became of a dark purple, heavy-looking, and with sharply defined edges. They then presented the appearance of clouds close to the horizon after sunset. It seemed as if a sun had set at the four points of the horizon. The sky was of a dark leaden blue, and the trees looked almost black. The faces of the observers looked dark, but not pallid or unnatural. The moment of *maximum* darkness seemed to be immediately before the total obscuration. For a few seconds nothing could be seen except objects quite close to the horizon. Suddenly there burst forth a luminous ring around the moon. This ring was composed of a multitude of rays quite irregular in length and in direction. From the upper and lower parts they extended in bands to a distance more than twice the diameter of the sun. Other bands appeared to fall to one side, but in this there was no regularity, for bands near them fell away apparently towards the other side. When I called attention to this, Lieutenant Ray said, " Yes, I see them ; they are like horses' tails," and they certainly resembled masses of luminous hair in complete disorder. I have said that these bands appeared to fall to one side, but I do not mean that they actually fell, or moved in any way, during the observations. If the atmosphere had not been perfectly clear, it is possible that the appearance they presented would lead to the supposition that they moved, but no optical illusion of the kind was possible under the circumstances. During the second when the sun was disappearing, the edge of the luminous crescent became broken up into numerous points of light.

The moment these were gone, the rays shot forth, and, at the same time, we noticed the sudden appearance of the rose-coloured protuberances. The first of these was about one-sixth of the sun's diameter in length, and about one-twentyfourth of the sun's diameter in breadth. It all appeared at the same instant, as if a veil had suddenly melted away from before it. It seemed to be a tower of rose-coloured clouds. The colour was most beautiful —more beautiful than any rose-colour I ever saw. Indeed, I know of no natural object or colour to which it can be with justice compared. Though one has to describe it as rose-coloured, yet in truth it was very different from any colour or tint I ever saw before. This protuberance extended from the right of the upper limb, and was visible for six minutes. In five seconds after this was visible, a much broader and shorter protuberance appeared at the left side of the upper limb. This seemed to be composed of two united together. In colour and

Q

aspect it exactly resembled the long one. The second protuberance gradually sunk down as the sun continued to fall behind the moon, and in three minutes it had disappeared altogether. A few seconds after it had sunk down there appeared at the lower corresponding limb—the right inferior corner—a similar protuberance, which grew out as the eclipse proceeded. This also seemed to be a double protuberance, and in size and shape very much resembled the second one; that is, its breadth very much exceeded its height. In colour, however, this differed from either of the former ones. Its left edge was a bright blue—like a brilliant sapphire with light thrown upon it. Next that was the so-called rose-colour, and, at the right corner, a sparkling ruby tint. This beautiful protuberance advanced at the same rate that the sun had moved all along, when suddenly it seemed to spread towards the left until it ran around one-fourth of the circle, making a long ridge of the rose-coloured masses. As this happened, the blue shade disappeared. In about 12 seconds the whole of this ridge vanished, and gave place to a rough edge of brilliant white light, and in another second the sun had burst forth again. In the meantime the long rose-coloured protuberance on the upper right limb had remained visible; and though it seemed to be sinking into the moon, it did not disappear altogether until the lower ridge had been formed, and had been visible for two seconds. This long protuberance was quite visible to the naked eye, but its colour could not be detected except through the telescope. To the naked eye it simply appeared as a little tower of white light, standing on the dark edge of the moon. The lower protuberance appeared to the naked eye to be a notch of light in the dark edge of the moon—not a protuberance, but an indentation. In shape the long protuberance resembled a goat's horn. Though the darkness was by no means so great as I had expected, I was unable to mark the protuberances in my note-book without the aid of a lantern, which the sailors lit when the eclipse became total. Those who were looking out for stars counted nine visible to the naked eye; one planet, Venus, was very brilliant. On board the *Rifleman* the fowls and pigeons went to roost, but the cattle showed no signs of uneasiness—they were lying down at the time."

Major J. F. Tennant, R.E., reports to the Astronomer-Royal: "Guntoor.—This morning was very promising, and if it had followed the course of its predecessor, we should have had a magnificent clear sky, but it clouded over the east with thin cumulostrati, which, while hardly stopping vision, interfere very much with the photographic energy; and the result was that every negative was under-exposed, and we have little more than very dense marks showing the protuberances. The six plates arranged for were duly exposed, but the heat so concentrated the nitrate of silver solution, that, besides showing but faint traces of protuberances and corona, they are all covered with spots. Still we

may make something of them, and will try. Captain Branfill reports the protuberance unpolarised, and the corona strongly polarized everywhere in a plane passing through the centre of the sun. Complementarily I have to report a continuous spectrum from the corona, and one of the bright lines from the prominence I examined. I am, I believe, safe in saying that three of the lines in the spectrum of the protuberances correspond to C, D, and b. I saw a line in the green near F, but I had lost so much time in finding the protuberance (owing to the finder having changed its adjustment since last night), that I lost it in the sunlight before measuring it, and I believe I saw traces of a line in the blue near G ; but to see them clearly involves a very large change in the focus of the telescope, which was out of the question then. I conclude that my result is that the atmosphere of the sun is mainly of non-luminous (or faintly luminous) gas at a short distance from the limb of the sun. It may have had faintly luminous lines, but I had to open the jaws a good deal to get what I could see at first, and consequently, the lines would be diffused somewhat; still I think I should have seen them. The prominence I examined was a very high narrow one, almost, to my eye like a bit of the sun through a chink, in brightness and colour (I could see no tinge of colour), and somewhat zigzagged like a flash of lightning. It must have been three minutes high, for it was on the preceding side of the sun near the vertex, and was a marked object, both in the last photoplate just before the sun reappeared, and to the eye. Captain Branfill saw the prominences coloured, as did two other gentlemen ; but one in my observatory (like myself) only saw it white. I should, however, say that for long I never saw an Orionis markedly red, nor Antares, and I may not catch red soon, though I cannot conceive this being so. In conclusion, I may note that the darkness was very slight, and the colour not half so gloomy as in the eclipse of 1857, which was partial at Delhi, where I was then."
—Monthly Notices of the Royal Astronomical Society.

As was anticipated, the eclipse of the sun was treated of at some length in General Sabine's address to the Royal Society, particularly with reference to the spectroscopic observations. In connexion with the most recent spectroscopic observations of the sun, a body of information has been gained which cannot fail to be of high importance in cosmical science. The Royal Society expended nearly £300 for instruments to send out to India for observation of the eclipse ; and it is gratifying to learn that, notwithstanding clouds and bad weather brought by the monsoon, so many of the interesting phenomena were really observed.

NEW PLANET.

THE newest of the New Planets, in our knowledge, is No. 98, *Ianthe*, discovered in America, April 18, by Professor Peters. There may be two more for aught we know. It is obviously a great difficulty to find names. Which Ianthe is it? The daughter of Oceanus, or the betrothed of Iphis, about whom Ovid tells a curious story? The two preceding were *Clotho* and *Ægle*. It will not do to go on giving names. No one can use them; he must go from the name to the number. It will become a question, if the thing goes on, whether knowledge of the positions is to be kept up. In process of time we may have a thousand—aye, ten thousand—of these little specks of planet-dust. It must be a small job, even now, when a new one is discovered, to be quite sure it is not one of the old ones; what will it be when the 100,001st is found? The astronomers are very patient, and in gradual accumulation are only surpassed by the coral-insect. When Francis Bailey died in 1845, the little outstanding jobs which he had nearly finished superintending were the new edition of the Astronomical Society's Catalogue (8,377 stars), the printing of Lacaille's Southern Catalogue (9,766 stars), and the superintendence of Lalande's Catalogue (47,390 stars). In 1846 appeared the reduction—only astronomers know what a job that is—of all the planetary observations made at Greenwich from 1750 to 1830—the work of the Greenwich Observatory. Some reader immediately remarks, "How absurd that the discoverer of a little comet should instantly be of European fame when works like these are unnoticed!" There is some truth in this remark, but not so much as may be supposed. The comet-finder may have been systematically watching, in a skilful way, which ensures no loss of labour, for many a night before he was repaid. William Herschel discovered Uranus, not by popping the telescope on it unawares, but as one fruit of a long examination of stars, for a purpose wholly unconnected with planet-searching. There is very little accident in these discoveries: those who look out have reasons for their particular courses. It was not by mere coincidence that Lassell in England, and Bond in America, discovered the 8th satellite of Saturn on the same night of 1848.—*Athenæum.*

TRANSIT OF VENUS.

IN view of the observations of the Transit of Venus that will doubtless be made in 1874 and 1882, Mr. E. J. Stone, of Greenwich Observatory, has re-discussed the various observations made in 1769, by Father Kell, Wales and Dymond, Captain Cook, and others; and he states that the investigation has led him to the "detection of several grave and fundamental errors which have previously been made in the discussion of these results, and to a value of the solar parallax entitled to be received with confidence." This value is 8″·91, which confirms the long-accepted conclusion that it was "about 8″·90," and gets rid of the serious discrepancies which have long perplexed astronomers.

TRANSIT OF MERCURY.

Mr. W. E. Denning reports from Bristol, November 5th:—"The Transit of the planet Mercury this morning was well observed here, the sky being nearly cloudless during the time of the phenomenon. A few minutes before sunrise I placed myself in readiness at the telescope, and soon the northern limb of the sun became visible. Before the whole disc could be perceived I distinctly saw the planet, but at this period it was impossible to obtain a satisfactory view. Several large spots were visible on the northern part of the sun's disc.

"As the sun attained a greater altitude, it became much better defined, and as the end of the transit approached, the planet was excellently seen. I noticed no bright spots on the planet's disc, neither did I recognize any ring of light surrounding it, like other observers appear to have done in former transits. The planet was quite black and perfectly round, and contrasted strongly with the irregular edges of the solar spots. One rather large spot was visible on the northern part of the sun, which was composed of four nuclei and a large uneven penumbra. Altogether I reckoned twenty spots, but most of these were very small, and could only be discovered with a powerful telescope. On the south-western edge a large group of faculæ was visible, which gave the disc a beautiful marbled appearance. Just before the planet commenced its egress it became elongated, and soon after notched the west-south-west edge of the disc, where it disappeared soon after 9 A.M.

"The planet Venus was shining brilliantly at the time, and could be easily discerned with the naked eye.

"In the above observations I employed a refracting telescope of 4¼ in. aperture and 7½ ft. focal length."

The Rev. Charles B. Gribble, Chaplain to the English Embassy at Constantinople, writes that the transit of Mercury was well seen there on the morning of the 5th November. The weather was very fine ; a slight haze hung before the sun soon after its rising. At 7.20 A.M., or 4 d. 19 h. 20 m., the dark body of the planet was clearly visible ; its movement was gradual but perceptible until it neared the sun's limb, when its motion seemed more rapid. Its disappearing from its first contact with the limb until it was lost to view occupied about 30 seconds. Its disappearance was complete at 10 h. 48 m. 8 s. A.M.

THE AURORA BOREALIS.

Prof. Christison, in a paper read to the Royal Society of Edinburgh, says :—" The phenomena of the Aurora Borealis in his country have often been minutely described on the occurrence of unusually fine displays of it. But no one, so far as I am aware, has studied carefully its prognostications. Thoroughly inquired into, however, these may prove practically valuable, as the following illustration will serve to show. Every one knows that when the aurora first begins to exhibit in the autumn, it is regarded as a sign of broken weather following. But at that period of the year it supplies a prognostic of far greater precision and importance. I have repeatedly mentioned to my friends the observation I have invariably made, that the first great aurora after autumn is well advanced, and following a tract of fine weather, is a sign of a great storm of rain and wind in the forenoon of the second day afterwards. I must have noticed this fact very early, because I applied it on the occasion of the first meeting of the British Association in Edinburgh on the 8th of September, 1834. There had been a long tract of very fine weather—for a fortnight and more—when on Saturday evening, the 6th of the month, there appeared the widest, brightest, and most flashing aurora I have ever seen. Next day the weather continuing remarkably fine, Professor Sedgwick described at breakfast at Dr. Alison's, in glowing language, the magnificent exhibition which the philosophers of Edinburgh had provided for their southern visitors. Presenting, then, to him the dark side of the picture, I told him that the Association meeting was to be inaugurated with a great storm. He was surprised at this, and appealed to the continuing cloudless sunny sky against me; but I told him the particulars of the prognostication, and that the storm would not begin till the middle of the following day. Next morning the weather was equally splendid; but soon after 11 the eastern sky began to be overcast; an ominous low north-easterly black cloud rose by degrees; at 12, as the offices of the Association opened, rain began to fall from that direction; and in a short time there commenced the most incessant and heavy fall of north-east rain I have ever witnessed, lasting without intermission till 1 o'clock on Wednesday, the 10th, when the fine weather was again restored to us.

I have often made the same prognostication since that time, and with invariable accuracy; and several friends to whom I have mentioned it have made the same observation, viz., that the first great aurora, occurring after a long tract of fine autumnal weather, foretells a storm commencing between 12 and 2 o'clock in the afternoon of the second day thereafter. I restrict the prognostication to these conditions. It is evident how valuable the knowledge of it may often be to agriculturists. Nevertheless, I never met with a farmer or farm-servant who knew it. On one occasion it was the means of saving the corn-crop of a friend in Dumfriesshire whose farm-steward was about to leave his corn half-led on the day after a very great aurora, and, deceived by the beauty of the weather, was on the point of taking his labourers to other work not at all pressing. His master, trusting to my positive assurances, ordered him to make haste in leading and thatching everything; and great was the steward's astonishment when a furious three days' storm set in on the fore-noon of the second day."

___

WATERSPOUTS.

Two of these phenomena were seen at the mouth of St. George's Channel by Captain Stevenson, of the ship *Barelaw*, arrived at Liverpool from Bombay :—" On Sept. 24, at noon, in lat. 50° N., long 8° W., two waterspouts crossed the *Barelaw's* stern from east to west, the wind at the time being light from the eastward, with rain, the barometer 29°·66 and thermometer at 64°·3. The sea and sky were in visible connection, a large column of cloud and vapour in the shape of a bow rising from west to north, the smaller end becoming lost in a huge raincloud. The larger end was connected with the water, causing the sea to boil and bubble tremendously, and apparently covering a circle the diameter of which would be 20 ft. The first one passed about one-eighth of a mile astern, and broke immediately after. The second continued its course to the westward as far as the eye could follow it. When nearest the ship a furious puff of wind heeled the ship over for a moment, and when it was gone the revolution of the circle of agitated water was from east round south to west, north, and east, and seemed a perfect cyclone, though on a very small scale. The wind during the passage of the second spout across the ship's stern shifted in an instant from east to south."

___

RETURN SHOCK OF LIGHTNING.

M. Decquerel brings to our notice a novel fact, in connection with Lightning, which he terms a "return shock." The occasion on which it occurred was during a thunder-storm on June 8th. An *employé* of a gas company felt a sinking within himself at the instant he perceived a flash of lightning. From this he felt a trembling that lasted two days. He found that the nails of his boots, which were almost new, had been lifted by the light-ning. M. Decquerel attributes this fact to what he terms a "return shock." When a person finds himself under the influ-ence of a storm-cloud, and the cloud bursts in the distance, he perhaps "shocks" himself by the immediate recomposition of the two opposite electricities—the one possessed by the earth, the other by himself. The accounts and stories we hear of relative to the vagaries of lightning are frequently marvellous. So many strange phenomena have come within our own knowledge, that we are by no means prone to doubt any new or strange fact that may be noticed. The facts mentioned above are quite compatible with

some of the extraordinary phenomena of lightning. The effects
are those of a person struck by lightning, with no further bodily
damage done than the continuous trembling, the mechanical effect
being exhibited by the action upon the boot-nails. We really
cannot see why the phenomenon of a return shock, which could
not be powerful enough for the effects produced, should be
brought forward as an explanation, when the ordinary shock
seems to be perfectly sufficient.—*Mechanics' Magazine.*

ACTION OF LIGHTNING IN FORFARSHIRE.

Sir David Brewster relates:—"In the summer of 1827, a
hay-stack, in the parish of Dun, in this county, was struck with
Lightning. The stack was on fire; but before much of the hay
was consumed, the fire was extinguished by the farm-servants.
Upon examining the hay-stack, a circular passage was observed
in the middle of it, as if it had been cut out with a sharp instrument. This circular passage extended to the bottom of the
stack, and terminated in a hole in the ground. Captain Thomson, of Montrose, who had a farm in the neighbourhood, examined the stack, and found in the hay-stack, and in the hole, a
substance which he described as resembling lava." A portion of
this substance was sent by Captain Thomson to Dr. Brewster, of
Craig, who forwarded it to Sir D. Brewster, with the preceding
statement. The substance found in the hole was a mass of silex,
obviously formed by the fusion of the silex in the hay. It had a
highly greenish tinge, and contained burnt portions of the hay.
Sir D. Brewster presented the specimen to the museum of St.
Andrew's.

AEROLITES.

None but the heaviest and largest ever succeed in reaching the
earth in a solid form, and even those usually burst into fragments
while still at a great distance from us. All the rest are first
ignited in the upper regions of air, then thoroughly consumed in
their passage through the next layers, and, lastly, they fall to the
earth's surface in the form of an impalpable powder. We believe
that Dr. Reichenbach was the first to point out that this powder
may be detected. He collected dust from the top of a high
mountain, which had never been touched by spade or pickaxe;
and he found this dust to consist of the very elements which
compose meteoric stones—nickel, cobalt, iron, phosphorus, &c.
Dr. Phipson also notes that "when a glass, covered with pure
glycerine, is exposed to a strong wind late in November, it receives a certain number of black angular particles, which can be
dissolved in strong hydrochloric acid, and produce yellow chloride
of iron upon the glass plate."—*Express.*

VOLCANOES AND LUNAR INFLUENCE.

PROFESSOR PALMIERI, of Naples, who has been studying the Eruptions of Mount Vesuvius, thinks the Volcano acts under Lunar Influence. In truth, the periods of its greatest eruptions get every day about half-an-hour later, coinciding with the movements of the moon. This observation, if trustworthy, confirms the theory that the interior of the earth is molten, in which case its substance would be as much subject to the laws of tides as the oceans.

THE METEORIC SHOWER.

THE brilliant Meteoric Shower of November, 1866, has had a second annual repetition, although in the present instance, in a comparatively modified and feeble form. The weather also was unfavourable, but the chief observations appear to have been made by Professor Phillips, at Oxford. The shower commenced about 2 A.M., on Saturday last, the 14th inst., the sky having been previously cloudy and overcast. A few stars were observed at intervals up to 3.45, when they became numerous, bright, and occasionally splendid, although their splendour—perhaps from the smallness of their number—was far behind that of those in 1866. The sky being favourable enabled accurate observations to be taken till 5.45, after which the heavens became obscured, except in patches, when still from time to time the asteroids were visible. The radiant point was nearer to Leonis than to any other star, and in the early part of the spectacle the constellation of Leo was often traversed in one direction or other, but towards the end of the observations most of the meteors seemed rushing westward, and rarely appeared near to Leo. In the greater part of the period they passed in all directions. Some went northward across the Great Bear, threatening to extinguish the Pole Star; others shot at Venus in the east, at the cloudy moon, not long risen, at Sirius, Procyon, Orion, and the Twins, while a few crossed the zenith. The very brightest left no trace, but flew in a moment across, and blazed with a surprising light. Others of considerable splendour left short traces, usually bluish, the meteoric head being red and fiery, with sparkling rays. The duration of the trace was 15 s., 30 s., and 1 m. The longest path (overhead) was about 60°—generally as little as 10° or 5°. No serpentine path. The number was not counted—rarely five in as many seconds, more frequently as many in five minutes, with occasional pauses and occasional crowding of meteors—probably 250 in each of the two hours from 3.45 to 5.45. There were three observers, not commanding more than half the sky-space in which the objects moved. The meteoric shower was observed in other places, and amongst others at Bridgend, Glamorganshire. The observations were made on the coast near that place by several persons from 6.45 a.m., Greenwich mean time, until daylight, in the direction W. and W.N.W. The meteors are described as falling frequently, of large apparent size and great brilliancy, notwithstanding the prevalence of a sea-fog of sufficient density to obscure the stars. The meteors were followed by long trains of light of different hues, but no noise accompanied their disappearance. The shower of 1867, it will be remembered, was well observed in America; the recurrence of the phenomenon this year at precisely the same time, and its increasing faintness, prove the correctness of astronomical calculations, and verify the assertion that we are gradually departing from their track, and shall not see them again in their full beauty for thirty-one years—that is, thirty-three years from the grand display of 1866.—*Mechanics' Magazine.*

The Meteoric display on the night of November 13th, was observed in all parts of the United States. In Philadelphia it continued from midnight until daylight, and many hundreds were seen. The weather was clear in nearly all parts of the country, so that there was no obstruction to the view. In New York the number of meteors seen is reported to exceed 1,500 by actual count. In Boston, between 2 and 5 A.M., on the 14th, 3,500 meteors were counted, many of them of great brilliancy. At Baltimore it is announced that the shower was most brilliant between 12 and 2, passing from the east towards the south-west. At Charleston, South Carolina, the shower was most numerous at 1 A.M. In New Orleans the display continued from midnight until daylight, and several of the meteors are said to have left trains visible for five minutes. At Nashville, Tennessee, the meteors began to be visible about 3 A.M., continuing until daylight. At San Francisco the display began at half-past 10 on the evening of the 13th, continuing until daylight, and the shower is said to have rivalled that of 1867, several of the meteors leaving brilliant trains. At Washington the meteors were observed by a corps of astronomers at the Naval Observatory, and Commodore Sands, the superintendent, makes a report of the result, in which he says that during the evening of the 13th no greater number of meteors was seen than are usually observed on clear nights, until about 11 P.M., when there seemed to be an increase in their number and brilliancy, though not sufficient to indicate the beginning of a shower. By midnight the number had considerably increased, and the prevalence of trains was generally noticed. At 12.35 the observers began to count the meteors. At 1.35 300 meteors had been counted, most of them quite brilliant, and nearly all leaving green, blue, or red trains. Thus far the display had not been confined to any particular portion of the sky. Many of the trains were visible several minutes, and one lasted ten minutes, while traces of another near Ursa Minor were seen for 30 minutes after the appearance of the meteor. The tracks of about 90 meteors had been sketched at 1.50 a.m., at which time 400 had been counted. The observations were continued until 6 a.m., when 5,078 had been counted.—*Times.*

ASTRONOMICAL PHOTOGRAPHY.

SINCE the first news reached England of the work of Major Tennant, R.E., in Photographing the total eclipse of the sun at Guntoor, he has made enlarged copies of some of the negatives then obtained. Although, as we have previously stated, the original negatives were faulty, still some of them not only show the flame prominences, but streaks in them, spiral in form. Dr. Vogel, who very successfully photographed the total phase at Aden, used a bromo-iodized collodion, which also contained an excessively large proportion of alcohol, to render evaporation less rapid in a tropical climate than if the collodion contained the usual amount of ether. He also used nothing but cadmium salts in the collodion. The developer consisted of ammonio-sulphate of iron 7 parts, acetic acid 5 parts, and water 102 parts. Of the three plates exposed, one gave a perfect image, the second was a little under-exposed, and the third was a failure because of the passage of a cloud at the moment. A long sliding back was used in the telescope camera, to allow two pictures to be taken on one plate, so that he really obtained six pictures of the eclipse, two of which were failures. The photographic parts of his apparatus were not directly attached to the telescope, because the insertion of the slides might otherwise set up vibrations, so the camera was fixed upon a separate stand, and connected with the telescope by a tube of india-rubber. The eclipse has been successfully observed in a clear sky, further east than India, by M. Stéphan, a French philosopher, who stationed himself at Wha-Wen, in the Malayan peninsula; also by Captain Reed, who observed it on the coast of Borneo. We can hardly expect news from a more eastern district still, as the island of Papua is too far from civilization, and it is not easy to effect a landing on large portions of its coast, because they consist of long banks of mud, almost perfectly flat.

Direct photographs of the sun and its spots are taken regularly at Kew Observatory, and preserved as records; some also, on a very large scale, have been taken by Mr. Warren De La Rue, F.R.S. Very beautiful fac-similes of the best of these pictures have been drawn upon steel plates, and are now published in a new and very good work on astronomy, by Mr. Norman Lockyer.[*] The same book contains descriptions of the apparatus used by Mr. William Huggins, F.R.S., in obtaining the spectra of the stars, with illustrations. Much care has been taken in getting

---

[*] *Elementary Lessons in Astronomy.* By J. Norman Lockyer, F.R.A.S. London: Macmillan and Co., 1868.

up this standard little work, the author having been aided by several leading philosophers experienced in the several branches of the science. The Royal Astronomical Society placed facilities at his disposal; so also did Mr. Warren De La Rue and Dr. Balfour Stewart, Superintendent of the Observatory of the British Association at Kew.

Some recent photographic experiments by Mr. George Dawson, M.A., of King's College, are of scientific interest. Three ounces of clippings from a bar of perfectly pure silver were dissolved in pure nitric acid, and the slight excess of acid removed by recrystallizing the salt many times, from the purest of distilled water, over a sand-bath. A negative bath was then made of this very pure nitrate of silver, which proved to be very slightly alkaline to test papers, a peculiarity always exhibited by this salt when it is absolutely pure. A plate was coated with an ordinary slightly acid bromo-iodized collodion, and sensitized in the bath. On trial in the camera, a foggy and scarcely visible picture was the result, yet the same collodion worked well in other baths. He then tried a very new and neutral sample of bromo-iodized collodion in the pure bath. This gave good and blooming negatives with an almost instantaneous exposure in the glass-house. That a neutral collodion and slightly alkaline bath should give good pictures is something entirely new to photography, though the principle was once laid down in a very unconvincing and complicated manner by Mr. M'Lachlan. The same new and neutral collodion did not give good pictures in the ordinary acid baths. If, after further testing, these facts should be verified beyond all doubt, the discovery will be valuable in astronomical photography, for the slight trace of acid usually given to secure clean pictures has a slight retarding influence, and necessitates a slightly longer exposure.

In rapid photography, where the exposure requires to be reduced to the smallest possible fraction of a second, slight influences like these have their weight, yet the value of these influences is not scientifically and accurately known. Celestial photography will not be brought to perfection till many obscure phenomena of this kind have their accelerating and retarding powers determined by accurate experiments. Under all ordinary circumstances, these slight influences are unnoticed and unknown, because they do not interfere appreciably with the results.

Many bad photographic varnishes constantly creep into the market, and some of them have a most disastrous effect upon the pictures, cracking valuable negatives all over, and rendering them worthless. We know an amateur photographer who, after a tour of a few weeks in Wales, taking pictures of beautiful seacoast scenery, had half his negatives cracked all over by a miserable sample of varnish. When such cracks are very fine, the negative will sometimes print well if a little plumbago be gently brushed into the cracks. In Kingham's work on photography, we find the following little-known remedy for the evil, and if it really answers in practice, it must be very valuable:—" First ascertain whether the solvent of the varnish on the plate be alcohol, chloroform, or benzole, by dropping on one corner a minute drop of each of these menstrua, to ascertain which dissolves the varnish. Next take a tin box somewhat larger than the picture, about 1 in. deep; at the bottom of this box solder a ring of tin, about ⅛ in. wide, of the same shape and nearly of the same size, as a support for the glass plate; pour a small quantity of the solvent on the outside of the support; place the plate, collodion side upwards, on the ring, cover the box as nearly airtight as possible with a piece of glass, and place it in a water bath; the vapour of the solvent will soon cause the varnish to swell, and the edges of the cracks to coalesce. As soon as this end in view is accomplished, the plate is carefully withdrawn and when cool is again varnished with a similar varnish."

A NEW WAY OF ESTIMATING THE MOTION OF THE STARS.

A REMARKABLE paper has been sent to the Royal Society by Mr. Huggins, one of the Fellows. It announces the application of a new and most promising method of inquiry to the determination of the Stars' Motions. Many of our readers are doubtless familiar with the fact that Sir W. Herschel was the first to point out the important results which may be gathered from the consideration of the stars' apparent motions on the celestial sphere. Just as a person travelling through a wood observes the trees in front of him to be opening out, while those behind him are closing in, and trees on either side of him apparently falling behind him, so, Herschel argued, if the solar system is really travelling in any direction through sidereal space, we ought to be able to detect a gradual opening out of the stars around that point towards which the sun is travelling, a corresponding closing in of the stars towards the opposite point, and a slow motion of all other stars from the former towards the latter point. He applied this principle to the examination of the motion of several stars, and obtained a result which has been confirmed by subsequent researches. He found that there is a certain point in the constellation Hercules towards which the sun, with all his attendant planets, is rushing with enormous velocity. Later astronomers have examined the motions of hundreds of stars in both hemispheres, and have proved beyond a doubt that the sun really has a motion in that direction. They have also determined the rate at which the sun is travelling which appears to be somewhere about 150,000,000 miles per annum.

This result has been obtained simply by considering the apparent motions of the stars as interpreted by the principle which Sir W. Herschel laid down. The fact that the stars themselves are probably also in motion has not been left out of con-

sideration; but it has been thought, and justly, that in the long
run, when a sufficient number of stars have been observed, the
effect due to their motion *inter se* will be eliminated, and may
therefore be neglected throughout the inquiry. Were it not for
this, the inquiry would have seemed altogether hopeless.

But, returning to our illustration, the traveller through a
wood has other means besides the one pointed out, of determining his rate of motion among the trees. For, as he draws nearer
to a tree, he observes an increase in its apparent size, and in the
clearness with which its various parts are seen. The same also
is true of other objects in the wood; and if any of these be themselves in motion he is able to detect the fact, and to determine
the nature of the motion, by noticing not merely their apparent
change of place, but the variations which take place in their
apparent magnitude and distinctness. Now, the astronomer on
earth seems to be wholly debarred from applying a method resembling the last to determine the nature of the stellar motions.
The distances at which the stars are placed from us are so enormous, that no motion we could reasonably imagine them to have
could appreciably affect their apparent brilliancy within hundreds
or even thousands of years. Nor is this the only difficulty in the
way of astronomers. If the stars shone always, or for very long
periods, with uniform brilliancy, it might be possible for the
astronomers of one age to judge of a star's motions by comparing
their own estimate of its brilliancy with that formed by the
astronomers of another age. But the stars are continually varying in brilliancy, some with a regular periodic change, others
without any semblance of law, and this peculiarity renders it
quite impossible to judge of their motions by means of any
observed changes of brilliancy.

Therefore it seems hopeless for astronomers to attempt to learn
anything respecting stellar motions, beyond the mere fact that
the sun is advancing towards a certain region of the sidereal
heavens with an assigned velocity. Yet Mr. Huggins has succeeded in obtaining an answer to the very questions we have
suggested above.

Many of our readers are doubtless aware that the solar spectrum is crossed by a multitude of dark lines. Now, what is the
meaning of a dark line in the solar spectrum? We do not ask
what is the physical interpretation of the phenomenon, that is,
what evidence it gives us as to the sun's physical state—but
what does the phenomenon signify in itself? The rainbow-
coloured spectrum signifies the presence in the solar light of light-
waves of various length from the longest rays, which produce
red light, to the shortest, which produce violet light. If there
were light-waves of every possible length between those limits,
the solar spectrum could not fail to be perfectly continuous. But
it is broken by dark lines. Hence it follows that each dark line
signifies that light-waves of a certain definite length are wanting

in the sun's light; a similar result holds, of course, in the case of a star.

Now, if a star is at rest—neither approaching the earth nor receding from it—it is perfectly clear that the waves of light which correspond to a given line (or rather, since those waves are wanting, those which correspond to the part of the spectrum just above and just below the line) will have exactly the same position in the spectrum as the corresponding waves in the solar spectrum. So that, if it were possible to bring the sun's spectrum side by side with that of the star, this line in the sun's spectrum would be exactly on a level with the corresponding line in the star's. But if the star is speeding from or towards us, the waves of light will seem to be somewhat lengthened or shortened, precisely as the waves crossed by a stalwart swimmer seem broader or narrower according as he swims with or against their course. Hence, in this case, if the sun's spectrum and that of the star could be brought side by side, there would no longer be observed that exact coincidence which we described as resulting in the former case. If the star were very rapidly approaching us, the lines in its spectrum would be shifted towards the violent end of the spectrum, whereas if the star were receding from us, the lines would be shifted towards the red end. The apparent change of place would be very minute indeed, because the rate of the star-motion would be very minute in comparison with the enormous velocity of light. But still there would exist a possibility of so magnifying the change of place as to make it measurable.

Now we cannot bring the spectrum of a star side by side with that of the sun. But we can do what serves our purpose just as well—we can bring it side by side with the spectrum of some terrestrial element which is known to exhibit the particular line whose change of position we wish to measure.

This has been done by Mr. Huggins in the case of the star Sirius. The spectrum of this star is crossed by a multitude of dark lines, and, amongst others, by one known to correspond to a bright line seen in the spectrum of burning hydrogen. The two spectra were brought side by side, and due care having been taken to magnify as much as possible any discrepancy which might exist, it was found that the dark line in the spectrum of Sirius was not exactly opposite the bright line in the spectrum of hydrogen, but was slightly shifted towards the red end of the spectrum. It followed from the amount of the displacement, that at the time of observation Sirius was receding from the earth at the rate of about forty miles per second. When due account is taken of the earth's orbital motion at the time of observation, it results that Sirius is receding from the sun at the rate of about twenty-eight miles per second, or upwards of nine hundred millions of miles per annum.

The new method of examining the stellar motions is a most

promising one. It will doubtless soon be extended to other stars. In fact, nothing but time and patience are required to enable astronomers to extend this method to all the visible stars, and even to many telescopic ones. For the latter purpose, however, an instrument of enormous light-gathering power will be required, and we hear with pleasure that Mr. Browning, F.R.A.S., the optician, is engaged in constructing a spectroscope to be used with the great six-feet mirror of the Parsontown reflector.—*Daily News.*

THE METEOROLOGICAL DEPARTMENT.

THE *Athenæum* describes the progress of the Department, from two different sources: first, the official blue-book just issued — *The Report of the Meteorological Committee of the Royal Society,*—and secondly, the Report of the Kew Committee of the British Association for 1867-68, presented at Norwich. The functions of the Committee, as stated at page 6 of the Report, are divided into three great branches: 1, Ocean Meteorology; 2, Telegraphic Weather Information; and 3, Land Meteorology of the British Isles.

With regard to Ocean Meteorology, it may be stated that the Committee have hit upon a good plan of recording observations, and are pursuing it steadily. The surface of the globe has been divided into spaces comprising 10 degrees of latitude and longitude, which are called ten-degree squares from their shape, which is rectangular, on a chart on Mercator's projection. Each of these spaces comprises 100 single-degree squares, and each opening of the data-book employed corresponds to one of the smaller divisions. Formerly, the observations dealt only with the monthly mean temperatures for each five-degree square, but in certain parts of the ocean the boundaries between currents are so sharply defined that, without more minute tabulation, the phenomena would be entirely masked. Operations have been commenced on that part of the Atlantic which lies between the parallels 20° N. and 10° S. The Report states that already we have a practical result from the tabulations. At certain seasons of the year ships bound south should avoid the coast of South America, or else their passage will be prolonged as much as it would be at other times by keeping too close to Africa.

The Directors of the various lines of Transatlantic steamers have also lent themselves to the good work of determining the temperature of the surface of the sea, with a view to throw light on the course of ocean currents.

In the matter of Telegraphic Weather Information, the Committee merit praise both for what they have done and what they have not done. They have agreed to telegraph to the outposts the actual presence within a certain radius of a meteorological disturbance, and to permit the drum to be hoisted, but they do not pretend to prophesy. This is as it should be. The Fitzroy drums and cones, however, are now considered inadmissible, as they foretold with a possible wind, while the Committee wish to indicate the wind blowing on a certain line of coast, and of the locality where that wind is blowing. Captain Toynbee has devised a semaphore to give this information, and several are on trial.

The principal point about the Committee's labours as to the Meteorology of the British Isles is the fact that observations are being taken at properly chosen stations by means of *self-recording instruments*, for the altogether admirable contrivance and results of which we are indebted to Mr. Balfour Stewart, whose long experience at Kew, together with the practical skill of Mr. Beckley, his mechanical assistant, have enabled the Kew Committee to present us with specimens of "barograms" and "thermograms," which will soon ensure the introduction of similar self-recording instruments even in second-rate meteorological observatories. We have here direct evidence of the great advantages of the Meteorological and Kew Committees working in unison. If the object of the Meteorological Committee had been merely to apply in a practical manner the meteorological information already acquired, such a union would not have been essential; but if they intend, as they doubtless do intend, to extend the boundaries of science, to draw their parallels continually nearer to the unknown, it is of the utmost importance that they should unite with a body that has already done so much in this direction. The description given of the instruments in this Report is so clear that all may understand it, and their excellence is not so much the subject of discussion as of proof. Several unbiassed meteorologists of standing have given their adhesion to the plan, and already, as stated in the Report, in addition to the sets of self-recording instruments on the new plan previously ordered for Melbourne and Sydney, a set has been ordered for Bombay. Mr. Crossley, of Halifax, has generously undertaken to fit up at his own expense an observatory, in connection with the Meteorological Committee. As far as the extension of science is concerned, these two Committees, with two such chairmen, and with a man of such scientific attainments as Mr. Stewart for scientific secretary to both, represent a scientific combination not only powerful but, as far as we know, unique; indeed, these self-recording instruments may be looked upon as one of the first-fruits of this combination.

The importance of the connection between the Meteorological and the Kew Committee may, however, be put on a higher and even more important ground. There is a class of observations which may be called Cosmo-Physical observations, of immense importance at the present moment; and there is moreover, a sort of preparatory scientific conviction gradually arising that we are on the eve of some grand generalization, which may co-ordinate many things which seem at present strangely diverse and unconnected. Among the phenomena that are likely to be embraced by such a generalization as the one to which we refer are especially those which have been either solely or most successfully studied and registered at Kew, by the Kew Committee and Mr. Stewart, by means of their excellent photographic processes. Among these phenomena we have the physics of sun-spots, magnetic and electric disturbances, and meteorological phenomena generally.

We may, indeed, liken Kew to the head-quarters of an invading army bent on attacking the realms of darkness; and the scientific authorities, by means of Kew and other stations working in concert with Kew, have accumulated vast supplies. In fact, a sort of mental Abyssinian Expedition is on foot. But here a word of warning is necessary. The supplies are perfect in kind and sufficient in quantity; the commanders are full of dash, energy, and skill; but as the front advances the stores must follow, and the reserves must ever feed the front through an ever-increasing chain of posts. Here, it seems to us, is the weakness of this scientific campaign, as at present arranged : the leaders will die for want of mental food, while the ideal Annesley Bay is daily more crammed with transports.

Nor is the Meteorological Committee alone in this dilemma ; in the problems that belong particularly to Kew, there is the same drawback to the scientific advance : the vanguard must halt. The Kew Report for 1866-67 informed us that, owing to a deficiency in the Kew funds, the Kew magnetic curves had been reduced at the Woolwich office up to January, 1865, by the kind assistance of General Sabine. Again, the expense of printing a very valuable series of sun-researches, now too well known to need special mention here, was undertaken by Mr. De La Rue.

It is absolutely necessary that this source of weakness, this bar to advancement, should be overcome, and at once. There can be no doubt that, were proper representation made by the proper persons to Government, it would be overcome. It is unfair to the scientific world, to their officers and to themselves; the Government should not leave the matter in its present state. Either the work is worth doing thoroughly or it is not worth doing at all. It is not a question of placing our scientific army on the Horse Guards' footing as regards pay,—we are not absurd enough to suppose that our most valued and world-renowned scientific men in the Government service should receive one-sixth of the pay of a third-rate general in that enviable locality,—but it is a question of adding some few privates to the ranks of our scientific pioneers, who are engaged in fighting the unknown, and doing battle for the most important branch of physical inquiry.—*Abridged from the Athenæum, Oct. 3, 1868.*

REMARKS ON THE WEATHER DURING THE YEAR 1868, BY JAMES
GLAISHER, ESQ., F.R.S., &c., PRESIDENT OF THE METEOROLO-
GICAL SOCIETY.

The weather was cold during the first 11 days of the year; for
this period the deficiency of temperature averaged 6¼° daily.
The wind was from N.E.; on the 12th day it changed to the
S.W.; the temperature increased and passed above the average,
and continued so for the most part till the end of the first
quarter, the exceptions from an excess of temperature being few
in number and small in amount. The average excess of temperature in the 80 days ending 31st March was rather more than
3¼° daily.

The weather during the whole quarter was remarkably fine
and warm; the temperature being nearly constantly in excess over
the average, the exceptions being few in number and small in
amounts. The average daily excess of temperature for the 91
days ending 30th of June was 3°·1, and for the 171 days (from
January 12th to June 30th) was daily more than 3¼° in excess.

The weather during the second quarter was of the same character
as in the preceding quarter, viz., remarkably fine and warm; the
temperature of the air was nearly constantly above and, at times in
the month of July and at the beginning of August, to very large
amounts. The average daily excess of temperature for the 92
days ending 30th September was very nearly 4°, and for the
263 days (from 12th January to 30th September) was 3¼° daily.
In no year, back to 1771, has the excess of temperature been so
large as this for so long a period.

The very long period of warm weather which had prevailed
from January 12th to September 30th changed to cold on October 1st, and during the months of October and November, the
temperature of the air, with the exception of a few slightly warm
days, was below its average for the season. For the 61 days
ending November 30th the average daily deficiency of temperature was 2°. On December 1st the mean temperature was that
of its average; on the 2nd day it passed above the average, and
so continued throughout the month (with the exception of the
11th day, when it was slightly in defect), and the excess of
mean daily temperature was as large as 5¼° over the average of
50 years. The month of December was, therefore, exceptionally
warm, its average temperature having been exceeded twice only in
the period of 98 years, viz., in the years 1806 and 1852. The month
was also distinguished by the lowest monthly reading of the barometer since the year 1841. The fall of rain was also exceptional
in this month, the amount collected was 5½ inches nearly, being
the greatest in any December as far back as the record at the
Royal Observatory extends.

The mean temperature of January was 37°·2, being 1°·0 higher
than the average of 97 years, and 3°·0 higher than the corresponding temperature of the preceding year.

The mean temperature of February was 43°·0, being 4°·5 higher than the average of the preceding 97 years, 1°·7 lower than the preceding year, and higher than the corresponding temperatures of any year, except 1859 and 1867, as far back as 1851.

The mean temperature of March was 44°·0, being 3°·1 higher than the average of the preceding 97 years, and higher than the corresponding temperature of any year since 1859.

The month of February was remarkably warm. There were less than the average of east winds, and compounds of east winds, in both the months of February and March. The weather in February was more like spring than winter, it caused vegetation to progress rapidly, and at the end of the month trees and shrubs were budding, and the accounts respecting autumn and winter-sown wheat were favourable.

The month of March, though less settled than February, was still favourable to agricultural pursuits ; good progress was made in plowing, sowing, and planting.

At the end of the quarter vegetation was in advance of ordinary seasons, and the prospects of the next harvest were favourable.

The reading of the barometer at Greenwich on the 1st of January was about 30 inches, and from this day to the 12th the readings were constantly above the average. A general tendency to decrease now followed, the readings fell to 28·8 inches by the 18th, and remained between 28·8 inches and 29·0 inches till the 20th, when violent gales were experienced. A rapid increase to 29·6 inches occurred by the morning of the 21st, followed by an equally rapid decrease to 28·86 inches by the following morning, then a steady increase took place, and the readings of the barometer were generally high till the 30th of January ; another rapid decline was then experienced accompanied by a gale of extreme violence on 31st January and 1st February. From February 3rd to 27th the readings of the barometer were generally high, at times reaching 30·4 inches; a period of depression below the average followed until the 12th of March, and from the latter day to the end of the quarter the readings were generally in excess of the corresponding averages.

The mean temperature of April was 48°·1, being 2°·2 higher than the average of 97 years, and 0°·9 lower than the corresponding temperature of the preceding year.

The mean temperature of May was 57°·3, being 4°·8 higher than the preceding 97 years, and higher than the corresponding temperatures of any year since 1848, when it was 50°·7.

The mean temperature of June was 62°·0, being 3°·9 higher than the averages of the preceding 97 years, and higher than has been recorded in the period 1841-1867, with the exceptions of 1842, 1846, and 1858, when they were respectively 62°·9, 65°·3, and 64°·9.

The mean high day temperatures were respectively 1°·3, 6°·0, and 5°·3 higher than the averages in April, May, and June.

The month of April was warm, but not remarkably so, for since the year 1771 there have been 24 Aprils of higher temperature.

The month of May was of higher temperature than any since the year 1818, when it was 59°·7, or 2°·4 warmer than in this year; the next and only other instance back to 1771 was in 1833, when the mean temperature of May was 59°·4. The mean temperatures of all the other Mays were less than 57°.

The month of June was of high temperature, but was greatly exceeded in the year 1846, when it was 65°·3, or 3°·3 warmer; the other instances in June of higher temperature than in June of the present year, back to the year 1771 were in the years 1812, 1822, 1818, 1781, and 1775. The highest of them was 62°·9 in 1812 and 1818, the lowest 62°·6, in the year 1781.

The mean temperature of the 3 months ending June was 55°·8; for the same period in 1775 it was 55°·5; in 1822 was 55°·0; in 1844 was 55°·1; in 1846 was 55°·7; in 1848 was 55·03; and in 1865 was 56°·2; so that the only instance in 98 years of higher temperature in the corresponding quarter than in the present year, was in 1865. In the latter year the temperature in April was 52°·3, being higher than any other April on record.

The other years since 1771, when the mean temperature of the three months ending with June exceeded 54° and less than 55°, were 1778, 1779, 1783, 1798, 1811, 1826, 1833, 1834, and 1859.

It will be interesting to compare the mean temperature of the longer period, viz., February to June. The mean temperature of this period for 1868 was 50°·9, the mean temperatures of the corresponding period of other years distinguished by high temperature, were as follow :—In the year 1775 it was 50°·0; in 1779 was 51°·1; in 1794 was 49°·4; in 1822 was 51°·1; in 1826 was 49°·5; in 1846 was 50°·8; in 1848 was 50°·6; and in 1859 it was 50°·1. The mean temperature of these five months for all the other years since 1771 was less than 50°·0. In two instances therefore, viz., in the years 1779 and 1822, have these five months been of higher temperature than in 1868, and in both by so small an amount only as the one-fifth part of a degree ; but if we compare the mean temperature of the 171 days ending 30th June with the corresponding period of other years, we find that the year 1822 is the only one distinguished by an excess of temperature over the present year.

GREENWICH METEOROLOGICAL ELEMENTS FOR THE YEAR 1862.

By James Glaisher, Esq., F.R.S.

| Date | Mean Reading of the Barometer | Temperature of the Air | | | | | | | Pressure from reading of 31 p.m. | Mean Temp. of the Dew-point. | Mean Elastic Force of Vapour | Weight of Vapour in a cubic foot of air | Mean Weight of Vapour required for Saturation | Mean Degree of Humidity, 100 | Mean Weight of a cubic foot of Air. | Relative Proportion of Wind | | | | Mean Amt. of Cloud 0 to 10 | Rain | |
		Highest by Day	Lowest by Night	Range in the Month	Mean of all Highest	Mean of all Lowest	Mean Daily Range	Mean for the Month								N.	E.	S.	W.		Amount Collected	No. of days on which it fell
	In.							°	In.	°	In.	Grs.	Grs.	Grs.								In.
Jan.	29.741	51.9	32.8		29.1		9.6	.372	−0.9	63.4	.701	7.2	6.1		80	9	5	9	10	6.4		4.10	21
Feb.	29.516	51.7	35.7	25.1	35.9		13.1	.475	+1.1	37.8	.223	2.9	0.8		83	8	10	8	17	6.3		1.93	11
March	29.854	55.5	33.1	30.3	32.9		16.1	.440	+2.3	33.3	.231	2.6	0.6		75	5	10	8	10	6.1		1.07	16
April	29.788	57.0	39.2	40.6			18.7	.431	+1.8	40.0	.261	3.6	1.4		63	6	4	14	8	6.4		1.39	10
May	29.830	70.0	44.7	53.1			23.5	.673	+3.2	49.5	.365	5.0	2.0		73	8	5	7	11	4.9		1.57	8
June	29.900	79.0	44.7	42.3			26.5	.629	+4.9	51.0	.370	5.1	2.7		65	5	11	7	12	6.6		1.08	6
July	29.941	80.6	45.3	45.2			23.5	.635	+5.9	54.9	.435	5.7	1.8		76	9	11	8	9	6.5		1.32	12
Aug.	29.755	80.5	47.9	42.7			21.5	.768	+2.4	54.3	.401	5.3	1.3		81	7	5	10	8	5.2		2.82	16
Sept.	29.696	69.4	43.0	40.5			20.6	.573	+2.6	53.8	.391	5.9	1.1		77	4	11	8	10	5.9		1.16	9
Oct.	29.792	60.5	37.8	37.9			16.7	.479	−2.5	38.0	.225	2.9	0.8		77	3	8	6	13	5.6		3.38	13
Nov.	29.436	57.1	36.1	31.9			10.9	.425	+4.5	37.0	.227	2.9	0.7		87	11	5	7	9	7.4		2.16	11
Dec.	29.370	57.8	39.3	38.5			9.4	.460	+5.9	42.9	.226	3.2	0.4		86	1	9	2	17	6.9		3.15	10
Means	29.726	77.1	34.3	39.8			17.5	.516	+2.2	44.6	.360	3.9	1.1		78	73	80	110	122	6.2		25.15	146

Note. — In column 10 the sign + implies above, and the sign − below: the average.

Explanation. — The cistern of the barometer is about 154 feet above the level of the sea, and its readings are coincident with those of the Royal Society's flint-glass barometer. The observations are taken daily at 9 a.m., noon, 3 p.m., and 9 p.m.; the mean of these readings are corrected for diurnal range by the application of Mr. Glaisher's corrections, in the Philosophical Transactions, Part I., 1848; and from the readings of the dry and wet bulb thermometers, thus corrected, the severally corrected deductions to columns 11 to 16 are calculated by means of Mr. Glaisher's Hygrometrical Tables, Fourth Edition.

The numbers in column 2 show the mean reading of the barometer every month, or the mean height of a column of mercury which balanced the whole weight of atmosphere of air and water; the numbers in column 12 show the length of a column of mercury balanced by the same stone; and if the numbers in this column be subtracted from those in column 2, the result will be the length of a column of mercury balanced by the air alone, or that reading of the barometer which would have been, had no water been mixed with the air.

These same five months (February to June) have been further distinguished by having an almost constant atmospheric pressure above the average; the mean monthly excess of pressure was more than 0·1 inch. They have also been distinguished by a deficiency of rain in each month, with the exception of April; the amount below the average in the five months ending June was 2·5 inches; but reckoning from 1st of January the fall of rain is very nearly the true fall for the period, the deficiency being only 0·1 inch. The period from 1st January has been distinguished by an unusual distribution of rain; in January it fell to the depth of 4·2 inches, being an excess for that month of 2·4 inches. The drought which was experienced towards the end of the quarter is not attributable, therefore, to a deficiency of rain since the beginning of the year up to the end of June, but to its unusual distribution over these months, there having been a great excess in January and a great deficiency in June, together with an unusual evaporation caused by continued high temperatures extending over a period of five months.

The highest temperature occurred on 19th June, when it was 87°, and on 13th and 14th June, when it was 85°. These temperatures were exceeded at some places in the Midland Counties.

It is very remarkable that notwithstanding the continuance of high temperatures, but one thunderstorm occurred at Greenwich during the quarter, that on the 29th of May, on which day the greater part of the rain for that month fell, and generally over the country there have been much less than the usual number of thunderstorms.

For agricultural pursuits the month of April was favourable, and at its end there was every prospect of an early and plentiful harvest.

The month of May was remarkable for brilliant sunshine, high temperature, the general forwardness of the season, and the promising appearance of the cereal crops.

The month of June was favourable to the ripening of the wheat crops, but injurious to grass lands, and to all spring and root crops.

The hay crop was housed in good condition at an unusually small expense; the quality is good, but the bulk is stated to be small.

The mean temperature of July was 67°·5, being 6°·1 higher than the average of 97 years, and 8°·1 higher than the corresponding temperature of the preceding year.

The mean temperature of August was 63°·6, being 2°·9 higher than the average of the preceding 97 years, 1°·6 higher than the preceding year, and higher than the corresponding temperatures of any year as far back as 1857.

The mean temperature of September was 60°·5, being 4°·0 higher than the average of the preceding 97 years, 2°·9 higher than the preceding year, and 4°·1 higher than the corresponding temperature in 1860.

The mean high day temperatures in July, August, and September were respectively 8°·4, 2°·4, and 4°·1 higher than the averages.

The month of July was remarkably warm, the temperature of the air on the 22nd day was as high as 90°·6, a higher temperature than has ever before been recorded at Greenwich; it reached 92° on two occasions, viz., on July 16th and 21st, and was 90° on two other occasions, viz., on July 20th and 28th. In 1859 the temperature once reached 93°, and in 1846 it was once 93°·3.

The mean temperature of the month was 67°·5, in the year 1859 it was 68°·1, and in the year 1778 was 67°; in all other years, back to 1771, it was less than 67°. The month was therefore very remarkable for its high mean temperature, being the highest, except one, of any month in the preceding 97 years.

The beginning of the month of August was of high temperature; on the 5th day the maximum temperature was 90½°. The mean for the month was high, but not remarkably so; it was 63°·6; in the year 1857 it was 65°·8; and back to 1771 there were nine other instances of as high or higher temperature than 63°·6, viz., in the years 1779, 1780, 1800, 1802, 1807, 1818, 1819 1842, 1856.

The month of September was warm throughout, particularly at the beginning; on the 7th day the maximum temperature reached 93°·1; in the warm month of 1846 the highest temperature reached was 86°·4, and in the warmest September on record, viz., that of 1805, it reached 86° only.

The mean for the month was 60°·5, being 3°·4 of lower temperature than in 1865, and 1°·8 lower than in 1815, and nearly the same as in the years 1779, 1793, 1818, 1840, and 1858, whilst in all other years the mean temperature has been below 60°.

The mean temperature of the three months ending September was 63°·9; for the same period in 1818 it was 63°·5, and in 1857 was 63°·3, in all other years it was below 63°; so that there is no instance in 98 years of so high a mean temperature in the corresponding quarter, as in the present year.

The warm period began about the middle of January. Comparing the monthly mean temperature of the eight months since, viz., February to September in this year, with the corresponding long period in other years as far back as we can, we find that of 1779 was 55°·6, in 1846 was 55°·2, in 1868 was 55°·7; in all other years, back to 1771, it was less than 55°. In no year, therefore, has the mean temperature of these eight months been so high as in this year; that of 1779 was, however, closely approximate.

The reading of the barometer at Greenwich on the 1st July was about 30·1 inches, and continued almost constantly above the average throughout that month; on the 1st August it began to fall, and continued steadily falling till the 7th, when the reading recorded was 29·52 inches. From the 10th to the 24th the readings were generally in defect of the average; a rapid fall from 29·84 inches on the 20th to 29·05 inches on the 22nd being accompanied by a violent gale on the 22nd. On the 25th August the readings rose above the average, and continued in excess till the 6th September, when a slight decrease took place, but was immediately succeeded by an increase to 30·21 inches on the 9th. On the 9th a steady fall commenced, and from the 11th the readings were constantly below the average. During the last few days of the month several gales were experienced.

At the end of the month of July the harvest was progressing in almost every part of the British Isles, and in some of the southern districts was brought near to completion. Harvest came in suddenly and nearly simultaneously in all parts of the country; and the crops proved to be in such a perfect state, that cutting, carrying, threshing, and grinding into flour followed in rapid succession.

There were many sudden deaths from sunstroke during the month. The want of water was severely felt, and this, combined with the great heat, acted injuriously both on animal and vegetable life to an extent unprecedented, so far as I can trace, in this country. Pastures and grass lands were generally brown and bare; in many places not a blade of grass was to be seen.

At the end of August the harvest was nearly completed; pastures and grass lands were of their ordinary verdant appearance, the rain-fall changing the appearance of the fields in a very short time, and root crops were benefited by the moisture.

The month of September was favourable to agricultural pursuits. Towards the end of the month heavy rains fell all over the country, and brought general relief to animal and vegetable life. Ponds and wells recommenced to yield their usual supply of water; streams and currents were filled. The rain loosened the ground for the plough; and the potato crop was, upon the whole, spoken of with satisfaction.

The mean temperature of October was 47°·9, being 1°·8 lower than the average of 97 years, and lower than the corresponding temperatures of any year as far back as 1852.

The mean temperature of November was 41°·5, being 0°·9 lower than the average of 97 years, 0°·1 higher than the preceding year, but 2°·8 lower than the corresponding temperature in 1866.

The mean temperature of December was 46°·0, being 6°·9 higher than the average of 97 years, and higher than any corresponding temperature in the period 1771–1867, with the sole exceptions of 1806 and 1852, when 46°·8 and 47°·6 respectively were recorded.

The mean temperature of the air in the three months ending November, constituting the three autumn months, was 50°·0, being 0°·8 higher than the average of the preceding 97 years.

The reading of the barometer on the 1st October was about 29·61 inches, and during the month oscillated above and below the average several times to a small extent; the movements in November were generally of the same character as in October, viz., at the beginning of the month below the average, in the middle, above, and at the end, alternately in excess and defect in periods of two or three days. The mean daily readings during December were, with one exception, below the average, on two occasions being over an inch in defect. From the 1st to the 8th the mean daily readings were below the average to the mean amount of 0·29 in. During the night of the 8th a sudden rise took place, and at 9 h. P.M. of the 9th the maximum for the month was attained, viz., 30·17 in.; this was followed by a no less sudden fall on the 10th and 11th to 29·08 in., and was accompanied by a destructive gale from the S.S.W.; the defect from the average on the 11th was 0·65 in.

From the 12th to the 20th, increase and decrease followed each other in rapid succession, the mean daily readings, however, being constantly below the average, and on 20th, 9 h. P.M., 29·52 in. was recorded. During the night of the 20th a fall set in, and with very slight fluctuations lasted till 2 h. P.M. of the 24th; the reading recorded was 28·53 in., being the absolute minimum in the year 1868, and the mean value for this day was as much as 1·22 in. in defect of the average. On the 26th the readings increased to 29·44 in. at noon; a rapid decrease then occurred, and the minimum (28·76 in.) was reached on the 27th at noon. Violent gales were experienced on the 27th and 28th, and pressures were recorded of 30 lbs. on the square foot. A rise then took place, and the reading at the end of the month was 29·85 in.

The mean reading of the barometer in December 1868, was 29·38 in.; the lowest mean reading in any month back to 1841 was 29·40 in. in January, 1865; the other instances of mean readings below 29·50 in., were 29·44 in. in October, 1841, and 1865; 29·47 in. in November, 1852, and January, 1856; and 29·49 in. in December, 1860.

The average mean reading for December is 29·83 in. The average pressure of the atmosphere for December was 29·379 in., being 0·45 in. below the average; and the range was 1·64 in.

Field Elm leaf-buds first appeared. On the 16th of February at Boston; on the 17th at Weybridge Heath; and on the 28th at Helston. On the 2nd of March at Eastbourne; on the 10th at Miltown; and on the 17th at Holkham.

Horsechestnut leaf-buds first appeared. On the 20th of February at Eastbourne; on the 24th at Boston; on the 25th at Miltown; and on the 26th at Helston. On the 14th of March at Guernsey and Strathfield Turgiss; on the 15th at Holkham; and on the 21st at Weybridge Heath.

Sycamore leaf-buds first appeared. On the 17th of February at Boston; on the 19th at Weybridge Heath; and on the 28th at Miltown. On the 10th of March at Strathfield Turgiss; on the 14th at Guernsey; on the 17th at Holkham; and on the 19th at Helston.

Hawthorn leaf-buds first appeared. On the 10th of February at Eastbourne and Weybridge Heath; on the 15th at Guernsey; on the 21st at Taunton; and on the 24th at Boston and Miltown. On the 9th of March at Holkham; on the 14th at Hull; and on the 21st at Strathfield Turgiss.

Horsechestnut in leaf. On the 21st of March at Ripon; on the 23rd at Strathfield Turgiss; on the 30th at Guernsey and Oxford; on the 31st at Eastbourne. On the 4th of April at Holkham; on the 5th at Wisbech; on the 16th at Miltown; on the 20th at Boston; on the 23rd at Culloden; on the 24th at Marlborough College; and on the 28th at Cockermouth. On the 3rd of May at Hull.

Sycamore in leaf. On the 21st of March at Ripon; and on the 30th at Guernsey. On the 5th of April at Tunbridge Wells and Holkham; on the 7th at Eastbourne; on the 9th at Weybridge Heath; on the 12th at Cockermouth; on the 20th at Miltown; on the 23rd at Wisbech; and on the 24th at Boston. On the 1st of May at Ripon; and on the 15th at Hull.

Hawthorn in leaf. On the 27th of February at Eastbourne; and on the 29th at Guernsey. On the 10th of March at Oxford; on the 11th at Streatley Vicarage; on the 12th at Ripon; on the 20th at Weybridge Heath; on the 23rd at Holkham; and on the 25th at Helston.

Apple in blossom. On the 31st of March at Guernsey and Wilton. On the 5th of April at Miltown; on the 14th at Wisbech; on the 15th at Eastbourne, Weybridge Heath, Strathfield Turgiss, Oxford, and Culloden; on the 16th at Marlborough College; on the 20th at Hull and Cockermouth; on the 22nd at Holkham; on the 25th at Grantham; and on the 30th at Cardington.

Pear in blossom. On the 14th of February at Guernsey. On the 7th of March at Helston; on the 25th at Wilton; on the 26th at Streatley Vicarage; on the 30th at Oxford, Ripon, and Miltown; on the 31st at Eastbourne. On the 5th of April at Wisbech; on the 6th at Weybridge Heath, Grantham, and Holkham; on the 7th at Cardington; on the 10th at Hull; on the 16th at Culloden; and on the 20th at Boston.

Peach in blossom. On the 25th of February at Guernsey; on the 27th at Helston and Taunton; and on the 28th at Eastbourne. On the 4th of March at Streatley Vicarage; on the 7th at Wilton; on the 8th at Oxford; on the 14th at Miltown; and on the 15th at Wisbech.

Plum in blossom. On the 25th of February at Guernsey. On the 11th of March at Streatley Vicarage; on the 12th at Strathfield Turgiss; on the 14th at Helston; on the 17th at Oxford; on the 18th at Ripon; on the 20th at Miltown; on the

25th at Weybridge Heath; on the 29th at Holkham; and on the
30th at Wilton. On the 3rd of April at Wisbech; on the 4th
at Strathfield Turgiss and Cardington; and on the 6th at Cockermouth.

Cherry in blossom. On the 10th of March at Helston; on the
27th at Holkham; on the 28th at Miltown; on the 30th at Wilton
and Oxford; on the 31st at Strathfield Turgiss and Hull. On
the 4th of April at Weybridge Heath; on the 5th at Marlborough
College; on the 13th at Wisbech; on the 15th at Kingsley
Parsonage, Cockermouth, and Culloden; and on the 28th at
Grantham.

Snowdrops in flower. On the 28th of January at Helston and
Streatley Vicarage; on the 29th at Eastbourne; on the 30th at
Sidmouth; and on the 31st at Lampeter. On the 3rd of
February at Boston; on the 6th at Taunton; and on the 7th at
North Shields.

Swallow arrived. On the 31st of March at Guernsey.

Wryneck arrived. On the 31st of March at Guernsey.

Woodcock departed. On the 20th of February at Helston.

Oak in leaf. On the 27th of April at Tunbridge Wells; and
on the 29th at Oxford and Miltown. On the 7th of May at Ripon;
on the 8th at Cockermouth; on the 10th at Culloden; on the
11th at Wisbech; and on the 16th at Hull.

Lime in leaf. On the 4th of April at Wisbech; on the 15th
at Oxford; on the 18th at Guernsey; on the 22nd at Miltown; on
the 24th at Boston; on the 25th at Culloden; on the 26th at
Strathfield Turgiss; and on the 30th at Marlborough College.
On the 4th of May at Ripon; and on the 9th at Hull.

Lilac in blossom. On the 4th of May at Helston; on the 10th
at Guernsey; on the 14th at Taunton; on the 18th at Holkham;
on the 20th at Bournemouth; on the 22nd at Weybridge Heath
and Battersea; on the 25th at Eastbourne; on the 28th at Wisbech and Miltown; and on the 29th at Hawarden. On the 1st
of May at Oxford; on the 2nd at Strathfield Turgiss, Lampeter,
and Boston; on the 3rd at Ripon and York; on the 7th at Hull;
on the 9th at Silloth; and on the 25th at Culloden.

Laburnum in blossom. On the 14th of April at Helston;
on the 20th at Bournemouth; on the 26th at Eastbourne; on
the 29th at Hawarden; and on the 30th at Wisbech. On the
1st of May at Battersea; on the 2nd at Taunton, Weybridge
Heath, Boston, and Miltown; on the 3rd at Oxford; on the 4th
at Strathfield Turgiss; on the 7th at Ripon; on the 8th at York
and Silloth; and on the 14th at Hull and Lampeter.

Wheat in ear. On the 24th of May at Royston; on the 26th
at Taunton and Boston; and on the 28th at Wisbech. On the
1st of June at Cardington; on the 2nd at Weybridge Heath; on
the 3rd at Helston; on the 7th at Grantham and Hull; on the
8th at Strathfield Turgiss; on the 10th at Miltown; on the 15th
at Cockermouth; and on the 27th at Culloden.

Wheat in flower. On the 5th of June at Boston; on the 8th at Cardington; on the 9th at Weybridge Heath and Marlborough College; on the 11th at Helston; on the 18th at Grantham; on the 26th at Hull; and on the 30th at Cockermouth.

Barley in ear. On the 5th of June at Weybridge Heath; on the 6th at Cardington; on the 14th at Miltown; on the 16th at Holston; on the 20th at Cockermouth; and on the 29th at Culloden.

Barley in flower. On the 10th of June at Cardington; on the 14th at Weybridge Heath; and on the 18th at Marlborough College.

Oats in ear. On the 7th of June at Weybridge Heath; on the 11th at Strathfield Turgiss; on the 16th at Helston; and on the 30th at Cockermouth, Miltown, and Culloden.

Wheat cut on the 7th July at Weybridge; on the 8th at Norwood; on the 10th at Worthing and Cardington; on the 13th at Strathfield Turgiss, Hawarden, and in Kent generally; on the 14th at Osborne and Eastbourne; on the 16th at Guernsey and Boston; on the 17th at Helston; on the 20th at Holkham; on the 24th at Hull; and on the 27th at Carlisle. On the 8th August at Miltown and Culloden.

Barley cut on the 3rd July at Worthing; on the 9th at Guernsey and Weybridge; on the 13th at Helston; on the 20th at Cardington; and on the 22nd at Carlisle. On the 6th August at Strathfield Turgiss; on the 10th at Culloden; and on the 12th at Miltown.

Oats cut on the 6th July at Weybridge; on the 9th at Guernsey; on the 11th at Boston; on the 13th at Helston; on the 24th at Carlisle; on the 27th at Strathfield Turgiss; and on the 31st at Culloden. On the 4th August at Miltown.

Apple ripe on the 20th July at Hull. On the 2d August at Strathfield Turgiss; on the 10th at Boston; on the 25th at North Shields; on the 27th at Helston; and on the 28th at Miltown and Culloden.

Pear ripe on the 28th July at Miltown. On the 11th August at Helston; on the 17th at Strathfield Turgiss; and on the 30th at North Shields.

Peach ripe on the 5th August at Helston; on the 7th at Strathfield Turgiss; on the 12th at Boston; on the 27th at Miltown; and on the 28th at Culloden.

Plum ripe on the 5th August at Helston; on the 7th at Weybridge; on the 10th at Strathfield Turgiss; on the 12th at Boston; on the 20th at Culloden; and on the 22nd at Miltown.

The Lime leafless on the 17th October at Strathfield Turgiss; on the 22nd at Boston and Hull; on the 24th at Wisbech and Helston; on the 25th at Marlborough College; and on the 30th at Oxford. On the 2nd of November at Weybridge Heath.

The Horsechestnut leafless on the 8th October at Helston; on the 16th at Hull; on the 20th at Oxford; on the 22nd at

Boston ; on the 24th at Strathfield Turgiss ; on the 25th at Marlborough College and Spital ; and on the 30th at Weybridge Heath.

The Common Poplar leafless on the 12th October at Helston; and on the 30th at Boston. On the 2nd November at Hull ; on the 3d at Weybridge Heath ; and on the 7th at Llandudno.

The Hawthorn leafless on the 7th October at Helston ; and on the 30th at Boston. On the 4th November at Weybridge Heath ; on the 5th at Llandudno ; and on the 24th at Hull.

Woodcock arrived on the 6th October at Helston; on the 15th at Cardington ; on the 24th at Hawarden ; and on the 29th at Strathfield Turgiss. On the 3rd November at Taunton ; and on the 5th at Guernsey.

GREENWICH OBSERVATIONS.

THE volume of Greenwich Observations for 1860 has been published by the Astronomer-Royal. It contains, perhaps, an unusual quantity of details, for it is thicker than ever—400 pages being taken up with the magnetical and meteorological observations alone. Any one who wishes to know how to conduct an observatory, and take account of magnetism and of the weather, has only to study this big book in order to qualify himself for the work. The arrangement of the different magnets—the compensations to be allowed—the method of preparing the self-registering photographic apparatus—the way to observe spontaneous terrestrial galvanic currents—to note and record the phenomena of temperature, fluctuations of moisture, of wind and radiation—how to read the rain-gauge, the actinometer and electrometer,—are all described with instructive fulness in this ponderous quarto.—*Athenæum.*

A MIRAGE

In May last was strikingly conspicuous on a Sunday afternoon and evening at Dover. The *Dover Chronicle* states that the dome of the Cathedral and Napoleon's Pillar at Boulogne were to be seen from the Crescent Walk by the naked eye, but with a telescope of ordinary power the entrance of the port, its lighthouse, its shipping, and the surrounding houses, the valley of the hillside of Capecure, and the little fishing village of Portel, were distinctly visible ; whilst on the eastern side the principal features of the country, the lighthouse of Cape Grinez, the adjacent windmill, numerous farms and villages, with their windows illuminated by the setting sun, stood out with extraordinary clearness. Whilst these were under observation, a locomotive was seen to leave Boulogne, and travel some miles in the Calais direction, by its puffs and wreaths of white steam. Shortly after sunset the mirage subsided.

Obituary.

JOHN CRAWFURD, F.R.S., the distinguished Oriental Scholar and Ethnologist. He studied for medicine. In 1803 he obtained a medical appointment in the Indian Service, embarked for India in April, and landed in Calcutta in September of the same year. For the first five years of his residence in India he was employed in his professional duties with the army, chiefly in the North-West Provinces, in the neighbourhood of Delhi and Agra. In 1808 the same duties took him to Penang, in the Straits of Malacca, where he began to devote himself to that study of the languages and manners of the Malay race which was destined to make him widely known. In 1811, having been brought under the notice of Lord Minto, then Governor-General of India, Mr. Crawfurd was invited to accompany him on the expedition which effected the conquest of Java. After that event, in consequence of his acquaintance with the Malay languages, he was appointed to represent the British Government at the Court of one of the native Princes, and for nearly six years he filled some of the principal diplomatic offices of the island. It was then that he collected the materials for the work which he afterwards published, entitled, *The History of the Indian Archipelago.* Java, and their other Indian possessions, having been restored to the Dutch, Mr. Crawfurd returned to England in 1817, and in 1820 published the work just mentioned. In 1821 he went back to India, and shortly after his return was appointed by the first Marquis of Hastings, at that time Governor-General, to the Diplomatic Mission to Siam and Cochin China. In 1823 Mr. John Adam, *ad interim* Governor-General, appointed him to administer the new settlement of Singapore, on the resignation of its founder, Sir Stamford Raffles. In that position he remained three years, and concluded with the native chiefs, to whom the settlement belonged, the convention by which we hold its sovereignty. In 1826 he returned to Bengal, and was forthwith appointed by the Governor-General, Lord Amherst, Commissioner in Pegu, and eventually, on the conclusion of peace, Envoy to the Burmese Court. In 1827 Mr. Crawfurd finally returned to England, and in the following year published an account of his mission to Siam and Cochin China, and in 1829 another of his mission to Burmah. After this period, long leisure, good health, and an inclination to study, and capacity for work enabled him to keep up and perfect his stores of Indian and Eastern information. He was an indefatigable contributor to the Press on matters relating to the East, and indeed on many other subjects. In 1852 he published a grammar and dictionary of the Malay languages, and in 1856 a descriptive dictionary of Malay, and the languages of the Philippine Archipelago, works which secured for their author the respect of the philological world. All members of the Geographical and Ethnological Societies will miss the tall form of the evergreen veteran, who scarcely ever failed to take part in their discussions, and who, while stoutly maintaining his own views, showed a forbearance and courtesy in listening to others which might well be imitated by all members of learned Societies. Of singularly simple and unostentatious bearing, few were more able, and certainly none more ready, to impart sound information to those who sought his advice and assistance. A self-made man, he showed none of that jealousy which sometimes makes self-made men believe that kind of creation ended when they were made.—*Abridged from the Times.*

MARTINUS DES AMORIE VAN DER HOEVEN, Professor of Law ot the Athenæum, Amsterdam.—He was born at Rotterdam, in 1824, and was a man of extraordinary learning. Hardly anything written exists by his hand. He took no notes. All he wanted to remember was locked up in his prodigious memory. He read books as fast as he could, and the number of volumes swallowed during his lifetime must be something tremendous.

HENRY, LORD BROUGHAM, F.R.S. (A Memoir of his Lordship's scientific labours appeared in *The Year Book of Facts*, 1861, with a portrait.)—Lord Brougham was elected a fellow of the Royal Society in 1803. His first paper in the *Philosophical Transactions*, " Experiments and Observations on the Inflection, Reflection, and Colours of Light," was published in 1796. For some years past the venerable Peer had become, by seniority of election, the " Father " of the Society, and was fond of the dignified title. By his decease, another Peer, the Earl of Lonsdale, who was elected F.R.S. in 1810—the year when he was a Lord of the Admiralty—succeeds to the parental distinction, and is now the Father of the Royal Society. Next to him came Sir Henry Ellis, who was elected in 1811 (since deceased) ; and, running our eyes down the list of Fellows, we find, in 1813, Sir John Herschel ; in 1814, Lord Broughton ; in 1815, Sir H. Holland and Sir W. E. Eliott ; in 1813, Dr. Roget ; in 1816, Sir F. Pollock, Mr. Babbage, and the Rev. H. H. Baber ; in 1817, Dr. Granville ; in 1818, General Sabine—the actual President of the Society of the same year was John Crawfurd, who has passed away ; then, in 1819, Sir Thomas Phillipps ; in 1820, Sir J. S. Lefevre and Commander Friend ; in 1821 comes the veteran geologist, Sedgewick ; followed by Sir Woodbine Parish in 1824, and in 1826 by the two now famous geologists, Sir Charles Lyell and Sir Roderick Murchison. Without exhausting the list of names of the same period, we have here a goodly file of veterans, many of whom had become men of mark in the Royal Society before the present generation was born. It is well for the Society that they thus combine the distinctions and traditions of the past and the present.

CHARLES JOHN KEAN, F.S.A., the celebrated Actor, who had been on the stage just forty years.—"Mr. C. Kean was the first manager who took a strolling company round the world. He obtained large profits, but he injured his health. Though hopeful of recovery, he was conscious of peril. Last March, when he took leave of Edinburgh, promising to begin his final visit on the 10th of this present month of February, his words were freighted with melancholy significance. All his old comrades there, of long, long ago, he remarked, had passed away; and, he added, 'the shades of evening are closing round me, the parting hour is rapidly approaching, and the *last* chapter of my theatrical history is about to be opened. One season more, and the dark curtain will descend on my professional existence.' The chapter *had* closed and the curtain *had* fallen, and there was little left for the sad yet hopeful player but to struggle for a while, and then to die. The disease which subdued him was albuminuria, or 'Bright's Disease,' with serous affection of the heart. With him, who was the noblest kinsman that poor relatives ever possessed, the name of Kean disappears from the bills, after being there in reference to various members of his family, Moses Kean, Carey and Kean, the travelling showmen, Edmund Kean, and others, some eighty or ninety years. In the last of them has gone, if not the best actor, certainly the worthiest man. One member of the family, however, survives, in an aged strolling actress, whose name at fairs and in barns is Mrs. Cuthbert, and who is said to be a half-sister of Edmund, their common mother being 'Nance Carey.'"—*Athenæum.* Mr. Kean, being of an over-sensitive nature, felt the severe treatment of a portion of the critical press; but the public pretty well understood the matter. Shortly before his leaving England on his great tour, he was entertained at a public dinner, and presented with a magnificent service of plate, purchased by subscription of those who really understood and appreciated his many excellent qualities.

PROFESSOR JULIUS PLÜCKER, at Bonn, where, with the exception of two years' occupation in Berlin and Halle, his life had been passed in professorial duties and scientific research. His industry was great. The titles only of his papers form a long list, embracing pure and applied mathematics, magnetism under various conditions, the optical and magnetic phenomena of crystals, and allied subjects. Among his latest works were three papers, published in the *Philosophical Transactions,* "On the Spectra of Gases and Vapours," "On a New Geometry of Space," and "Fundamental Views regarding Mechanics," all of which are marked by great insight and originality of view. Mathematicians will the more deplore their loss as he was engaged in a continuation of his geometrical researches to within a short timo of his decease. He was born at Elberfeld in 1801, was elected a Foreign Member of the Royal Society, 1855, and 1866 was honoured by the gift of their Copley Medal.—*Athenæum.*

ROBERT PORRETT, F.R.S.—To the younger class of chemists it will scarcely appear possible that the discoverer of ferrocyanic acid, an event to be referred to the very infancy of their science, has only just passed away from among us. Mr. Porrett had attained the age of eighty-six, and to the last was a steady attendant of the principal scientific Societies, and participated in their business. His era of successful activity was quite fifty years ago, when he established a working laboratory at his residence in the Tower, and contributed his historical essays to the *Transactions* of the Royal Society. These embrace a perilous investigation of the explosive chloride of nitrogen, in addition to the careful study of ferrocyanic acid. Mr. Porrett spent his life in the Civil Service in connection with the Tower.—*Ibid.*

CAPTAIN CLAXTON, R.N., who has rendered signal assistance to our engineers on many important occasions.—Captain Claxton distinguished himself during the Peninsular war. Later on in life he held the post of Secretary to the Great Western Steamship Company, and when their vessel, the *Great Britain*, was stranded in Dundrum Bay, Captain Claxton was selected to protect the ship during winter, which he did most effectually under Mr. Brunel. Subsequently, Captain Claxton was engaged in the critical operations of floating the tubes of the Conway and Britannia bridges, which he successfully accomplished, to the satisfaction of Mr. Robert Stephenson. He has also assisted Mr. Brunel in similar operations at Chepstow.

M. BOUCHER DE PERTHES, the founder of the new science called Archæogeology. M. Boucher de Crèvecœur de Perthes was the first to call the attention of the learned world to those remarkable relics of the earliest ages, the flint implements used by man before the discovery of metals. At first ridiculed as a visionary, then, by slow degrees, listened to with increasing interest, he at length succeeded in proving to archæologists that there had been in Europe an age of stone; nay, he went further, and conquered the incredulity of geologists by producing the first human jaw ever found in the undisturbed Alpine drift, showing thereby that man had been coeval with the extinct races of large carnivora that peopled Europe before the commencement of the present geological period. His valuable collection of flint implements now forms an important part of the Gallo-Roman Museum at St. Germain.—*Galignani's Messenger.*

JEREMIAH CARHART, the inventor of the Melodeon, the leading principles of which instrument have been adopted by our present makers of parlour or cabinet organs.

THE REV. J. G. CUMMING, formerly Warden of Queen's College, Birmingham, well known among scientific men for geological and antiquarian researches ; and the author of several valuable historical works upon the Isle of Man.

MICHAEL LANE, Engineer-in-Chief to the Great Western Railway Company. The deceased gentleman was engaged on the the works of the Thames Tunnel in 1825, which was the commencement of a career of usefulness and honour. A few years later he became the Resident Engineer of the Bristol Docks, under Mr. Brunel. He was afterwards engaged on the Monkwearmouth Dock, Sunderland (north shore), the works of which he carried out. In 1840 Mr. Lane commenced his duties on the Great Western Railway, being engaged on the western division of that line. He was afterwards appointed Resident Engineer of the Hull Docks, but, in 1845, he rejoined the Great Western Company as Superintendent of Permanent Way. At the close of 1860 Mr. Lane was made Engineer-in-Chief to the Company, which office he retained up to the time of his death.—*Mechanics' Magazine.*

SIR JOHN SCOTT LILLY, C.B., Lieutenant-Colonel Grenadier Guards, and Major-General in the Portuguese Service, expired in London on Monday. The deceased officer was one of the few survivors of the army which sailed to Portugal with Sir Arthur Wellesley. Sir John was present at almost every action from Roleia to Toulouse. He had the honour of beginning the battle of Salamanca by occupying the Arapiles, which afterwards became the pivot of the battle. The French simultaneously tried to seize it, but were repulsed after a severe fight, in which Sir John captured the colours of the 110th French Regiment of the Line with his own hand. At Albuera, Sir John took part in Sir Lowry Cole's famous charge ; but perhaps the most memorable event of his life occurred in the Pyrenees, on the 30th of June, when Wellington, after four days' hard fighting on the road to Ostiz, determined to seize a hill in the very centre of the French position as a preliminary to a decisive movement. This hill was captured from the French by Sir John, with the 6th Portuguese Cacadores, although defended by two battalions.—*Ibid.*

SIR CUSACK RONEY, whose name is intimately associated with railway history both in the old and the new world. The deceased knight first became known as the Secretary of the Eastern Counties Line, an office which he filled while Mr. George Hudson was still the leading potentate of railways. He was afterwards closely connected with the Grand Trunk of Canada Railway, and with Irish and Continental railway enterprise, his long experience of all matters pertaining to railways rendering his services valuable when any new scheme had to be launched or any grand plan of operations to be accomplished. Sir Cusack received his knighthood for his very able services to the Dublin Great Exhibition.

NATHANIEL BAGSHAW WARD, the inventor of Wardian Cases by which many most interesting exotic plants have been introduced into this country, and which have enabled so many to cultivate ferns in their rooms with delight and instruction, died, at the age of seventy-seven, at St. Leonards, on the 4th inst. He was a Botanist of considerable reputation, Fellow of the Royal Society and other scientific Societies; lately Master of the Apothecaries' Company, one of the Examiners, and took considerable interest in the education and examination of females for the medical profession. His house in Wellclose Square, and afterwards at Clapham Rise, was one of the scientific curiosities of London, showing how many and how well plants might be cultivated in a small space in or near a large city.—*Athenæum.*

Mr. PETIT, the accomplished and laborious Archæological Writer. His *Lectures on Architectural Studies, Lectures on Architectural Principles, Remarks on Architectural Characters,* and still more broadly, his characteristic mode of sketching are known to modern students.

SIR DAVID BREWSTER, K.H., of the University of Edinburgh, and one of the first Natural Philosophers of his time. He was born at Jedburgh, on the 11th of December, 1781. His father, who was Rector of the Grammar-school there, destined him for the ministry; and he was accordingly sent to the University of Edinburgh, and maintained there for several sessions, during which his performances as a student were promising and even brilliant. He passed through the theological classes, and took licence as a Preacher of the Church of Scotland; but he was strongly attached during his college career towards the study of science and the observation of natural phenomena. He had received the honorary degree of M.A. in 1800; and at and after that period he enjoyed the acquaintance, and assistance in his scientific studies—in which he already gave evidence of surpassing powers of observation,—of Robison, the Professor of Natural Philosophy, and of Playfair and Dugald Stewart. He had already so far improved upon the instructions he had received, that, in maturely examining the bases of Newton's theory of Light, he succeeded in discovering a novel and important fact in optics—that of the influence of the condition of the surfaces of bodies on the "inflection," or change of direction of the rays of light, which had been formerly accepted as a consequence of the nature of the bodies themselves. He had already devoted himself principally to the science of Optics, in which he was destined to attain so distinguished a reputation. In 1807, a number of honours poured in upon him. He was made LL.D. of Aberdeen University; Oxford conferred on him the degree of D.C.L., and Cambridge that of A.M. Next year Dr. Brewster was elected a member of the Royal Society of Edinburgh, of which he subsequently filled the offices of Secretary, Vice-President, and President—holding the latter office at his death; and in the same year he took in hand the task of editing the *Edinburgh Encyclopædia,* a work to which he

made a number of important and interesting scientific contributions, and which he did not complete till 1830. This considerable undertaking, however, was far from occupying the whole of
Dr. Brewster's almost marvellous working energy. In 1813,
under the title of a *Treatise on New Philosophical Instruments,
&c.*, he presented to the public some of the results of his optical
researches during the preceding twelve years. In 1811 he had bestowed some attention upon the experiments prosecuted by
Buffon, with the purpose of discovering the nature and emulating
the effects of the burning-mirrors of Archimedes; and these experiments suggested to him the construction of what he styled
"polyzonal" lenses. Lighthouses at that time were usually
fitted up with plain parabolic reflectors; Dr. Brewster proposed
instead the use of lenses built up of zones of glass, each of
which might be composed of several circular segments, arranged
concentrically round a central disc, with the effect of strengthening the light and transmitting it to a greater distance. The
invention, or adaptation of Buffon's invention, excited a good
deal of interest at the time, as it promised to lead to an improvement in the illumination of our lighthouses and the safe
conduct of our coast navigation; but it was not then practically
taken up in this country, though it was in France. In 1815, at
the desire of the Corporation of Edinburgh and of Professor
Playfair, he undertook to take the place of the latter in deliver-
ing the lectures on natural philosophy; but he did not long per-
sist in this task, grudging every moment and every effort that did
not lead him further in the investigation and knowledge of his
favourite subject. In the same year he sent again to the Royal
Society of London a paper "On the Polarization of Light by
Reflection," and the Society elected him a Fellow, and voted him
their Copley Medal for his discoveries and researches. In 1816
he had the honour to receive from the French Institute, half of
the prize of 3,000 f. awarded for the two most important discoveries made in Europe in physical science during the two
years preceding. In that year also he achieved the invention
which has rendered his name most popular—that of the kaleido-
scope. Thenceforward honours continued to flow in rapidly
on him, and in 1831 he received the decoration of the Guelphic
Order of Hanover. The year following he was knighted by King
William IV. In 1833 he was a candidate for the Chair of Natural
Philosophy in the University of Edinburgh, but was defeated by
Mr. James D. Forbes, now Principal Forbes, of St. Andrew's.
To the distinctions we have enumerated as falling to his share
the King of Prussia added (in 1817) the Order of Merit. In
1849 he was elected one of the Foreign Associate Members of
the Institute of France, and the Emperor Napoleon (in 1855)
conferred upon him the Cross of the Legion of Honour. The
list of Sir David Brewster's contributions to scientific and
general literature is very extensive.

ANTOINE VECHTE, one of the most remarkable Artists of this or any age,—the Cellini of modern times,—the "art-workman" whose productions—especially his designs and chasings in repoussé oxydised silver, executed principally for the eminent firm of Messrs. Hunt and Roskell—have won an European reputation. M. Vechte's career was in many respects extraordinary, and affords a noble lesson of perseverance to workmen of all countries.

DR. WILLIAM BIRD HERAPATH, of Bristol, a son of the late Mr. William Herapath, so eminent as an Analytical Chemist, and, like his father, had attained to a high degree of knowledge and skill in the same science. Dr. Herapath's name has also been associated with some useful discoveries in the microscope. On passing his M.D. examination, in 1844, at the London University, he took honours in no fewer than six branches of medical knowledge. He subsequently became an M.D. of the same Institution, and his rapid and brilliant succession of chemical and toxicological discoveries was rewarded by the Fellowships of the Royal Societies of Edinburgh and London, and corresponding membership of most of our learned bodies. Among a mass of scientific communications to various periodicals, we may mention his papers on "The Optical and Chemical Characters," "Sulphate of Soda Quinine," on "The Iodo-Sulphate of the Cinchona Alkaloids," "Discovery and Manufacture of Artificial Tourmalines," "Address on Chemistry in its Relation to Medicine and the Collateral Sciences," on "A New Method of Detecting the Hydrogen, Arsenic, and Phosphorus when in company as Mixed Gases," &c. Although suffering from an exhausting and painful disease, his zeal for science remained until the last, and within a few days of his decease he was engaged in laborious researches with spectrum analyses, more especially as to bloodstains and the chlorophilis of plants. His early death, at forty-eight years of age, is deeply regretted by a large circle of professional and other friends.—*Times.*

CHRISTIAN FREDERIC SCHONBEIN, who, besides acquiring scientific celebrity, made himself universally known as the discoverer of Ozone and the inventor of Gun-cotton, died on August 2 at a friend's house in Sauersburg, where he was stopping *en route* for Bâsle. He forsook mere practical work and strove to prepare himself for the theoretical research, by studying at the Universities at Tubingen and Erlangen. After finishing his college course, he taught chemistry and physics in a school at Keilhau, near Rudolstadt. This employment he subsequently relinquished, and visited England and France, where he continued his scientific pursuits. In 1835 the university of Bâsle appointed him Professor of Physics and Chemistry. He held the Chair of Chemistry until his death; the Chair of Physics, in 1852, being made a separate one. Schönbein's studies were neither confined to chemistry or physics, but liberally embraced both. To testify this, we need but mention a few of his works. His researches on iron and other metals; the influence of temperature in changing the colour of substances; the chemical action of the rays of the sun; and, finally, the theory of voltaic electricity. On this last subject Schönbein has thrown considerable light. In the well-known debate on the contact and chemical theories, he impartially investigated the opposing opinions, and demonstrated what was faulty in both. He attributed the origin of voltaic electricity to chemical affinity, at the same time establishing a positive difference between the development of electricity in the open pile, and the production of current, accompanied by chemical decomposition, which occurs when the circuit is closed. Schönbein was married in 1835, but it is said that he was true to the advice of his teacher at Erlangen, the celebrated Schelling, and ever regarded science as his *fiancée.*—*Basler Nachrichten.* Quoted in *Scientific Opinion.*

DR. JOHN DAVY, the younger brother of Sir Humphry Davy. —He was able to continue his important Chemical researches nearly to the time of his death, and communicated papers to the Royal Society in the course of last year. It was to his observations on the effect of cold on fishes that Australia is indebted to the introduction of salmon, and the possibility of moving the ova and fish from place to place has been proved. The Australian agents had so little faith in his experiment, that when they acceded to the recommendation of taking some eggs out, packed in ice, they forgot or did not think it worth while to look after the box in the ice-house on their arrival; but when the ice-house was cleared out to be refilled on the vessel's return to England, the box was discovered, and the greater part of the eggs were found to be alive. These were the first eggs of fish that had ever survived the voyage and been hatched in Australia: now the plan is in general use. By the death of Dr. Davy, the service of plate presented to Sir Humphry Davy, in 1817, by several coal-mine proprietors, for the discovery of his celebrated safety-lamp, reverts to the Royal Society. Sir H. Davy, in his will (made two years before his death), left the plate in question " to Lady Davy, to revert to his brother in case of his surviving her, and if not, to any child of his who may be capable of using it; but if he bo not in a situation to use or enjoy it, I wish it to be melted and given to the Royal Society to found a medal, to be given annually for the most important discovery in chemistry anywhere made in Europe or Anglo-America." Dr. Davy directs by his will that " the service of plate formerly belonging to Sir Humphry, in accordance with his wishes, after the decease of testator's wife, or earlier, if she thinks proper, is to be given to the Royal Society." The value of the plate is £2,500.

CAPTAIN BLAKELEY, R.A., the inventor of the strengthened Gun bearing his name; his death occurred on May 3, in Lima, Peru, whither Captain Blakeley had gone for the purpose of examining the condition of the guns furnished by him to the Peruvian Government, that did such good service during the engagement of May 2 with the Spanish fleet.

ROBERT GRIER.—He entered the army in February, 1810, as Ensign in the 44th Regiment, and shortly afterwards went on service to the Peninsula. At the battle of Fuentes d'Onor, he was wounded in the right shoulder; at the siege of Badajos he led the advance of the feigned attack which ultimately became the successful one, and commanded the "forlorn hope" when the place was carried. He was in the campaign of 1815; severely wounded in the ankle at Quatre Bras. Captain Grier had the War Medal with three clasps, and the Waterloo Medal. He was placed on half-pay in March, 1817, as Lieutenant, and was appointed a Captain of Invalids in September last.

LIEUTENANT LE SAINT, the French African Traveller, in exploring the country about the White Nile. He had already overcome many difficulties, and reached Abon-Konka, within some sixty leagues north of Gondokoro, when he fell a victim, at the age of thirty-five, to the insalubrity of the climate.

JOHN STEVELLY, of Belfast, a man of high attainments in science, for many years the Chief Secretary of Section A. in the British Association. He died at an advanced age, after a life of modest usefulness.

EYRE EVANS CROWE, the experienced Writer and Politician. Mr. Crowe's most important work is the *History of France*, the last volume of which has recently been published.

G. WALKER ARNOTT, the well-known Professor of Botany in the University of Glasgow.

WILLIAM HENRY BARTON, late Deputy Master and Comptroller of Her Majesty's Mint. The deceased gentleman was connected with that establishment for the long period of thirty-eight years, and since 1851 had held the combined offices named above. Mr. Barton was always distinguished for urbanity and kindness, and his decease leaves a vacancy in the Mint which is not likely to be more ably or honourably filled than it was by himself. The late officer was an excellent amateur mechanic, and several useful inventions owe their origin to his knowledge and skill.

THOMAS COOKE, the English Fraunhofer, whose science and skill have restored to England the pre-eminent position she held a century ago in the time of Dolland. His loss will be greatly felt in the astronomical world; his mathematical attainments and large scientific mind insured the admiration and respect of all who knew him, and as an Artist he has never been excelled. It is, we believe, now a little over ten years since his first large equatorial telescope, the one in the possession of Mr. Fletcher, was completed. In that period his genius has left its mark on every important astronomical instrument. The equatorial he found comparatively clumsy: it is now perfect. Nothing can exceed the simplicity and completeness of the means by which he has supplied all the wants of the observer. Called to make a transit for the Indian Survey, he at once thought out an entirely original plan of mounting, which will certainly be the basis, if not the actual form, of all future important instruments of this class. A dividing engine, which is entirely automatic and greatly superior to any existing form, is one of his least triumphs. Had a wonderful mechanical talent been the only strong side of Mr. Cooke's mind, he would probably never have been an optician. An early acquaintance with, and love of, mathematics, however, led him to the study of optics, and his success as an optician is due to this combination. After commencing the construction of object-glasses, he was soon dissatisfied with the method of hand-polishing, and in his perfected arrangements the hand is scarcely called into play. The introduction of steam-power, as arranged by him, not only ensured perfect accuracy of figure, but it has enabled a number of object-glasses to be made, which seems almost fabulous, if we compare it with what was formerly considered the maximum rate of manufacture. Further, it has enabled the size to be increased almost in the same ratio as the rate of construction. Telescopes, nearly all of Mr. Cooke's manufacture, of six up to ten inches aperture, are now as plentiful as five feet achromatics were formerly; and it is in this particular that Mr. Cooke, in his all too short life, earned the gratitude of every lover of astronomy. At the time of his death, a 25-inch telescope, a triumph of skill, required only a few touches to make it complete; and we believe that other glasses, varying from 10 to 16 inches, are also in hand. When it is remembered that a few years ago 16½ inches in aperture was the largest size of object-glass, and that German, and that this has only been quite recently exceeded by Mr. Alvan Clark in America, it is not difficult to appreciate how Mr. Cooke's labours in England have succeeded in restoring our ancient pre-eminence, and it is well that Mr. Cooke has left sons behind him upon whom the duty now devolves, and we do not doubt their training to sustain their father's reputation. Mr. Cooke was a Fellow of the Royal Astronomical Society. It is certain that, had he lived, the rewards for which scientific men generally care would have been bestowed upon him, although with his modesty and retiring disposition he would never have expected them.—*Athenæum*.

SIR JAMES BROOKE, K.C.B., late Governor of Labuan, and Rajah of Sarawak. After more than three years' sailing and cruising in the Mediterranean and other European seas, during which he was training and "educating" them for greater things —he left the Thames on the 27th of October, 1838, and steered straight for those Eastern seas of which he had read as a child, and which he now resolved to penetrate again. He had heard much, too, of the wretched condition of the natives of some of those Eastern islands; of their habits of plunder, piracy and murder; of their discontent under the rule of native chiefs almost as savage and lawless as themselves; and of the gradual cessation of trade and commerce, which threatened to plunge them deeper in the gloom of barbarism. He passed the southern shores of India and Ceylon, crossed the Indian Ocean, and speedily landed at Singapore. This was in July, 1839, and he reached Sarawak, which lies a few leagues up country from the sea-coast of Borneo, in the following month. On reaching the coast of Borneo, he found the Sovereign or Sultan of that island engaged in a long and almost hopeless attempt to subdue one of the rebellions which so frequently happen among the rival rulers of subordinate districts. What he could not do in four years, Brooke helped him to do in as many months, if not in as many weeks. His aid was solicited by and given to the Rajah Muda Hassim; and it secured the triumph of authority and law. It appears that Muda soon afterwards being called to the post of Prime Minister, recommended the Sultan to make the English adventurer his successor as Rajah of Sarawak. The advice thus tendered was accepted, and the honour and dignity of Rajah was laid at the feet of the Englishman. When the news came to England that he had taken an active and successful part in the suppression of the Malay pirates, and that the Prince had ceded to him the territory of Sarawak as the representative of England, James Brooke became a popular idol. This was in 1841; and his official proclamation as Governor of Sarawak dates from the 21st of September, 1841, on which day the British flag was hoisted there. The result of these expeditions has been published. The rest of the ex-Rajah's story is soon told. In 1858 he returned to England, but he had been in this country only a few months when his health received a serious shock in the shape of a paralytic attack. Towards the close of 1861 he paid Borneo a visit, accompanied by Mr. Spencer St. John; but he had the mortification of finding the north-west part of the island in rebellion. As soon as this outbreak was suppressed he returned to England, but was again recalled to the East by fresh complications which had arisen in the internal administration of Borneo. These diffi- culties, however, he had the satisfaction of seeing arranged on his farewell visit to the island about five years since. From that date the fortunes of Borneo and of Sarawak have been, on the whole, peaceful and quiet. Brooke, though himself placed on a sort of honorary retired list, saw the independence of his favourite settlement recognized by the British Government, and a British Consul being established there, it may be assumed that, in spite of the indifference of the Home Government, henceforth our rule in that portion of the Eastern seas is established upon a safe and sure basis.—*Abridged from the Times.*

SIR EDMUND WALKER HEAD, Bart.; in Literature chiefly known by his *Handbook of Spanish Painters.*

CHRISTOPHER BENSON, once the most popular Preacher in London, and Master of the Temple.

PIERRE ANTOINE BERRYER, the glory of the Paris Bar. He was the oldest and ablest Advocate of his time, and his powers as an orator were enhanced by the virtues of his character, and the splendid consistency of his career.

The Very Rev. HENRY HART MILMAN, Dean of St. Paul's, Poet and Historical Writer.

GIOACCHIMO ANTONIO ROSSINI, the celebrated Musical Composer.

His Grace the Most Reverend CHARLES THOMAS LONGLEY, D.D., Archbishop of Canterbury, and Primate of England.

GEORGE PRYME, M.A., Professor of Political Economy.

DR. RICHARD HOBSON, Naturalist, who formed one of the best collections of stuffed birds, mosses, lichens, in the West Riding of Yorkshire.

SAMUEL LUCAS, M.A., Journalist.

WILLIAM EDWARD SHUCKARD, Entomologist.

T. DUNCAN, Water-Engineer to the Liverpool Corporation.

EDWARD JESSE, Deputy-Surveyor of the Royal Parks and Palaces. In this capacity his knowledge of natural history enabled him to effect many useful and permanent improvements in the Royal residences and gardens, more especially at Windsor and Hampton Court Palace.

SIR WILLIAM NEWTON, the Miniature Painter.

MICHAEL WALKER, Hydrographer.

J. R. LEON FOUCAULT, Natural Philosopher.

PROFESSOR NANDER HOEVEN, Zoologist.

JOHN BURNET, Engraver.

J. W. CARMICHAEL, Marine Landscape Painter.

GEO. HOUSEMAN THOMAS, Artist.

G. R. BURNELL, Engineer.

GEORGE CATTERMOLE, Artist.

DR. JOHN ELLIOTSON, Physician

JOHN DOUGLAS COOK, Journalist.

E. H. WEHNERT, Artist.

HENRY LE KEUX, Engraver.

JOHN WATERER, Botanist.

CAUSES OF DEATH.

THE Registrar-General's Report for 1866 contains the usual letter from Dr. W. Farr on the causes of death in England in the year. Five great gates of death are distinguished. In round numbers, 24 in 100 of those who quitted the world died from zymotic diseases; 18 from constitutional diseases; 39 from local diseases ; 16 from developmental diseases; and 3 in 100 (3·43) were violent deaths. The fact of the occurrence of more than a fifth of the deaths from zymotic diseases shows what

great defects of conservancy still prevail in England, and how
in unclean towns preventible diseases scourge the negligent
population. Dr. Farr has to state that there is no apparent
evidence of decline in the rate of death from fever. He considers it exceedingly probable that typhoid fever is sustained by
the increasing contamination of the waters, and typhus by the
increased density of population. Hydrophobia, which had been
fatal to one, two, three, or four persons in the seven years from
1857 to 1863, became fatal to 12, 19, 36 persons in the three
years ending with 1866. The amount of mortality from constitutional diseases varies little from year to year. "Local
diseases" include all the recognized inflammations which give
rise to general derangements both of the nerve-force and of the
circulating blood, but produce specific changes of particular
organs, and may hence be called monorganic. The deaths from
phthisis and the respiratory order of diseases have increased
from 5,580 per annum to a million persons living in the five
years 1850-54, to 5,803 in the five years 1855-59, 5,976 in
1860-64, 5,956 in 1865, and 6,331 in 1866. Dr. Farr notes
that the smoke in towns, the dust of the atmosphere in shops,
and the shut-up life in chambers are increasing as towns increase;
and he suggests that the smoke from private houses should be
diminished by the simple practice of lighting fires from the top.
The violent deaths were 16,915 in 1866; 12 were public execu-
tions; 480 were, according to the findings of coroners' juries,
murders and manslaughters, 183 of the victims being infants
not a year old. There were 1,329 suicides; the number of
suicides fluctuates little, from 64 (the proportion in 1866) to 70
in a million living. The number of deaths by accident or
negligence in 1866 was 14,886; the increasing number measures
to some extent the increase of the chemical and mechanical
forces in use, as it is some time before workmen and others learn
to manage them and keep them under control. The deaths by
burns and scalds have diminished to 2,533; possibly a substitu-
tion of woollen and worsted dresses for cotton has had some
effect on the deaths by burns. The deaths by poison have also
decreased, probably owing to the better regulations for the sale
of poisons, and the greater precautions of chemists; but still, 406
deaths by poison—278 by accident or negligence, and 128 by
suicides—are a large number. Twenty deaths by lightning were
registered in 1866, mostly of persons engaged in outdoor pursuits. These 15 were the chief causes of death in 1866:—
Phthisis killed 55,714; bronchitis, 41,334; atrophy and debility,
31,097; old age, 28,546; convulsions, 27,431; pneumonia,
25,155; heart disease, 21,197; typhus, 21,104; diarrhœa,
17,170; whooping-cough, 15,764; cholera (epidemic that year),
14,378; scarlatina, 11,685; measles, 10,940; paralysis, 10,504;
apoplexy, 10,297.

GENERAL INDEX.